Medical Biofilms

Medical Biofilms

DETECTION, PREVENTION AND CONTROL

Edited by

JANA JASS
Department of Microbiology and Immunology,
The University of Western Ontario, and
The Lawson Health Research Institute,
London, ON, Canada

SUSANNE SURMAN
London Food, Water & Environmental Microbiology Laboratory,
CPHL, London, UK

JAMES WALKER
CAMR, Porton Down, Salisbury, UK

WILEY

Telephone (+44) 1243 779777

Email (for orders and customer service enquiries): cs-books@wiley.co.uk
Visit our Home Page on www.wileyeurope.com or www.wiley.com

Other Wiley Editorial Offices

John Wiley & Sons Inc., 111 River Street, Hoboken, NJ 07030, USA

Jossey-Bass, 989 Market Street, San Francisco, CA 94103-1741, USA

Wiley-VCH Verlag GmbH, Boschstr. 12, D-69469 Weinheim, Germany

John Wiley & Sons Australia Ltd, 33 Park Road, Milton, Queensland 4064, Australia

John Wiley & Sons (Asia) Pte Ltd, 2 Clementi Loop #02-01, Jin Xing Distripark, Singapore 129809

John Wiley & Sons Canada Ltd, 22 Worcester Road, Etobicoke, Ontario, Canada M9W 1L1

Wiley also publishes its books in a variety of electronic formats. Some content that appears in print may not be available in electronic books.

British Library Cataloguing in Publication Data

A catalogue record for this book is available from the British Library

ISBN 0 471 98867 7

Typeset in 10/12 point Palatino by Dobbie Typesetting Ltd, Tavistock, Devon
Printed and bound in Great Britain by Biddles Ltd, Guildford, Kings Lynn
This book is printed on acid-free paper responsibly manufactured from sustainable forestry in which at least two trees are planted for each one used for paper production.

Dedications

I would like to dedicate this book to my parents, Jan and Marie, and sister Irena.
Jana Jass

With thanks to John, and my daughter and son, Nicola and David, and my stepsons Philip and Richard.
Susanne Surman

I would like to dedicate this book to my mother and father, Hector and Rosina, as well as to my sisters, Catherine, Mary and Winnie, and my brothers Thomas and Hector.
Jimmy Walker

Contents

Contributors

Peter Cadieux, Department of Microbiology and Immunology, The University of Western Ontario, London, Ontario, N6A 4V2, Canada.

Howard C. Ceri, Biofilm Research Group, Department of Microbiology and Infectious Diseases, Department of Biological Sciences, University of Calgary, Calgary, Alberta, T2N 1N4, Canada.

John Denstedt, Lawson Health Research Institute and Department of Surgery, The University of Western Ontario, London, Ontario, N6A 4V2, Canada.

Rodney M. Donlan, Division of Healthcare Quality Promotion, National Center for Infectious Diseases, Center for Disease Control, Mailstop C-16, Atlanta, GA 30333, USA.

David Dymock, Department of Oral and Dental Science, Dental School, Lower Maudlin Street, Bristol, BS1 2LY, UK.

Peter Gilbert, School of Pharmacy and Pharmaceutical Sciences, University of Manchester, Oxford Road, Manchester, M13 9PL, UK.

Jana Jass, Department of Microbiology and Immunology, University of Western Ontario, and The Lawson Health Research Institute, Grosvenor Campus, London, Ontario, N6A 4V2, Canada.

Andrew J. McBain, School of Pharmacy and Pharmaceutical Sciences, University of Manchester, Oxford Road, Manchester, M13 9PL, UK.

Yoshimitsu Mizunoe, Department of Bacteriology, Faculty of Medical Sciences, Kyushu University, Fukuoka, 812-8582, Japan.

Douglas W. Morck, Biofilm Research Group, Department of Microbiology and Infectious Diseases, Department of Biological Sciences, University of Calgary, Calgary, Alberta, T2N 1N4, Canada.

Merle E. Olson, Department of Microbiology and Infectious Diseases, Faculty of Medicine, University of Calgary, Calgary, Alberta, T2N 4N1, Canada.

Georg Peters, Institute of Medical Microbiology, University of Münster Hospital and Clinics, Domagkstaße 10, D-48149 Münster, Germany.

Gregor Reid, Lawson Health Research Institute and Department of Microbiology and Immunology, The University of Western Ontario, 268 Grosvenor Street, London, Ontario, N6A 4V2, Canada.

Alexander H. Rickard, School of Pharmacy and Pharmaceutical Sciences, University of Manchester, Oxford Road, Manchester, M13 9PL, UK.

Rachel Sammons, University of Birmingham, School of Dentistry, St Chad's Queensway, Birmingham, B4 6NN, UK.

Sarah R. Schooling, School of Pharmacy and Pharmaceutical Sciences, University of Manchester, Oxford Road, Manchester, M13 9PL, UK.

David Spratt, Departments of Microbiology and Conservative Dentistry, Eastman Dental Institute for Oral Health Care Science, University College London, 256 Gray's Inn Road, London, WC1X 8LD, UK.

Susanne Surman, London Food, Water & Environmental Microbiology Laboratory, Central Public Health Laboratory, London, NW9 5HT, UK.

Christof von Eiff, Institute of Medical Microbiology, University of Münster Hospital and Clinics, Domagkstraße 10, D-48149 Münster, Germany.

Sun N. Wai, Department of Molecular Biology, Umeå University, SE-901 87 Umeå, Sweden.

James T. Walker, CAMR, Porton Down, Salisbury, SP4 0JG, UK.

James Watterson, The University of Western Ontario, London, Ontario, N6A 4V2, Canada.

Preface

Biofilms are a complex heterogeneous consortium of microorganisms associated with surfaces and interfaces and have been shown to play an important role both in causing disease and for maintaining health. Microorganisms growing in biofilms may cause or prolong infections through, colonisation of implants or prosthetic devices and problems resulting from dental plaque formation. Modern medical practices and implant technology have alerted clinicians to the implications of biofilm-associated infections due to their persistence and resistance to antimicrobial treatment. Biofilms are also shown to be important in maintaining health by supporting commensal microflora that may assist in preventing pathogen infectivity.

This book is the first to deal specifically with biofilms associated with different medical areas including: the contamination of medical devices such as catheters, orthopaedic prostheses, renal dialysis, shunts, pacemakers and drug delivery systems; infection of tissue surfaces as in the lungs of cystic fibrosis patients, on damaged tissue surfaces (i.e. burns and surgery), bone (osteomyelitis), cardiac tissue (endocarditis) and genitourinary tract; and dental plaque, the cause of caries and periodontal disease. For each of these topics the book provides an overview of current research in medical biofilms focusing on detection and monitoring the problems associated with biofilm and current strategies for control and eradication. To fully understand infectious biofilms, a current summary of the basic concepts in biofilm research and future prospects are included.

The editors intend that the book be used as an aid in teaching and research. Persons with an interest in medical diseases will find this book fascinating to read and the format is aimed at complimenting many hospital teaching courses for clinicians as well as medical and dentistry students. The publication came about due to the increased awareness of the importance of adherent microbial populations in human health and disease, yet lacking a comprehensive text on information and research investigating the problems of medical biofilms, how to detect them and ultimately how to control their presence.

Industrial environments have been ahead of the medical profession on their understanding and research progress on biofilms and biofouling of surfaces by microorganisms. The companion volume ''Industrial Biofouling'' (2000) discusses biofilms as a persistent problem and how to control them in potable water systems, industrial water systems and the food industry.

Glossary

Aerobes Bacteria that require oxygen for growth and are dependent on a respiratory metabolism to generate energy, with molecular oxygen usually serving as the terminal electron acceptor.

Alloplastic Inert material suitable for implanted prostheses, generally made of metal, ceramic or polymeric synthetic substances.

Amylolytic Organisms with enzymes that reduce starch.

Anaerobes Microorganisms that only grow in the absence of molecular oxygen and that generate energy by fermentative reactions that do not involve molecular oxygen.

Antagonists Opposing actions or processes. E.g. microbial antagonists inhibit the growth or presence of another; drug/biochemical antagonists inhibit or produce opposite effects to each other.

Apoptosis Programmed cell death.

Atherosclerosis A disease of the arteries in which plaque deposits of cholesterol, lipoid substances, and lipophages are formed within large and medium-sized arteries.

Atomic force microscope (AFM) A form of scanning probe microscopy that enables visualization of a surface at atomic (nanometre to micrometre) resolution.

Autochthonous flora Usually non-pathogenic, *commensal*, naturally colonizing microorganisms.

Autolysis Self-digestion or automatic dissolution of cells or tissue by the enzymes contained within them, occurring upon death or under certain pathological conditions.

Bacteraemia Bacterial infection of the blood.

Bactericidal Substances that kill bacteria; includes many antibiotics.

Bacteriocin Exotoxin produced by bacteria to kill other bacteria; often plasmid coded.

Bacteriostatic Substances that inhibit bacterial growth without killing them.

Bioacoustic Application of ultrasound in conjunction with exposure of the biofilm to antibiotics.

Bioelectric Electric fields used to enhance the efficacy of charged biocides and antibiotics in killing biofilm bacteria.

Biofilm Microorganisms attached to an interface and growing within a matrix including exopolymeric substances.

Biosurfactants Any surface-active agent or substance that modifies the nature of surfaces, often reducing the surface tension of water, and is produced by a living organism.

Bronchiectasis Chronically enlarged bronchi with inflammation, most commonly occurring in the lower portion of the lung.

Calculus A mass of crystallized or precipitated salts or other material, such as cholesterol, that is formed within a body chamber, such as the gallbladder, kidney, or urinary bladder.

Caries Decomposition and decay of teeth, causing discoloration, softening, and porosity.

Cariogenic Producing or promoting the development of tooth decay.

Catalase Tetrameric enzyme, which breaks down hydrogen peroxide.

Category 1 devices Those that are totally implanted in the tissues of the body and intended to remain in place for the life of the patient. Examples include large joint replacements, prosthetic heart valves, and hydrocephalus shunts.

Category 2 devices These are partially implanted, and intended to remain *in situ* for long time periods (e.g. central venous catheters, external ventricular drains).

Category 3 devices These are not true implants, and include urinary catheters and voice prostheses.

Cellulolytic The ability to hydrolyse cellulose (protozoans and certain bacteria).

Cementum Modified bone surrounding roots of teeth, beneath the gum in vertebrates, binding the teeth to the jaw by the periodontal ligament.

Checkerboard hybridization The extraction and labelling of total genomic DNA from culturable microorganisms for use as a probe in hybridization experiments with DNA.

Chemotaxis The movement of cells or microorganisms towards or from a chemical substance.

Cholangitis Inflammation of the bile duct.

Coaggregation When microorganisms bind to each other in suspension to form aggregates.

Coadhesion When microorganisms from suspension bind to surface-adherent cells.

Commensal Microorganisms, which are usually non-pathogenic, naturally occur on the host surface and give protection against pathogens.

Conjugation Union between two bacterial cells that leads to a transfer of genetic material.

Cytotoxic Producing toxin lethal to cells.

Debridement The removal of foreign matter or dead/infected tissue from a wound or lesion to leave healthy tissue.

Endotoxin Heat-stable polysaccharide toxins produced by Gram-negative bacteria that is responsible for many of their virulent effects.

Epididymus Convoluted duct on the posterior surface of the testicle.

Exotoxin Toxins released by either Gram-negative or Gram-positive bacteria into growth medium or tissue during growth phase.

Extracellular polymeric matrix Material produced by cells and secreted into the surrounding medium applied to non-cellular portions of animal tissues and to biofilms.

Extracellular polysaccharide (EPS) Polymeric material produced by cells and secreted into the surrounding medium and is primarily composed of sugar residues.

Fibrinogen Soluble plasma protein (340 kDa) composed of six peptide chains.

Fibronectin Glycoprotein of high molecular weight that occurs in an insoluble fibrillar form in the extracellular matrix of animal tissues. Fibronectin has multiple domains and specific membrane receptors.

Fibrosis Connective tissue that occurs normally during scar tissue formation.

Fluoroplastic A plastic composed of linear polymers with some or all of the hydrogen atoms replaced by fluorine.

Fluoroquinolone Synthetic antibiotics that inhibit bacterial DNA gyrase, which is necessary for the synthesis of bacterial DNA. They are active against a wide range of Gram-negative and Gram-positive organisms.

Furanones Natural biochemicals involved in cell–cell signalling.

Furcation Branch/fork; in dentistry, bifurcations or trifurcations are conditions in which a bifurcation of a molar tooth root is denuded because of periodontal disease.

Genomics The study of genomes, which includes genome mapping and gene sequencing.

Gingivitis An inflammation of the gingiva; when associated with bony changes it is referred to as periodontitis.

Glycocalyx A carbohydrate-rich cell coat on the extracellular side of the plasma membrane that may be involved in cellular recognition and confers a unique identity upon the cell.

Glycoprotein A membrane-bound protein that has attached branching carbohydrates. These may function in cell–cell recognition, such as in the immune system response, as well as in resisting compression of cells.

Glycosaminoglycan (GAG) Polysaccharide side chains of proteoglycans forming a hydrated space-filling polymer found in the extracellular matrix.

Gram-negative Microorganisms, with thin peptidoglycan walls bounded by an outer membrane, that do not retain the crystal violet stain during the Gram staining process.

Gram-positive Microorganisms with thick cell walls containing teichoic or lipoteichoic acid complexed to peptidoglycan that retain the crystal violet stain during the Gram staining process.

Haemagluttinin (HA) Substance that causes agglutination of erythrocytes.

Haematogenous Produced by or derived from blood; or disseminated by the circulation or by the blood stream.

Homeostasis The ability of an organism to maintain a constant internal environment, such as body temperature or fluid content, by regulating its physiological processes and by making adjustments to the internal/external environment by feedback mechanisms.

Human leucocyte antigen (HL-A) A set of genes that code for the most important histocompatibility and related markers, occurring on human nucleated cells, including lymphocytes.

Iatrogenic Describing any adverse condition that is a reaction to medical treatment, especially to infections transmitted during therapy.

Immunogenic Producing immunity.

Ionotophoresis The use of DC fields to generate a bioelectric effect; can be used to reduce biofilms.

Isogenic Having the same genotype.

Keratinocytes Epidermal cells that produce keratin.

Lactoferrin An iron-binding/transport protein found in tears, bile, milk and saliva.

Laminin Link proteins of the basal lamina which induces adhesion and spreading of many cell types.

Lethal photosensitization Use of substances such as the dye Toluidine blue to increase the sensitivity of a cell to light, resulting in cell death due to light exposure.

Leucocidin A substance produced by some pathogenic bacteria that is toxic to some leucocytes, killing the cells with or without lysis.

Leukotrienes A member of the family of lipoxygenase metabolites of arachidonic acid; that can act as a mediator of an allergic response or as a chemotactic factor. (From *leuko*cytes + *triene*, indicating three double bonds.)

Ligand A functional group, atom, or molecule, of a non-metallic substance that combines with another substance in solution by a coordinate bond, in most cases to a metallic central atom or ion.

Lithotripsy The use of sound waves to disperse kidney stones *in situ*.

Metastatic disease This is a disease that has spread from its original source to a distant part of the anatomy.

Micelle A spherical arrangement formed by a group of lipid molecules in an aqueous environment with a hydrophilic interior and a hydrophobic exterior.

Minimum biofilm eradication concentration (MBEC) The minimum concentration of a substance, chemical/biochemical, required to kill microorganisms growing within a biofilm.

Minimum inhibitory concentration (MIC) Minimum concentration of a substance, chemical/biochemical, that prevents growth of microorganisms.

Miswak Twigs of certain trees that have been used on a regular basis by Muslims for centuries as a natural toothbrush; also called siwak.

Mucolytics A dissolving agent for mucin.

Myringotomy Removal of fluid (usually infected) from the middle ear space by incising the eardrum.

Necrotizing enterocolitis The severe ulceration and necrosis of the ileum and colon, which can be caused by perinatal intestinal ischaemia and bacterial invasion.

Nosocomial An infection originating in the hospital, which was neither present nor incubating prior to admittance to the hospital.

Operons A unit consisting of adjacent cistrons (the nucleotide coding for a single polypeptide, excluding regulators and terminators) that function coordinately under the control of an operator gene.

Opsonization (Hsl) A process in which an antigen is combined with opsonin, to make it more susceptible to the engulfing action of phagocytes.

Osteomyelitis Inflammation of the bone marrow and adjacent tissue.

Otitis media Inflammation of the middle ear and eardrum.

Pathogenicity The ability to give rise to morbid tissue changes or to a pathological disorder or disease.

Pericarditis Heart inflammation, specifically of the pericardium.

Periodontitis Inflammation of the gingiva associated with bony changes.

Phenotype Appearance of an organism determined by interactions affecting the genotype during development or growth as a result of environmental factors. Different phenotypes may be derived from the same genotype.

Phospholipase An enzyme that acts as a catalyst during the hydrolysis of a phospholipid.

Planktonic Microorganisms that exist in the aqueous phase of a system as opposed to sessile microorganisms, which are attached to surfaces.

Plasmid A small, closed entity of double-stranded extra-chromosomal DNA forming a self-replicating genetic element that occurs in many bacteria and some eukaryotes. It often carries genetic sequences for resistance to antibiotics; widely used in genetic engineering as a cloning vector.

Polymerase chain reaction (PCR) A process for amplifying a DNA molecule by up to 10^6- to 10^9-fold following heat denaturation of DNA strands.

Polysaccharide intercellular adhesin (PIA) A polysaccharide cell-surface appendage or extracellular macromolecular substance that facilitates adhesion of a cell to a surface or to other cells.

Porin pumps Active transport mechanism found in the outer membrane of Gram-negative bacteria that, grouped as dimers or trimers, form transmembrane channels for the entry of certain molecules into the cell.

Prebiotic Nutrients not utilized by the body, used to promote growth of normal flora.

Probiotic The use of safe living organisms taken to promote health in the user.

Prophylaxis Preventive measures or treatment taken to prevent disease.

Prostatitis Prostate inflammation.

Protamine sulphate Mixture of sulphates of basic peptides.

Protease Any enzyme, such as pepsin or trypsin, that catalyses the hydrolysis of a protein during the first stage of its degradation to a simpler substance.

Proteomics The study of the function of all the proteins in an organism.

Pyrogenic A substance that induces fever.

Quorum-sensing Cell–cell communication by extracellullar signals produced by bacteria at high cell densities. The quorum-sensing system has been shown to coordinate/regulate the expression of virulence factors in a number of organisms and has also been implicated in the formation, development, differentiation and maturation of biofilms.

Sessile Microorganisms attached to surfaces.

Shunts Channels or passages, natural or artificial, to allow fluid to pass between two natural channels, as in a bypass between two arteries to divert the blood flow from one part of the body to another.

Sialidases Virulence factors produced by pathogens.

Sigma (σ) factor A subunit of bacterial RNA polymerase that does not take an active role in the catalytic activity of the enzyme, but is necessary for the recognition of and binding to specific sites during the initiation of RNA transcription.

Siwak In Islam, a piece of a branch or root of a tree that is used as a toothbrush; may also be called a miswak.

Stent A metal or plastic tube inserted into a vessel or lumen to maintain patency. A moulded device used to maintain a skin graft in position or to provide support for tubular structures for anastomosis.

Struvite A crystalline mineral formed from hydrous phosphate of magnesia and ammonia.

Sub-inhibitory A substance that is below the concentration that is required to inhibit growth of microorganisms.

Sub-lethal A substance that is below the concentration that is sufficient to cause death.

Substantivity With respect to oral health, the retention of a substance, such as antiplaque agents in the mouth, for long enough to be effective.

Sulcus A groove or furrow found in the body.

Synergism The cooperative action of two or more organisms so that the effect of their collective effort is greater than it would be by their individual effect.

Teichoic acid A cell wall component of some bacteria formed from ribitol or glycerol phosphate and other compounds, such as glucose.

Thrombospondin A glycoprotein found in the extracellular matrix of certain cells; may be involved in autocrine growth regulation during platelet aggregation.

Total parenteral nutrition (TPN) Feeding the body by some means other than through the intestine.

Transposon mutants A mutation in which a purine/pyrimidine base pair is replaced by a pyrimidine/purine or *vice versa*, e.g. GC with TA.

Tympanostomy tube A tube inserted into the eardrum to facilitate drainage.

Urolithiasis Stone formation causing disease, e.g. kidney and bladder stones.

Van der Waals forces A general term for those forces of attraction between atoms or molecules that are not the result of chemical bond formation or simple ionic attraction; e.g. the relatively brief and weak interactions that neutral, chemically saturated molecules experience, such as dipole–dipole forces.

Vitronectin A protein that promotes adhesion; also called serum spreading factor.

Von Willebrand factor Necessary for the adhesion of platelets to vascular elements; a deficiency results in a prolonged bleeding time, as seen in von Willebrand's disease.

Xenobiotic A synthetic chemical compound that does not occur naturally.

1 Microbial Biofilms in Medicine

JANA JASS,[1] SUSANNE SURMAN[2] and JAMES T. WALKER[3]
[1]*Department of Microbiology and Immunology, University of Western Ontario and The Lawson Health Research Institute, London, ON, Canada*
[2]*London Food, Water & Environmental Microbiology Laboratory, Central Public Health Laboratory, London, UK*
[3]*CAMR, Porton Down, Salisbury, UK*

INTRODUCTION

Bacteria in nature do not generally grow in nutrient-rich suspensions as in the laboratory, but thrive on surfaces or interfaces (Bar-Or 1990). For many years, microorganisms have been viewed as simple life forms, growing as individual cells suspended in liquid when the required nutrients are present and surviving as dormant organisms or in spores when environmentally stressed. Although this view has been useful, to a limited extent, for both characterizing and studying bacteria, it is not their natural state of growth and care needs to be taken when extrapolating any results from such studies to growth in their natural state. After a number of early observations that microorganisms were found on surfaces, it became apparent that they prefer to grow in surface-associated communities or microcosms, now commonly called biofilms (ZoBell 1943; Geesey *et al.* 1978; Gristina *et al.* 1984; Costerton *et al.* 1987). Within these biofilms, bacterial species demonstrate cooperative behaviour (Kolenbrander 1993) and can subsequently differentiate further to exhibit complex multi-cellular behaviour (Shapiro 1998). Microorganisms may be susceptible to harsh environmental conditions, and growing within complex communities has been shown to offer protection (Rowbotham 1999).

The biofilms most often encountered include dental plaque and the slime on surfaces within both natural and man-made water systems, including domestic water supplies and drains. It is only recently that biofilms have been implicated in many medical conditions and infections (Gristina *et al.* 1984). With the increasing use of invasive medical procedures, infections involving biofilms form an important consideration as a risk factor for

Medical Biofilms. Edited by J. Jass, S. Surman, J. Walker

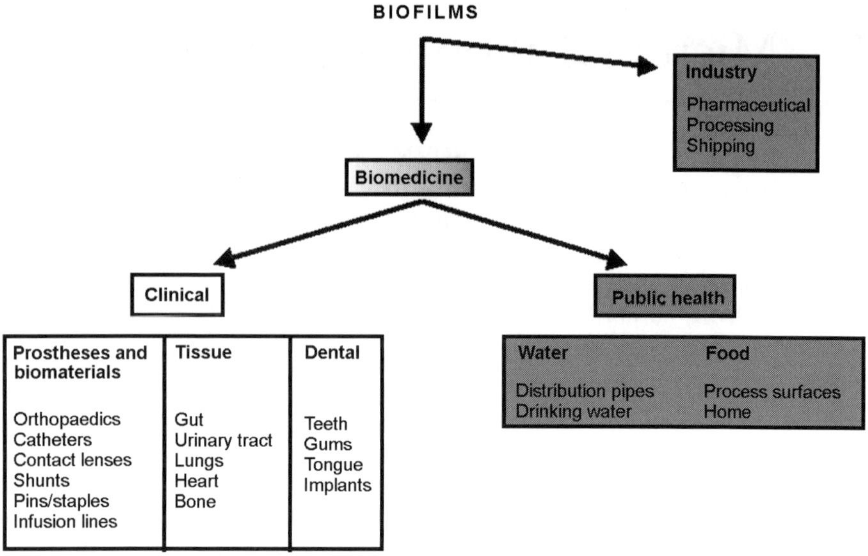

Figure 1.1. The many diverse environments that man is directly associated with that harbour biofilms.

complications postoperatively. Many persistent and chronic infections, such as endocarditis, osteomyelitis, periodontitis, otitis media and biliary tract infections, have also been attributed to bacterial biofilms (Costerton *et al.* 1999). The modern-day lifestyle provides more opportunity for biofilms to cause problems to mankind (Figure 1.1). Although we often only focus on the detrimental biofilms, commensal microorganisms within biofilms of the gut, skin and oral cavity can also have positive effects by inhibiting the colonization and establishment of pathogenic microorganisms, and so assist in maintaining health (Tannock 1994). Therefore, it is important for us to develop an understanding of biofilm formation to help us prevent and/or control its formation when desired, and subsequently remove detrimental biofilms. Here, we introduce what biofilms are and describe the current status of knowledge.

A BIOFILM DEFINITION

A number of working definitions have evolved over the years, with the increased understanding of biofilm structures and how they form. The definition must be broad, encompassing the numerous environmental and nutrient conditions and diverse microbial populations. One definition based on the biofilm morphological structure is:

> Complex communities of microorganisms attached to a surface or interface enclosed in an exopolysaccharide matrix of microbial and host origin to produce a spatially organized three-dimensional structure (Costerton *et al.* 1995).

This definition emphasizes the complexity of microbial composition and structure and may well describe the infections associated with implants, such as catheters and stents. However, a number of infections where bacteria form large aggregates on tissue surfaces have also been considered as biofilms, including *Pseudomonas aeruginosa* infections in cystic fibrosis (CF) lungs or microbial plaques on heart valves (Costerton *et al.* 1999). Although biofilms are generally perceived as a complex consortium of microorganisms attached to a surface or interface, it is difficult to be too prescriptive as, in the latter example, a biofilm may also consist of a monolayer or layers comprised of a single species. These microbial structures have been identified as biofilms based on phenotypic characteristics and properties. Consequently, the definition should also include the phenotypic variation created within a biofilm. A definition that includes the different phenotypic aspects between biofilm and planktonic bacteria may more accurately describe the important features without specifically defining all the physical properties that may vary between biofilm structures.

There are multitudes of research groups investigating biofilm structure and, as such, an actual consensus may be difficult to achieve. This reflects the varying methods used to investigate biofilm growth, and many factors have to be taken into consideration, such as types of cell and their growth phase, type and quantity of nutrients, type of substratum and its affect on the cells and nutrients. A simple definition may assist those who are new to the field, but the biofilm is complex and an agreed definition will fuel debate for many decades and may never be achieved.

BIOFILM STRUCTURE AND PHENOTYPE

There is a wide diversity of biofilm structures and architectures. These are influenced by the available physical (surface properties, pH, charge) and environmental conditions (temperature, humidity, etc.), nutrient and physiological status of the microorganisms and, certainly, microbial content (Stoodley *et al.* 1997). Regardless of this diversity, biofilm structure has a number of common features that have been used for identification. The primary and common features of a biofilm are illustrated in Figure 1.2 and include: a substratum to which the bacteria attach; a conditioning film; the biofilm matrix; and the liquid or gas phase. The substratum can be an abiotic material such as plastics (catheters and shunts), titanium metal

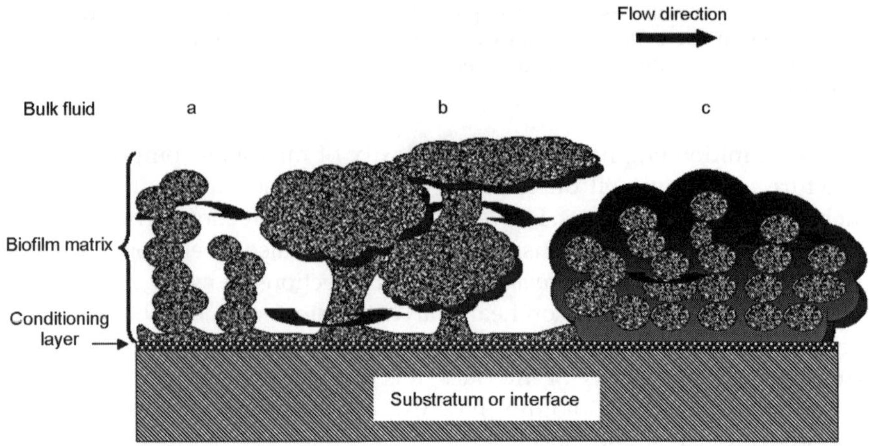

Figure 1.2. Illustration of the three structural variants of the biofilm matrix. (a) The heterogeneous mosaic is characterized by a basal layer and stacks of microcolonies extending up into the aqueous phase. (b) The porous biofilm is illustrated with mushroom-like structures interdispersed with water channels. (c) The dense-confluent biofilm appears more tightly packed, often containing multiple species of bacteria with regions of lower density that may act as transport channels within the biofilm.

alloys or ceramics (dental or orthopaedic implants) and hydrogels (contact lenses). Alternatively, the substratum can be of biotic origin, for instance tissue cell surfaces colonized by bacterial biofilms such as the lungs of CF patients, cardiac tissue in endocarditis and epithelial cells of the bladder in cystitis (Costerton *et al.* 1999).

In the first instance, a conditioning layer often composed of glycoproteins and lipids, is formed on the surface of any material placed into a liquid environment (Characklis 1990). For example, the pellicle on dental surfaces (Busscher *et al.* 1989; Bradshaw *et al.* 1997; Hannig 1999) and proteins from urine on catheter surfaces (Reid *et al.* 1998) create a conditioning film to which bacteria will subsequently adhere. The nature of the conditioning film is dependent on the substrate properties and chemical composition of the liquid medium, and this influences the mechanisms involved in early attachment events (Bradshaw *et al.* 1997; Shahal *et al.* 1998). The conditioning layer will influence which bacterial strains will act as primary colonizers and adhere to the surface first. This has been shown in the formation of dental plaque, where the pellicle (conditioning layer) aids in the attachment of primary colonizers (streptococci and actinomycetes) and these, through coadhesion, then adhere to other planktonic bacteria and subsequently build up a thick dental biofilm (Kolenbrander 1993; Bos *et al.* 1995).

The biofilm matrix is an important part of the biofilm, containing the microbial cells, exopolysaccharide (EPS) and water (Costerton *et al.* 1995; Sutherland 2001). The major component of the biofilm matrix is water and is believed to constitute approximately 95–99% of the biofilm (Costerton *et al.* 1995; Sutherland 2001). The microbial content is only approximately 2–5%, surrounded by EPS that may reach up to 2% of the total matrix (Sutherland 2001). Other substances often found in the biofilm matrix include DNA, RNA, proteins and enzymes reaching levels of approximately 2% in total.

The EPS is a highly hydrated, gel-like biopolymer that immobilizes bacteria creating the three-dimensional structures characteristic of biofilms and microbial aggregates (Flemming *et al.* 2000). The EPS composition is important not only for adhesion and biofilm matrix stabilization, but also for creating heterogeneity and increasing nutrient availability within the biofilm. The EPS contains microenvironments of local positive or negative charge and hydrophobicity in a non-uniform distribution, thus creating specialized niches within the biofilm (Flemming *et al.* 2000). In high-nutrient environments, it appears that microorganisms tend to increase their EPS production along with an increase in cell numbers, which may lead to denser biofilm structures. Under low-nutrient conditions, however, the biofilms are less dense and inter-dispersed with water channels, thus creating an opportunity for increased mass transfer from the bulk fluid (deBeer *et al.* 1994a; Stoodley *et al.* 1999). Stoodley *et al.* (1999) indicated that, when the flow velocity was increased at low-nutrient levels, a decrease in the biofilm density would follow, with the formation of streamers and ripple-like structures increasing the biofilm surface area. These structures were not present at higher nutrient concentrations. The heterogeneous environments created within the biofilm also aid in the growth and survival of diverse populations of microorganisms that arrange themselves for optimum nutrient exchange, waste product maintenance and microenvironment stability (Kolanbrander 1993; Møller *et al.* 1998).

Three structural variants of the biofilm matrix have been identified: the heterogeneous mosaic, the porous biofilm inter-dispersed with water channels and the dense-confluent biofilm (Figure 1.2). The heterogeneous mosaic is a biofilm where thin layers of bacteria form a dense basal layer of approximately 5 μm, from which microcolonies extend to form large pillars of up to 100 μm (Walker *et al.* 1995). These biofilms are believed to prevail in low-nutrient environments and may contain pathogens such as *Legionella pneumophila*, which form microcolonies within a diverse biofilm population (Rogers and Keevil 1992). Costerton and colleagues (Costerton *et al.* 1994) have visualized the hydrated biofilm using scanning confocal laser microscopy as being highly porous: mushroom-like structures are formed, whereby large microcolonies encompassed in EPS are attached to the surface by thinner EPS columns, thus creating water channels (Lawrence *et al.* 1991). This structure is believed to increase the mass transfer of materials and nutrients from the surrounding medium to the individual cells by having the bulk phase flow through channels within the biofilm matrix. Both of the structures described are based on biofilms formed by environmental microorganisms, predominantly Gram-negative bacteria such as *Pseudomonas* spp. in both mono- and mixed-populations. Mixed populations within biofilms may contain a very diverse consortium of microorganisms, including protozoan species, fungi and diatoms. The third type of architecture is called the dense-confluent biofilm, and is found in dental plaque. This type of biofilm is generally very thick, hosting a large number of different organisms existing in a cooperative environment with interdependence for nutrients between the different bacterial populations (Bradshaw *et al.* 1994; Kinniment *et al.* 1996). It may be suggested that such biofilms would have severe nutrient limitations; however, with this system of cooperative nutrient exchange between the different bacterial populations the biofilms can build up to become very thick and dense (Bradshaw *et al.* 1994).

The different structures described above are the result of a combination of physical factors, nutrient availability and population diversity. Although the three biofilm structures described are morphologically and structurally different, they do, in fact, share similarities, in that they each contain microorganisms, EPS and less-dense regions that may act as transport channels. To date, there is no formal structural description of medical biofilms. Bacteria within the body may be found both associated with biomaterials or on tissue surfaces, and the local environment will influence the structures formed. Biofilms on fully implanted devices exposed to blood will include a large portion of host clotting proteins and immune cells intended to isolate the infection from the rest of the body. Alternatively, biofilms on catheters and contact lenses will comprise of local host proteins and a few clotting factors, thus resulting in very different structures. In medical biofilms, it may be more appropriate to describe a biofilm by the microbial phenotypic and physiological properties rather than by the structure.

PROPERTIES OF A BIOFILM

Bacteria growing within biofilms have a number of properties that clearly distinguish them from planktonic populations. These include: protection, where the biofilm structure or phenotype protects the bacteria from different environmental conditions and substances; differences in phenotypic expression and growth characteristics; competition and exchange of nutrients affecting availability of nutrient acquisition; and microbial communication (Table 1.1).

Table 1.1. General features and advantages of microbial growth as a biofilm

Feature	Description	References
Protection	• From host defences and predators	Anwar *et al.* 1992; Jensen *et al.* 1993
	• From antimicrobial agents * slow growth rate * poor penetration * altered phenotype	Nickel *et al.* 1985; Allison *et al.* 2000; Brown *et al.* 1988; Gilbert *et al.* 1990; Chen and Stewart 1996; Suci *et al.* 1994; deBeer *et al.* 1994b; Huang *et al.* 1995; Brown and Williams 1985
	• From desiccation	Nielsen and Jahn 1997
	• From fluid hydrodynamic and mechanical forces	Sanin and Vesilind 1994
Nutrient acquisition	• Elevated concentrations of nutrients * surface phenomenon * nutrient trapping	Kjelleberg *et al.* 1982; Samuelsson and Kirchman 1990; Marshall 1996
	• Microbial and environmental heterogeneity for metabolic cooperation	Kinniment *et al.* 1996; Møller *et al.* 1998; Bradshaw *et al.* 1994; Morisaki 1983
	• Spatial heterogeneity to optimize transport of by-products and increase nutrient influx	Nielsen *et al.* 2000; Xu *et al.* 1998
New traits	• Phenotypic plasticity—novel gene expression and bacterial phenotype	Sauer *et al.* 2002; Davies *et al.* 1993; Prigent-Combaret *et al.* 1999; Otto *et al.* 2001
	• Plasmid or genetic transfer between organisms	Roberts *et al.* 1999
	• Mutation due to selection	Mathee *et al.* 1999
Intercellular communication	• Quorum sensing/density-dependent communication	Davies *et al.* 1998; Stickler *et al.* 1998
	• Interspecies communication	Riedel *et al.* 2001

Protection from the Environment

Growth within the biofilm matrix imparts protection to the individual cells from often extreme conditions in the surrounding environment (Stickler 1999). Cells growing within biofilms are able to evade the host immune system (Anwar *et al.* 1992; Jensen *et al.* 1993) and are frequently 1000 times more resistant to antimicrobial agents than are the planktonic cells (Nickel *et al.* 1985). The biofilm matrix itself may create a physical barrier to both exposure of the immunogenic epitopes and to the immune response. Researchers found that *P. aeruginosa* within a biofilm survived the action of normal human serum (Anwar *et al.* 1992) and had a lower activation of the complement cascade than the planktonic population (Jensen *et al.* 1990, 1993). Complement activation can be initiated by lipopolysaccharides (LPSs) of Gram-negative bacteria and the peptidoglycan of Gram-positive bacteria (Høiby *et al.* 1995). In biofilm-associated infections, complement activation would not only be from the planktonic bacteria shed at the infected site, but also from cell fragments and debris. The biofilm matrix then protects the majority of the cells from the polymorphonuclear leucocytes (PMNs) and antibodies. Biofilms are thus associated with chronic infections where there is an accumulation of PMNs, resulting in inflammation and ultimately leading to local tissue damage (Pedersen *et al.* 1992; Brown *et al.* 1995; Hull *et al.* 1997). A prime example of where a biofilm causes tissue damage is in the lungs of CF patients chronically infected with mucoid *P. aeruginosa*. This is due to the activation of the oxidative burst of the PMNs (Jensen *et al.* 1990) and the complement system (Jensen *et al.* 1993) resulting in tissue deterioration (Baltimore *et al.* 1989).

The mechanisms by which biofilm bacteria show increased resistance to antimicrobial agents are not fully understood (reviewed by Mah and O'Toole 2001). The biofilm matrix is thought to provide a physical barrier to some antimicrobial agents by reducing penetration into the biofilm, however, this alone cannot explain the high level of resistance often observed (Suci *et al.* 1994; deBeer *et al.* 1994b; Chen and Stewart 1996). The EPS is a polyanionic matrix that may be able to bind cationic compounds, such as aminoglycosides (Hoyle *et al.* 1992; Huang *et al.* 1995). Interestingly, resistance is not limited to positively charged antimicrobial agents, therefore, other mechanisms must be involved. An increased concentration of antibiotic degrading enzymes, such as β-lactamases, may aid in resistance to some penicillins (Giwercman *et al.* 1990). However, it appears that the matrix may only delay the transport of antibiotics to the cells rather than prevent their access (McBain and Gilbert 2001). Alternatively, others have suggested that free anions are impeded from entering and moving through the biofilm matrix (Chamberlain 1997).

An important factor is that microbial cells within a biofilm are slower growing and phenotypically different from planktonic cells (Brown *et al.* 1988; Gilbert *et al.* 1990). The biofilm matrix contains a multitude of different microenvironments where the bacteria may experience nutrient gradients and waste by-products that may generate resistant phenotypes and slow growth rates (Brown and Williams 1985; Evans *et al.* 1990). Furthermore, it is established that slow growth or no growth produces an increased resistance to some antimicrobial compounds, and biofilms contain a range of growth rates and physiological states (Evans *et al.* 1990; Lewis 2001). Decreased growth rate is often due to nutrient limitation that may lead to an early onset of the general stress response. *RpoS* encodes for the σ^{s}, which regulates a number of genes required for survival in stationary phase in *Escherichia coli*. Bacteria lacking *rpoS* were unable to form biofilms and thus were unable to gain the protective nature of the biofilm (Adams and McLean 1999). Biofilm formation may select for biofilm-specific phenotypes that are also antibiotic resistant (Allison *et al.* 2000). Bacteria within biofilms experience nutrient and gas gradients (oxygen and carbon dioxide) and heterogeneous local microenvironments (Dillon and Fauci 2000) that can result in changes of gene expression and physiology, and alter membrane permeability to antimicrobial substances. These gradients may also select for resistant clones; for example, only a single point mutation can render *Enterobacteriaceae* resistant to triclosan (McMurray *et al.* 1998). Furthermore, multi-drug-resistant operons, such as the *mar* and efflux pumps (*acrAB*), may contribute towards bacterial protection. Constitutive production of the *acrAB* genes provided low levels of protection against ciprofloxacin and the expression of these genes was inversely affected by growth rate, therefore, together they may add to the protective nature of the biofilm phenotype (Maira-Litran *et al.* 2000).

The EPS has been shown to have other important protective functions. The EPS matrix is primarily composed of water that is tightly bound within the matrix and protects the cells from rapid desiccation (Nielsen and Jahn 1997). Water bound in such a manner is often more difficult to release, and thus evaporates more slowly. The EPS and other adhesins hold the biofilm matrix in place, immobilizing the organisms sufficiently to prevent them from being washed out. Thus, biofilm bacteria are able to persist under severe hydrodynamic conditions and host clearance mechanisms (Sanin and Vesilind 1994). It also protects the cells within the biofilm from UV light, pH fluctuations and oxidative and osmotic shock (Flemming 1991).

Nutrient Acquisition

The EPS matrix has been described as an anionic exchange column that may trap cationic substances. This ionic nature of the EPS has been shown to bind metal cations and enzymes (Kepkay *et al.* 1986; Ferris *et al.* 1987). The physical property of the EPS matrix is a highly hydrated gel-like material that can physically trap particles, and some of these may be a nutrient source to the resident bacteria.

The biofilm architecture, with its water channels, provides a mode of transport for soluble nutrients into the inner regions of the biofilm. Similarly, metabolites and waste products can be transported out of the biofilm matrix. This is of particular consequence in mono-species biofilms or those consisting of a few species. The more dense biofilms, such as dental plaque, have additional means by which nutrients are made available and waste products are removed. These dense biofilm structures do not appear to have the open water channels and pores as in monoculture biofilms, however, there are regions of less dense material and a very dynamic population of microorganisms. The cooperation between organisms in mixed populations in utilizing nutrients is of primary importance (Bradshaw *et al.* 1994), however, this is more than just sharing nutrients, this also prevents the build up of toxic by-products as neighbouring organisms utilize them as nutrient sources or mineralize them. The biofilm matrix thus provides a unique environment where cooperative populations of bacteria with differing growth requirements (nutrient, mineral and oxygen concentrations) can be maintained in close proximity to each other. In turn, the nutrient gradients created within these dense biofilms lead to microenvironments supporting a diverse microbial community (Xu *et al.* 1998; Nielsen *et al.* 2000). For example, anaerobic bacteria are able to survive in aerated situations within biofilms as a result of the gradient formed where the oxygen is rapidly used in the uppermost regions of the biofilm creating anoxic conditions at the base (deBeer *et al.* 1994a). Additionally, anaerobic organisms can also closely associate themselves with aerobic bacteria that quickly use up the oxygen, thereby creating a local anoxic microenvironment (Kinniment *et al.* 1996). Another example of community cooperation within the biofilm environment is within the gut. Here, one group of microorganisms can degrade complex compounds that can then serve as nutrients for another group of microorganisms or the host.

Phenotypic Variation

Biofilms are characterized by their heterogeneous microbial populations that rapidly adapt to new environments and by exhibiting a wide range of phenotypes. Biofilm bacteria may acquire new traits by either attaining a different phenotype within the biofilm due to heterogeneous growth conditions (Prigent-Combaret *et al.* 1999; Sauer *et al.* 2002) or at the genetic level by gene exchange or mutations (Mathee *et al.* 1999). Phenotypic plasticity, or the ability of bacteria to alter their phenotype in response to their immediate surroundings, is understood to occur in biofilms. That is, cells growing within a biofilm express different genes and are therefore phenotypically different from planktonic cells that grow in homogeneous environments (Prigent-Combaret *et al.* 1999; Sauer *et al.* 2002). The diversity of growth conditions during the stages of biofilm development result in multiple phenotypic expressions of traits required for survival (Sauer *et al.* 2002). Sauer *et al.* (2002) have determined that more than 800 proteins have altered expression from the corresponding planktonic population; greater than 50% of the proteome. Furthermore, Prigent-Combaret *et al.* (1999) found altered transcription of 38% of the genes following attachment of *E. coli*.

Alternatively, bacteria can survive the different conditions by genetic mutation of specific genes. Mathee *et al.* (1999) showed that exposure of *P. aeruginosa* to H_2O_2 induced a mutation in the *mucA* gene, resulting in a mucoid phenotype. This mucoid variant is often isolated from CF lungs and is caused by oxygen radicals released by PMNs (Mathee *et al.* 1999). Additional changes as a result of this mutation include a decrease in elastase activity, LasA protease synthesis and slight reduction in β-lactamase production (Mathee *et al.* 1999).

Horizontal gene transfer between bacteria is another way that bacteria can attain new traits. Genetic transfer within biofilms is less understood and has been investigated with differing results. In some cases, it is believed that the bacteria are held at a distance that prevents conjugation and plasmid transfer (Hausner and Wërtz 1999). Alternative theories are that the bacteria are held close to each other, thus favouring conjugation, or that bacteria may be able to move within the biofilms to overcome this distance and to conjugate under specific conditions. With the extensive use of antibiotics and the current emergence of multi-resistant microorganisms, plasmid transfer within biofilms has become a major concern. Roberts *et al.* (1999) investigated gene transfer in dental biofilms by forming a *Streptococcus* biofilm *in vitro* and introduced a *Bacillus subtilis* possessing a tetracycline-resistance conjugative transposon. Subsequently, tetracycline-resistant *Streptococcus* were isolated, indicating that genetic transfer between unrelated species was possible and, furthermore, that it was possible in organisms commonly associated with humans (Roberts *et al.* 1999).

Intercellular Communication

Under some environmental conditions, bacteria are able to display a collective response to the environment and demonstrate the same behaviour, which is indicative of communication among the population individuals (Riedel *et al.* 2001). One form of communication is cell-density-dependent signalling, otherwise called 'quorum-sensing' (Fuqua *et al.* 1994; de Kievit and Iglewski 2000). This type of communication was first observed in *Vibrio fischeri*, where the bacterium fluoresces when the population reaches a critical mass in the light organ of a marine fish. The signals were identified as *N*-acylhomoserine lactones (HSLs), small diffusible organic molecules produced at basal levels at low population densities and that accumulate at high cell densities. These signalling molecules, called autoinducers, monitor cell density and, at critical concentrations, induce or repress target genes. Many Gram-negative bacteria use HSL as a signalling molecule, whereas others have different molecules that have yet to be identified (Surette and Bassler 1998). Gram-positive organisms use post-transcriptionally processed peptides or γ-butyrolactones (Kleerebezem *et al.* 1997).

Quorum-sensing signals are important in coordinating multicellular behaviour in bacteria, and they regulate a number of physiological processes, including swarming, bioluminescence, antibiotic synthesis, conjugated plasmid transfer and the expression of virulence factors; see Van Delden and Iglewski (1998) and de Kievit and Iglewski (2000) for reviews. Two quorum sensing systems have been identified in *P. aeruginosa*, the *las* and *rhl* systems arranged in a cascade, where the *las* products positively regulate the *rhl* genes. In addition, they regulate the expression of exoenzymes (elastase and alkaline protease), secondary metabolites (pyocyanin, hydrogen cyanide and pyoverdin) and toxins (exotoxin A). Studies using animal models have demonstrated that bacterial mutants defective in quorum-sensing are less virulent (Tang *et al.* 1996; Rumbaugh *et al.* 1999). A number of different studies have shown that autoinducers are indeed produced *in vivo* and may be associated with biofilm-related infections (McLean *et al.* 1997; Stickler *et al.* 1998) and control the expression of some virulence factors *in vivo* (Erickson *et al.* 2002). Furthermore, Riedel *et al.* (2001) revealed unidirectional signalling between *P. aeruginosa* and *Burkholderia cepacia* during co-infection within a biofilm and mouse lung model.

The fact that bacteria do not respond to the autoinducer signals at low densities suggests that they are not important in the initial stages of biofilm formation but are in the later stages of biofilm formation and differentiation (Davies *et al.* 1998). Davies *et al.* (1998) observed that mutants of *P. aeruginosa* lacking *lasI*, the gene involved in the synthesis of the long-chain homoserine lactone, produced a dense thin biofilm that was only 20% of the thickness of the wild-type biofilm. The biofilm structure was recovered to the level of the wild type with the addition of the homoserine lactone, suggesting that cell–cell communication is involved in later stages of biofilm formation and maturation (Davies *et al.* 1998).

BIOFILM FORMATION

Biofilm formation is a continual dynamic sequence of events that has been divided into distinct developmental stages. As illustrated in Figure 1.3(a), we have generally divided biofilm formation into four developmental stages, finally returning to the planktonic stage in a cyclical scheme. The first stage is bacterial growth as planktonic cells. These are then transported to a surface or interface, leading to the second stage, where the bacteria become associated with a conditioned surface and form a monolayer. During this stage, the bacteria initially attach to the surface in a reversible manner, so that they can easily detach and move along the surface. This surface-associated motility has been visualized by O'Toole and Kolter (1998) using time-lapse phase-contrast microscopy of *P. aeruginosa* biofilms,

(a) **Biofilm formation**

Structure	Planktonic →	Monolayer →	Microcolonies →	Biofilm →	Planktonic
Activity	Suspended in bulk fluid	Association & reversible attachment	Irreversible attachment & cell division	Cell division, recruitment & 3-D structure formation	Dispersal & detachment

(b) **Bacterial features required for the progression of biofilm formation**

	→	→	→	→
Gram-negative bacteria	***Motility & surface adhesins***	***Motility EPS***	***EPS***	***EPS destabilizing compounds***
P. aeruginosa	Flagella LPS OMPs	Type IV pili Flagella Alginate Crc	Alginate Acyl-HSLs	Alginate lyase ***Acyl-HSL analogues***
E. coli	Flagella Type I pili Ag43	Flagella	Colanic acid	Altered surface properties at cell division ?
Gram-positive bacteria	***Adhesins Surface & extracellular proteins***	***Adhesins EPS***	***Adhesins EPS***	***Surface releasing enzymes***
S. epidermidis	Surface & extracellular proteins PS/A AtlE	PIA ?	PIA ?	Altered surface properties at cell division ?

Figure 1.3. Biofilm formation has been divided into distinct structures that require specific events and bacterial properties. (a) Planktonic bacteria become associated with a surface, adhere and begin to divide to form microcolonies. Once attached, the bacteria divide and produce EPS, which helps to cement the biofilm matrix together to create the characteristic three-dimensional structure. The biofilm expands until the growth and attachment equals the death and detachment thus called the mature biofilm. Environmental or genetic signals may be presented for cells to detach from the biofilm and return to the planktonic state. (b) The genetic requirements for biofilm formation are listed for *P. aeruginosa*, *E. coli* and *Staphylococcus epidermidis*. Many aspects of biofilm formation and detachment are still unanswered and are identified by a question mark. For further detail, see the text. (Constructed from information in Davey and O'Toole (2000) and O'Toole *et al.* (2000a)).

demonstrating the formation and dispersal of microcolonies. Eventually the bacteria become irreversibly attached and form microcolonies in the third stage of biofilm formation, through specific (adhesins) and non-specific interactions (hydrogen bonds, van der Waals forces and hydrophobic interactions) with the surface (Characklis 1990; Busscher and Van der Mei 1997). For mature biofilm formation, the fourth stage, EPS is essential for irreversible attachment and the development of the three-dimensional structure characteristic of the mature biofilm. Finally, the bacteria eventually return to the planktonic phase through dispersal and detachment from the biofilm. Though this is not a developmental stage of biofilm formation, it is important in maintenance of the mature biofilm and bacterial growth.

The different stages of biofilm formation have been described for Gram-negative species such as *P. aeruginosa*, *Pseudomonas fluorescens*, *E. coli* and *Vibrio cholerae*, however, less is known about the biofilm forming processes for Gram-positive bacteria (for reviews, see Davey and O'Toole 2000; O'Toole *et al.* 2000a). This is of particular interest, as the majority of implant-related infections are associated with Gram-positive strains, predominantly coagulase negative staphylococcus, *Staphylococcus aureus* and enterococcus. Though each stage is characterized by bacterial activity that, in part, dictates the structural features, the bacteria must be able to express particular genes to be able to progress to the next biofilm developmental stage. Some of these genetic requirements, in addition to other factors that influence biofilm formation, such as environmental and physical conditions, have been identified for a limited number of bacterial species (Stoodley *et al.* 1999; Davey and O'Toole 2000).

Environmental Factors Influencing Biofilm Formation

Environmental factors, including nutrient sources and local conditions such as pH, osmolarity, temperature, oxygen, surface properties and hydrodynamic conditions, greatly influence what species will be able to colonize to form biofilms and the maximum biofilm thickness and density (Fletcher and Pringle 1986; van Loosdrecht *et al.* 1995; Stoodley *et al.* 1999). Different nutrients and environmental conditions influence biofilm formation by signalling the bacteria to express different adhesins and EPS (Davies *et al.* 1993; Fletcher 1996). It is expected that diverse conditions are required for different organisms; there is also substantial diversity in requirements between strains of the same bacterium. For instance, *E. coli* O157:H7 form biofilms under low nutrient or starvation conditions (Dewanti and Wong 1995), some strains of enteroaggregative *E. coli* require high sugar and osmolarity (Sheikh *et al.* 2001), while K12 strains of *E. coli* require minimal medium supplemented with amino acids for biofilm formation (Pratt and

Kolter 1998). Alternatively, *P. aeruginosa*, the principal organism used for biofilm studies, and *P. fluorescens* will form biofilms under most nutrient and environmental conditions (O'Toole and Kolter 1998). Furthermore, Kolter's laboratory has identified that bacteria initiate biofilm formation through different genetic pathways depending on their environmental conditions; therefore, a single strain can achieve biofilm phenotype under different conditions by different mechanisms (O'Toole *et al.* 2000a). Using mutagenesis, they created a set of *sad* (surface attachment defective) mutants of *P. fluorescens* that were unable to form biofilms, however, biofilm formation was restored in some mutants by switching carbon sources from minimal medium and glucose to citrate or glutamate (Pratt and Kolter 1998).

Additional factors that affect biofilm formation are the physical conditions, such as hydrodynamics and surface physico-chemical characteristics (Stoodley *et al.* 1999, 2000). Consequently, biofilms will differ between catheter-associated infections that undergo intermittent urine flow, infections of orthopaedic prosthesis without strong liquid forces and those in the mouth that are continuously compacted by chewing. All of these affect the biofilm structure and formation: the stronger the forces are that are placed on the biofilm during development, the more adherent the initial colonizers must be, and these factors then limit the size and constitution of the biofilm (Characklis 1990; Brading *et al.* 1995). Physical surfaces also play a role in the formation of biofilms. Surface roughness, in particular the scale of surface topographical features, is the most important physico-chemical surface characteristic determining the distribution of the microflora (Quirynen and Bollen 1995). In the oral cavity, rough surfaces and stagnation will promote plaque formation and maturation, and high-energy surfaces are known to collect more plaque, to bind the plaque more strongly and to select for specific bacteria (Quirynen and Bollen 1995). Although both variables interact with each other, the influence of surface roughness overrules that of the surface free energy. However, the influence of surface roughness and surface free energy on supragingival plaque justifies the demand for smooth surfaces with a low surface free energy in order to minimize plaque formation, thereby reducing the occurrence of caries and periodontitis (Quirynen and Bollen 1995). Whilst studying urinary catheters, Brisset *et al.* (1996) provided evidence that the more hydrophobic the bacteria, the more they were able to colonize hydrophobic materials, whereas hydrophilic cells were able to colonize hydrophilic materials more easily.

Genetic Requirements for Biofilm Formation

Genetic requirements for biofilm formation have been identified for monocultures of different organisms. Figure 1.3(b) summarizes some of the requirements for selected bacteria. More detailed characterization of the genetic requirements for Gram-negative bacteria have been established, since only recently have Gram-positive strains been studied genetically for biofilm formation.

The primary requirements for initial attachment and early biofilm formation events of Gram-negative bacteria involve motility and adhesins. Initial attachment of *P. aeruginosa* is mediated by the presence of flagella involved in bacterial transport and Type IV pili involved in twitching motility along the surface (Ottemann and Miller 1997; O'Toole and Kolter 1998). Additionally, LPS has been found to decrease adhesion to hydrophilic surfaces and increase attachment to hydrophobic surfaces (Makin and Beveridge 1996). Adhesins and specific receptors necessary for *P. aeruginosa* adhesion to tissue surfaces may also alter the cell surface characteristics to mediate binding to biomedical surfaces (Prince 1992). O'Toole *et al.* (2000b) have observed that the catabolite repression control locus (*crc*) is important in biofilm formation not only by regulating the biogenesis of Type IV pilus but also possibly through recognition of environmental signals such as carbon availability. As biofilm formation progresses towards a three-dimensional structure, an increased synthesis of EPS is observed (Davies *et al.* 1993; Danese *et al.* 2000). In *P. aeruginosa*, an up-regulation of alginate synthesis and a down-regulation of flagella genes was observed upon adhesion to a surface (Davies *et al.* 1993; Garrett *et al.* 1999).

The genes involved at the different stages of Gram-positive biofilm formation are less well understood and believed to involve only two steps: adhesion to the surface followed by cell–cell adhesion (Cramton *et al.* 1999). This suggests that the same genes required for a mature biofilm are also required for the formation of microcolonies. The *ica* locus found in *S. epidermidis* and *S. aureus* encodes for two adhesin molecules: a capsular polysaccharide, PS/A, that mediates adhesion to biomaterial surfaces, and a polymer of <30 000 kDa, polysaccharide intercellular adhesin (PIA), that is involved in cell aggregation (McKenney *et al.* 1998; Ziebuhr *et al.* 1999). PIA is important in cell aggregation that contributes to the formation of the three-dimensional biofilm structure. However, the surfaces of most implanted materials are rapidly coated with host plasma molecules, such as fibrinogen/fibrin and blood clotting proteins, that act as receptors to specific adhesins (Vaudaux *et al.* 1994). The clumping factors ClfA and ClfB, in *S. aureus*, mediate binding to fibrinogen-coated haemodialysis tubing (Eidhin *et al.* 1998). Alternatively, *S. aureus* may adhere directly to tissue surfaces, as in endocarditis and osteomyelitis, in which case collagen adhesion is an important virulence factor that mediates biofilm formation (Patti *et al.* 1994). The *S. epidermidis AtlE* gene encodes for a cell surface autolysin that has vitronectin binding activity and will promote specific adhesion to implant and tissue surfaces (Heilmann *et al.* 1997).

Detachment is an important part of biofilm maintenance and has always been identified as a stage in biofilm formation, however, it is the least understood phenomenon. Detachment occurs through chemical destabilization of the biofilm matrix or hydrodynamic shear forces (Moore *et al.* 2000). Physical shear forces are often associated with the detachment of large aggregates from a biofilm through high liquid velocity and particle concentration or frequent changes in fluid velocity (Chang *et al.* 1991; Stoodley *et al.* 1999). Chemical or biological factors affecting detachment from a biofilm can be caused by enzymatic or chemical destabilization of the matrix so that the bacterial cells leave as planktonic bacteria. Allison *et al.* (1990) suggested that cell division has a role in releasing planktonic cells from a biofilm through altered bacterial surface characteristics, and this could be achieved by changes in nutrient availability (Stoodley *et al.* 1999). Sawyer and Hermanocicz (1998) observed that *Aeromonas hydrophila* exhibited increased cell detachment under nutrient-limiting conditions which could also be a result of changes to bacterial surface properties (Allison *et al.* 1990; Gilbert *et al.* 1991). A number of organisms also produce enzymes that degrade the EPS matrix, thus releasing bacteria from the biofilm (Boyd and Chakrabarty 1994; Allison *et al.* 1998). *P. aeruginosa* produces alginate lyase (Boyd and Chakrabarty 1994); similarly, *P. fluorescens* produces a lyase under starvation conditions (Allison *et al.* 1998), and *Streptococcus mutans* produces a protein-releasing enzyme that enhances cell detachment (Lee *et al.* 1996). Quorum-sensing signals have also been implicated in cell detachment from a biofilm (Lynch *et al.* 1999) and have therapeutic potential. To move into the planktonic phenotype, the individual bacteria must also receive signals that activate genes for this mode of growth. Regulation of this aspect of detachment is still unknown.

MIXED-CULTURE BIOFILMS

Clinical infections are the only environments where monoculture biofilms are found; most other environments are open systems where multi-species biofilms exist. However, for developing an understanding of the processes by which bacteria tend to form biofilms, monoculture systems have been useful (Kuehn *et al.* 1998). Mixed-culture biofilms, such as those found in natural environments or in the mouth and gut, proceed through additional developmental stages and can contain diverse microbial populations, including bacteria, yeast and fungi, viruses, diatoms and protozoa (Shu *et al.* 2000). Clinical biofilms, however, primarily contain diverse bacterial species and yeast, specifically *Candida albicans* (Chandra *et al.* 2001; Sbarbati *et al.* 2001). The best-described and characterized mixed population biofilm is dental plaque, where biofilm formation occurs through a build up of different bacterial communities (Kolenbrander 2000). Though the primary colonizers follow the initial adhesion stages described for monoculture biofilms, the subsequent stages involve recruitment to the biofilm through coaggregation and coadhesion with other organisms (Kolenbrander 2000; Marsh and Bowden 2000). Coaggregation, where bacteria bind to each other in suspension, and coadhesion, where cell–cell interactions are between an attached organism and a planktonic cell, are sensitive to environmental conditions of pH, temperature, ionic strength and divalent cation concentrations (Busscher and van der Mei 2000). The primary colonizers are able to attach to the pellicle-coated surface through specific receptor-mediated binding and this precedes the coadhesion of the bridging organisms that mediate the binding of the primary colonizers and late colonizers (Kolenbrander 2000; Marsh and Bowden 2000). Thus, thick biofilm communities are formed with stratification of bacterial species based on physiological cooperation and the establishment of microenvironments.

The mature biofilm is highly dynamic and continually undergoes reorganization in response to nutrient availability and the production of essential growth factors modifying local environments. Bradshaw *et al.* (1998) investigated the coaggregation and coadhesion of *Fusobacterium nucleatum* in an *in vitro* mixed-culture population containing obligate anaerobes and oxygen-tolerant species. *F. nucleatum* is able to coaggregate with both anaerobic and oxygen-tolerant bacterial species that would otherwise not be associated with each other. Interestingly, they observed that coaggregation and survival of anaerobic species in an oxygen-rich environment were enhanced in the presence of *F. nucleatum*, probably due to the removal of oxygen in the immediate surroundings by the aerobic bacteria so that the anaerobes were able to grow (Bradshaw *et al.* 1998).

Mixed bacterial populations have the potential for pathogenic synergism: individually, the bacteria may not be pathogenic, however, together they cause disease (Baumgartner *et al.* 1992). Cooperation may be through the production of essential nutrients, as in the production of succinate by *Klebsiella pneumoniae* for the growth of *Bacteroides asaccharolyticus* (Mayrand and McBride 1980). Cooperation may be through protection from antibiotics, where an antibiotic-sensitive pathogen may reside in a mixed population with a strain that produces β-lactamase, preventing its eradication through antibiotic treatment (Brook 1989). Determining the genetic requirements of mixed-culture biofilms is complex, and it is only recently that the molecular techniques have become available to achieve this.

CLINICAL BIOFILMS

Clinical biofilms have been primarily regarded as implant-related infections, however, clinicians have realized that tissue-associated infections are also caused by bacteria growing as biofilms. That is, they contain microcolonies of bacteria enclosed in an EPS matrix and are resistant to a normal course of antibiotic treatment. It is estimated that approximately 65% of all bacterial infections in humans are caused by biofilms (Costerton and Stewart 2000). These include many nosocomial infections associated with implants, which are often caused by *S. epidermidis* and coagulase-negative staphylococcus, *S. aureus* and Enterococci and less frequently by *E. coli, P. aeruginosa* and *C. albicans* (Costerton *et al.* 1999). Infections caused by biofilms not associated with implants include dental caries, periodontitis, otitis media, biliary tract infections, osteomyelitis, native valve endocarditis and prostatitis (Costerton *et al.* 1999). These infections are often caused by a variety of different organisms that are either translocated into a normally sterile site or the person is immunocompromised by a predisposing genetic factor or disease. Medical biofilm formation may indeed proceed through different developmental stages and genetic requirements. Consequently, many *in vitro* characterizations and studies must be validated through *in situ* and *in vivo* studies. This contrast is most obvious in antimicrobial sensitivity testing, where biofilm cells cannot be eradicated with the same concentration of antibiotics as log-phase planktonic cells, the standard protocols for clinical testing (Nickel *et al.* 1985).

In a world where there is an increasing number of immunocompromised, elderly and long-term hospitalized patients, there is an ever-increasing likelihood of clinical infections or hospital-acquired infection (Marone *et al.* 1992). Of particular concern are multiple-antibiotic-resistant organisms that appear to be hospital specific, in that they are common in hospitals but only rarely encountered in infections in the community (Abudu *et al.* 2001). Predominantly surface associated, these nosocomial biofilm infections create a huge economic burden. It has been calculated that the socio-economic burden of hospital-acquired infections has created an additional cost to the health system in England and Wales of £986 million annually (Plowman *et al.* 1999), and similarly in the rest of the developed world. Hospital-acquired bacteraemia rates vary between specialities, with the higher rates occurring in intensive-care units, haematology and oncology units; nearly half of all isolates are staphylococci, and 50% of the *S. aureus* are methicillin-resistant *S. aureus* (MRSA).

Bacteria are able to display multi-cellular properties through the formation of biofilms, thus, some have compared a biofilm to the prokaryotic version of multi-cellular organisms (Shapiro 1998). These multi-cellular properties include special organization, metabolic cooperation and cell communication (Marsh and Bowden, 2000). Thus, biofilm microbial communities coordinate their activities to provide extraordinary advantages for survival in extreme environments. The extent to which biofilms are involved in medicine is now becoming clear, and vigorous investigations into detection, prevention and control of clinical biofilms are under way. Although significant progress has been made towards these goals, novel approaches to control or prevention of biofilm-associated infections are crucial.

REFERENCES

Abudu, L, Blair, I, Fraise, A and Cheng, KK (2001) Methicillin-resistant *Staphylococcus aureus* (MRSA): a community-based prevalence survey. *Epidemiology and Infection* **126**:351–356.

Adams, JL and McLean, RJ (1999) Impact of *rpo*S deletion on *Escherichia coli* biofilms. *Applied and Environmental Microbiology* **65**:4285–4287.

Allison, DG, Evans, DJ, Brown, MRW and Gilbert, P (1990) Possible involvement of the division cycle in dispersal of *Escherichia coli* from biofilms. *Journal of Bacteriology* **172**:1667–1669.

Allison, DG, Ruiz, B, Sanjose, C, Jaspe, A and Gilbert, P (1998) Extracellular products as mediators of the formation and detachment of *Pseudomonas fluorescens* biofilms. *FEMS Microbiology Letters* **167**:179–184.

Allison, DG, McBain, AJ and Gilbert, P (2000) Biofilms: problems of control. In: Community Structure and Co-operation in Biofilms (Eds. Allison, D, Gilbert, P, Lappin-Scott, HM and Wilson, M), Cambridge University Press, Cambridge, pp. 309–327.

Anwar, H, Strap, JL and Costeron, JW (1992) Susceptibility of biofilm cells of *Pseudomonas aeruginosa* to bactericidal action of whole blood and serum. *FEMS Microbiology Letters* **92**:235–242.

Baltimore, RS, Christie, CD and Smith, GJ (1989) Immunohistopathologic localization of *Pseudomonas aeruginosa* in lungs from patients with cystic fibrosis. Implications for the pathogenesis of progressive lung deterioration. *American Review of Respiratory Diseases* **140**:1650–1661.

Bar-Or, Y (1990) The effects of adhesion on survival and growth of microorganisms. *Experientia* **46**:823–826.

Baumgartner, JC, Falkler, WA and Beckerman, T (1992) Experimentally induced infection by oral anaerobic microorganisms in a mouse model. *Oral Microbiology and Immunology* **7**:253–256.

Bos, R, van der Mei, HC and Busscher, HJ (1995) A quantitative method to study co-adhesion of microorganisms in a parallel plate flow chamber. II: analysis of the kinetics of co-adhesion. *Journal of Microbiological Methods* **23**:169–182.

Boyd, A and Chakrabarty, AM (1994) Role of alginate lyase in cell detachment of *Pseudomonas aeruginosa*. *Applied and Environmental Microbiology* **60**:2355–2359.

Brading, MG, Boyle, JD and Lappin-Scott, HM (1995) Biofilm formation under laminar flow using *Pseudomonas fluorescens* EX101. *Journal of Industrial Microbiology* **15**:297–304.

Bradshaw, DJ, Homer, KA, Marsh, PD and Beighton, D (1994) Metabolic cooperation in oral microbial communities during growth on mucin. *Microbiology* **140**:3407–3412.

Bradshaw, DJ, Homer, KA, Marsh, PD and Beighton, D (1997) Effects of conditioning films on oral microbial development. *Biofouling* **11**:217–226.

Bradshaw, DJ, Marsh, PD, Watson, GK and Allison, C (1998) Role of *Fusobacterium nucleatum* and coaggregation in anaerobic survival in planktonic and biofilm oral microbial communities during aeration. *Infection and Immunity* **66**:4729–4732.

Brisset, L, Vernet-Garnier, V, Carquin, J, Burde, A, Flament, JB and Choisy, C (1996) *In vivo* and *in vitro* analysis of the ability of urinary catheter to microbial colonization. *Pathology and Biology (Paris)* **44**:397–404.

Brook, I (1989) Direct and indirect pathogenicity of beta-lactamase-producing bacteria in mixed infection in children. *Critical Reviews in Microbiology* **16**:161–180.

Brown, MRW and Williams, P (1985) The influence of environment on envelope properties affecting survival of bacteria in infections. *Annual Reviews of Microbiology* **39**:527–556.

Brown, MRW, Allison, DG and Gilbert, P (1988) Resistance of bacterial biofilms to antibiotics: a growth-rate related effect. *Journal of Antimicrobial Chemotherapy* **22**:777–780.

Brown, RK, McBurney, A, Lunec, J and Kelly, FJ (1995) Oxidative damage to DNA in patients with cystic fibrosis. *Free Radicals in Biology and Medicine* **18**:801–806.

Busscher, HJ and van der Mei, HC (1997) Physico-chemical interactions in initial microbial adhesion and relevance for biofilm formation. *Advances in Dental Research* **11**:24–32.

Busscher, HJ and van der Mei, HC (2000) Initial microbial adhesion events: mechanisms and implications. In: Community Structure and Co-operation in Biofilms (Eds. Allison, DG, Gilbert, P, Lappin-Scott, HM and Wilson, M), Cambridge University Press, Cambridge. pp. 25–36.

Busscher, HJ, Uyen, HMW, Stokroos, I and Jongebloed, WL (1989) A transmission electron microscopy study of the adsorption patterns of early developing artificial pellicles on human enamel. *Archives of Oral Biology* **34**:803–810.

Chamberlain, AHL (1997) Matrix polymers: the key to biofilm processes. In: Biofilms, Community Interactions and Control (Eds. Wimpenny, J, Handley, P, Gilbert, P, Lappin-Scott, H and Jones, M), BioLine, Chippenham, pp. 41–46.

Chandra, J, Kuhn, DM, Mukherjee, PK, Hoyer, LL, McCormick, T and Ghannoum, MA (2001) Biofilm formation by the fungal pathogen *Candida albicans*: development, architecture, and drug resistance. *Journal of Bacteriology* **183**:5385–5394.

Chang, HT, Rittman, BE, Amar, D, Heim, R, Ehkinger, O and Lesty, Y (1991) Biofilm detachment mechanisms in a liquid-fluidized bed. *Biotechnology and Bioengineering* **38**:499–506.

Characklis, WG (1990) Biofilm process. In: Biofilms (Eds. Characklis, WG and Marshall, KC), Wiley, New York, pp. 195–231.

Chen, X and Stewart, PS (1996) Chlorine penetration into artificial biofilm is limited by a reaction–diffusion interaction. *Environmental Science and Technology* **30**:2078–2083.

Costerton, JW and Stewart, PS (2000) Biofilm and device-related infections. In: Persistent Bacterial Infections (Eds. Nataro, JP, Blaser, MJ and Cunningham-Rundles, S), ASM Press, Washington, pp. 423–439.

Costerton, JW, Cheng, K-J, Geesey, GG, Ladd, TI, Nickel, JC, Dasgupta, M and Marrie, TJ (1987) Bacterial biofilms in nature and disease. *Annual Reviews of Microbiology* **41**:435–464.

Costerton, JW, Lewandowski, Z, deBeer, D, Caldwell, D, Korber, D and James, G (1994) Biofilms, the customized microniche. *Journal of Bacteriology* **176**:2137–2142.

Costerton JW, Lewandowski, Z, Caldwell, D, Korber, DR and Lappin-Scott, HM (1995) Microbial biofilms. *Annual Reviews in Microbiology* **49**:711–745.

Costerton, JW, Stewart, PS and Greenberg, EP (1999) Bacterial biofilms: a common cause of persistent infections. *Science* **284**:1318–1322.

Cramton, SE, Gerke, C, Schnell, NF, Nichols, WW and Götz, F (1999) The intercellular adhesion (*ica*) locus is present in *Staphylococcus aureus* and is required for biofilm formation. *Infection and Immunity* **67**:5427–5433.

Danese, PN, Pratt, LA and Kolter, R (2000) Exopolysaccharide production is required for the development of *Escherichia coli* K-12 biofilm architecture. *Journal of Bacteriology* **182**:3593–3596.

Davies, DG, Chakrabarty, AM and Geesey, GG (1993) Exopolysaccharide production in biofilms: substratum activation of alginate gene expression by *Pseudomonas aeruginosa*. *Applied and Environmental Microbiology* **59**:1181–1186.

Davies, DG, Parsek, MR, Pearson, JP, Iglewski, BH, Costerton, JW and Greenberg, EP (1998) The involvement of cell-to-cell signals in the development of bacterial biofilm. *Science* **280**:295–298.

Davey, ME and O'Toole, GA (2000) Microbial biofilms: from ecology to molecular genetics. *Microbiology and Molecular Biology Reviews* **64**:847–867.

deBeer, D, Stoodley, P, Roe, F and Lewandowski, Z (1994a) Effects of biofilm structures on oxygen distribution and mass transfer. *Biotechnology and Bioengineering* **43**:1131–1138.

deBeer, D, Srinivasan, R and Stewart, PS (1994b) Direct measurement of chlorine penetration into biofilms during disinfection. *Applied and Environmental Microbiology* **60**:4339–4344.

de Kievit, TR and Iglewski, BH (2000) Bacterial quorum sensing in pathogenic relationships. *Infection and Immunity* **68**:4839–4849.

Dewanti, R and Wong, ACL (1995) Influence of culture conditions on biofilm formation by *Escherichia coli* O157:H7. *International Journal of Food Microbiology* **26**:147–164.

Dillon, R and Fauci, L (2000) A microscale model of bacterial and biofilm dynamics in porous media. *Biotechnology and Bioengineering* **68**:536–547.

Eidhin, DN, Perkins, S, Francois, P, Vaudaux, P, Höök, M and Foster, TJ (1998) Clumping factor B (ClfB), a new surface-located fibrinogen-binding adhesin of *Staphylococcus aureus*. *Molecular Microbiology* **30**:245–257.

Erickson, DL, Endersby, R, Kirkham, A, Stuber, K, Vollman, DD, Rabin, HR, Mitchel, I and Storey, DG (2002) *Pseudomonas aeruginosa* quorum-sensing systems may control virulence factor expression in the lungs of patients with cystic fibrosis. *Infection and Immunity* **70**:1783–1790.

Evans, DJ, Brown, MRW, Allison, DG and Gilbert, P (1990) Growth rate and resistance of Gram-negative biofilms towards centrimide USP. *Journal of Antimicrobial Chemotherapy* **26**:473–478.

Ferris, FG, Schultze, S, Witten, TC, Fyfe, WS and Beveridge, TJ (1987) Metal interactions with microbial biofilms in acidic and neutral pH environments. *Applied and Environmental Microbiology* **55**:1249–1257.

Flemming, H-C (1991) Biofilms as a particular form of microbial life. In: Biofouling and Biocorrosion in Industrial Water Systems (Eds. Flemming, HC and Geesey, GG), Springer, Heidelberg, pp. 5–11.

Flemming, H-C, Wingender, J, Mayer, C, Körstgens, V and Borchard, W (2000) Cohesiveness in biofilm matrix polymers. In: Community Structure and

Co-operation in Biofilms (Eds. Allison, D, Gilbert, P, Lappin-Scott, HM and Wilson, M), Cambridge University Press, Cambridge, pp. 87–105.

Fletcher, M (1996) Bacterial attachment in aquatic environments: a diversity of surfaces and adhesion strategies. In: Bacterial Adhesion: Molecular and Ecological Diversity (Ed. Fletcher, M), Wiley & Sons, New York, pp. 1–24.

Fletcher, M and Pringle, JH (1986) Influence of substratum hydration and adsorbed macromolecules on bacterial attachment to surfaces. *Applied and Environmental Microbiology* **51**:1321–1325.

Fuqua, WC, Winans, SC and Greenberg, EP (1994) Quorum sensing in bacteria: the LuxR–LuxI family of cell density-responsive transcriptional regulators. *Journal of Bacteriology* **176**:269–275.

Garrett, ES, Perlegas, D and Wozniak, DJ (1999) Negative control of flagellum synthesis in *Pseudomonas aeruginosa* is modulated by the alternative sigma factor AlgT (AlgU). *Journal of Bacteriology* **181**:7401–7404.

Geesey, GG, Mutch, R, Costerton, JW and Green, RB (1978) Sessile bacteria: an important component of the microbial population in small mountain streams. *Limnology and Oceanography* **23**:1214–1223.

Gilbert, P, Collier, PJ and Brown, MRW (1990) Influence of growth rate on susceptibility to antimicrobial agents: biofilms, cell cycle and dormancy. *Antimicrobial Agents and Chemotherapy* **34**:1865–1868.

Gilbert, P, Evans, DJ, Evans, E, Duguid, IG and Brown, MRW (1991) Surface characteristics and adhesion of *Escherichia coli* and *Staphylococcus epidermidis*. *Journal of Applied Bacteriology* **71**:71–77.

Giwercman, B, Lambert, PA, Rosdahl, VT, Shand, GH and Høiby, N (1990) Rapid emergence of resistance in *Pseudomonas aeruginosa* in cystic fibrosis patients due to *in-vivo* selection of stable partially derepressed beta-lactamase producing strains. *Journal of Antimicrobial Chemotherapy* **26**:247–259.

Gristina, AG, Costerton, JW and McGanity, PLJ (1984) Bacterial-laden biofilms: a hazard to orthopaedic prosthesis. *Infections in Surgery* (**September**):655–662.

Hannig, M (1999) Transmission electron microscopy of early plaque formation on dental materials *in vivo*. *European Journal of Oral Science* **107**:55–64.

Hausner, M and Wërtz, S (1999) High rates of conjugation in bacterial biofilms as determined by quantitative *in situ* analysis. *Applied and Environmental Microbiology* **65**:3710–3717.

Heilmann, C, Hussain, M, Peters, G and Gotz, F (1997) Evidence of autolysin mediated primary attachment of *Staphylococcus epidermidis* to a polystyrene surface. *Molecular Microbiology* **24**:1013–1024.

Høiby, N, Fomsgaard, A, Jensen, ET, Johansen, HK, Kronborg, G, Pedersen, SS, Pressler, T and Kharazmi, A (1995) The immune response to bacterial biofilms. In: Microbial Biofilms (Eds. Lappin-Scott, HM and Costerton, JW), Cambridge University Press, Cambridge, pp. 233–250.

Hoyle, BD, Wong, CK and Costerton, JW (1992) Disparate efficacy of tobramycin on Ca(2+), Mg(2+), and HEPES-treated *Pseudomonas aeruginosa* biofilms. *Canadian Journal of Microbiology* **38**:1214–1218.

Huang, CT, Yu, FP, McFeters, GA and Stewart, PS (1995) Nonuniform spatial patterns of respiratory activity within biofilms during disinfection. *Applied and Environmental Microbiology* **61**:2252–2256.

Hull, J, Vervaart, P, Grimwood, K and Phelan, P (1997) Pulmonary oxidative stress response in young children with cystic fibrosis. *Thorax* **52**:557–560.

Jensen, ET, Kharazami, A, Garred, P, Kronborg, G, Fomsgaard, A, Mollnes, TE and Høiby, N (1993) Complement activation by *Pseudomonas aeruginosa* biofilms. *Microbial Pathogenesis* **15**:377–388.

Jensen, ET, Kharazami, A, Lam, K, Costerton, JW and Høiby, N (1990) Human polymorphonuclear leukocyte response to *Pseudomonas aeruginosa* grown in biofilm. *Infection and Immunity* **58**:2383–2385.

Kepkay, PE, Schwinghamer, P, Willar, T and Bowen, AJ (1986) Metabolism and metal binding by surface colonising bacteria: results of microgradient measurements. *Applied and Environmental Microbiology* **51**:163–170.

Kinniment, SL, Wimpenny, JW, Adams, D and Marsh, PD (1996) Development of a steady-state oral microbial biofilm community using the constant-depth film fermenter. *Microbiology* **142**:631–638.

Kjelleberg, S, Humphrey, BA and Marshall, KC (1982) Effect of interfaces on small, starved marine bacteria. *Applied and Environmental Microbiology* **43**:1166–1172.

Kleerebezem, M, Quadri, LE, Kuipers, OP and de Vos, WM (1997) Quorum sensing by peptide pheromones and two-component signal transduction systems in Gram-positive bacteria. *Molecular Microbiology* **24**:895–904.

Kolenbrander, PE (1993) Coaggregation of human oral bacteria: potential role in the accretion of dental plaque. *Journal of Applied Bacteriology* **74**(Supplement):79S–86S.

Kolenbrander, PE (2000) Oral microbial communities: biofilms, interactions, and genetic systems. *Annual Reviews in Microbiology* **54**:413–437.

Kuehn, M, Hausner, M, Bungartz, HJ, Wagner, M, Wilderer, PA and Wuertz, S (1998) Automated confocal laser scanning microscopy and semiautomated image processing for analysis of biofilms. *Applied and Environmental Microbiology* **64**:4115–4127.

Lawrence, JR, Korber, DR, Hoyle, BD, Costerton, JW and Caldwell, DE (1991) Optical sectioning of microbial biofilms. *Journal of Bacteriology* **173**:6558–6567.

Lee, SF, Li, YH and Bowden, GH (1996) Detachment of *Streptococcus mutans* biofilm cells by an endogenous enzyme activity. *Infection and Immunity* **64**:1035–1038.

Lewis, K (2001) Riddle of biofilm resistance. *Antimicrobial Agents and Chemotherapy* **45**:999–1007.

Lynch, MJ, Swift, S, Kirke, D, Dodd, CER, Keevil, CW, Stewart, GSAB and Williams, P (1999) Investigation of quorum sensing in *Aeromonas hydrophila* biofilms formed on stainless steel. In: Biofilms—The Good, the Bad and the Ugly (Eds. Wimpenny, J, Gilbert, P, Walker, J, Brading, M and Bayston, R), BioLine, Cardiff, pp. 209–223.

Mah, T-F and O'Toole, GA (2001) Mechanisms of biofilm resistance to antimicrobial agents. *Trends in Microbiology* **9**:34–39.

Maira-Litran, T, Allison, DG and Gilbert, P (2000) Expression of the multiple antibiotic resistance operon (*mar*) during growth of *Escherichia coli* as a biofilm. *Journal of Applied Microbiology* **88**:243–247.

Makin, SA and Beveridge, TJ (1996) The influence of A-band and B-band lipopolysaccharide on the surface characteristics and adhesion of *Pseudomonas aeruginosa* to surfaces. *Microbiology* **142**:299–307.

Marone, WJ, Jarvis, W, Culver, D and Haley, R (1992) Incidence and nature of endemic and epidemic nosocomial infections. In: Hospital Infections (Eds. Bennet, JV and Brachman, PS), Little, Brown and Company, Boston, pp. 577–596.

Marsh, PD and Bowden, GHW (2000) Microbial community interactions in biofilms. In: Community Structure and Co-operation in Biofilms (Eds. Allison, DG, Gilbert, P, Lappin-Scott, HM and Wilson, M), Cambridge University Press, Cambridge. pp. 167–198.

Marshall, KC (1996) Adhesion as a strategy for access to nutrients. In: Bacterial Adhesion: Molecular and Ecological Diversity (Ed. Fletcher, M), Wiley–Liss, New York, pp. 59–87.

Mathee, K, Ciofu, O, Sternberg, C, Lindum, PW, Campbell, JIA, Jensen, P, Johnsen, AH, Givskov, M, Ohman, DE, Molin, S, Høiby, N and Kharazmi, A (1999) Mucoid conversion of *Pseudomonas aeruginosa* by hydrogen peroxide: a mechanism for virulence activation in cystic fibrosis lung. *Microbiology* **145**:1349–1357.

Mayrand, D and McBride, BC (1980) Ecological relationships of bacteria involved in a simple, mixed anaerobic infection. *Infection and Immunity* **27**:44–50.

McBain, AJ and Gilbert, P (2001) Biocide tolerance and the harbingers of doom. *International Biodeterioration & Biodegradation* **47**:55–61.

McKenney, D, Hübner, J, Muller, E, Wang, Y, Goldmann, DA and Pier, GB (1998) The *ica* locus of *Staphylococcus epidermidis* encodes production of the capsular polysaccharide/adhesin. *Infection and Immunity* **66**:4711–4720.

McLean, RJ, Whiteley, M, Stickler, DJ and Fuqua, WC (1997) Evidence of autoinducer activity in naturally occurring biofilms. *FEMS Microbiology Letters* **154**:259–263.

McMurray, LM, Oethinger, M and Levy, SB (1998) Triclosan inhibits lipid synthesis. *Nature* **394**:531–532.

Møller, S, Sternberg, C, Andersen, JB, Christensen, BB and Molin, S (1998) *In situ* gene expression in mixed-culture biofilms: evidence of metabolic interactions between community members. *Applied and Environmental Microbiology* **64**:721–732.

Moore, GF, Dunsmore, BC, Jones, SM, Smejkal, CW, Jass, J, Stoodley, P and Lappin-Scott, HM (2000) Microbial detachment from biofilms. In: Community Structure and Co-operation in Biofilms (Eds. Allison, D, Gilbert, P, Lappin-Scott, HM and Wilson, M), Cambridge University Press, Cambridge, pp. 107–127.

Morisaki, H (1983) Effect of solid–liquid interface on metabolic activity of *E. coli*. *Journal of General and Applied Microbiology* **29**:195–204.

Nickel, JC, Wright, JB, Ruseska, I, Marrie, TJ, Whitfield, C and Costerton, JW (1985) Antibiotic resistance of *Pseudomonas aeruginosa* colonizing a urinary catheter *in vitro*. *European Journal of Clinical Microbiology* **4**:213–218.

Nielsen, PH and Jahn, A (1997) Conceptual model for production and composition of exopolymers in biofilms. *Water Science and Technology* **36**:11–19.

Nielsen, T, Barken, KB and Molin, S (2000) Role of commensal relationships on the spatial structure of a surface-attached microbial consortium. *Environmental Microbiology* **2**:59–68.

O'Toole, G and Kolter, R (1998) Flagellar and twitching motility are necessary for *Pseudomonas aeruginosa* biofilm development. *Molecular Microbiology* **30**:295–304.

O'Toole, G, Kaplan, HB and Kolter, R (2000a) Biofilm formation as microbial development. *Annual Reviews in Microbiology* **54**:49–79.

O'Toole, G, Gibbs, KA, Hager, PW, Phibbs, PV Jr and Kolter, R (2000b) The global carbon metabolism regulator Crc is a component of a signal transduction pathway required for biofilm development by *Pseudomonas aeruginosa*. *Journal of Bacteriology* **182**:425–431.

Ottemann, KM and Miller, JF (1997) Roles for motility in bacterial–host interactions. *Molecular Microbiology* **24**:1109–1117.

Otto, K, Norbeck, J, Larsson, T, Larlsson, K-A and Hermansson, M (2001) Adhesion of Type I–fimbriated *Escherichia coli* to abiotic surfaces leads to altered composition of the outer membrane. *Journal of Bacteriology* **183**:2445–2453.

Patti, JM, Bremell, T, Krajewska-Peitrasik, D, Abdelnour, A, Tarkowski, A, Ryden, C and Höök, M (1994) The *Staphylococcus aureus* collagen adhesin is a virulence determinant in experimental septic arthritis. *Infection and Immunity* **62**:152–161.

Pedersen, SS, Moller, H, Espersen, F, Sorensen, CH, Jensen, T and Høiby, N (1992) Mucosal immunity to *Pseudomonas aeruginosa* alginate in cystic fibrosis. *Acta Pathologica, Microbiologica et Immunologica of Scandanavia* **100**:326–334.

Plowman, R, Graves, N, Griffin, M, Roberts, J, Swan, A, Cookson, B and Taylor, L (1999) The socio-economic burden of hospital-acquired infection. *London: Public Health Laboratory Service and the London School of Hygiene and Tropical Medicine* 12.

Pratt, LA and Kolter, R (1998) Genetic analysis of *Escherichia coli* biofilm formation: roles of flagella, motility, chemotaxis and type I pili. *Molecular Microbiology* **30**:285–293.

Prigent-Combaret, C, Vidal, O, Dorel, C and Lejeune, P (1999) Abiotic surface sensing and biofilm-dependent regulation of gene expression in *Escherichia coli*. *Journal of Bacteriology* **181**:5993–6002.

Prince, A (1992) Adhesion and receptors of *Pseudomonas aeruginosa* associated with infection of the respiratory tract. *Microbial Pathogenesis* **13**:251–260.

Quirynen, M and Bollen, CM (1995) The influence of surface roughness and surface-free energy on supra- and subgingival plaque formation in man. A review of the literature. *Journal of Clinical Periodontology* **22**:1–14.

Reid, G, Van der Mei, HC and Busscher, HJ (1998) Microbial biofilms and urinary tract infections. In: Urinary Tract Infections (Eds. Brumfitt, W, Hamilton-Miller, JMT and Bailey, RR), Chapman & Hall, London, pp. 111–116.

Riedel, K, Hentzer, M, Geisenberger, O, Huber, B, Steidle, A, Wu, H, Høiby, N, Givskov, M, Molin, S and Eberl, L (2001) *N*-Acylhomoserine-lactone-mediated communication between *Pseudomonas aeruginosa* and *Burkholderia cepacia* in mixed biofilms. *Microbiology* **147**:3249–3262.

Roberts, AP, Pratten, J, Wilson, M and Mullany, P (1999) Transfer of a conjugative transposon, Tn5397 in a model oral biofilm. *FEMS Microbiology Letters* **177**:63–66.

Rogers, J and Keevil, CW (1992) Immunogold and fluorescein immunolabelling of *Legionella pneumophila* within acquatic biofilms visualised using episcopic differential interference contrast microscopy. *Applied and Environmental Microbiology* **58**:2326–2330.

Rowbotham, TJ (1999) Current views on the relationships between amoebae, legionellae and man. *Israeli Journal of Medical Science* **22**:678–689.

Rumbaugh, KP, Griswold, JA, Iglewski, BH and Hamood, AN (1999) Contribution of quorum sensing to the virulence of *Pseudomonas aeruginosa* infections in cystic fibrosis. *Infection and Immunity* **67**:5854–5862.

Samuelsson, MO and Kirchman, DL (1990) Degradation of adsorbed protein by attached bacteria in relationship to surface hydrophobicity. *Applied and Environmental Microbiology* **56**:3643–3648.

Sanin, FD and Vesilind, PA (1994) Effect of centrifugation on the removal of extracellular polymers and physical properties of activated sludge. *Water Science and Technology* **30**:117–127.

Sauer, K, Camper, AK, Ehrlich, GD, Costerton, JW and Davies, DG (2002) *Pseudomonas aeruginosa* displays multiple phenotypes during development as a biofilm. *Journal of Bacteriology* **184**:1140–1154.

Sawyer, LK and Hermanocicz, SW (1998) Detachment of biofilm bacteria due to variations in nutrient supply. *Water Science Technology* **37**:211–214.

Sbarbati, A, Fanos, V, Bernardi, P and Tato, L (2001) Rapid diagnosis of fungal infection of intravascular catheters in newborns by scanning electron microscopy. *Scanning* **23**:376–378.

Shapiro, JA (1998) Thinking about bacterial populations as multicellular organisms. *Annual Reviews of Microbiology* **52**:81–104.

Shahal, Y, Steinberg, D, Hirschfeld, Z, Bronshteyn, M and Kopolovic, K (1998) *In vitro* bacterial adherence onto pellicle-coated aesthetic restorative materials. *Journal of Oral Rehabilitation* **25**:52–58.

Sheikh, J, Hicks, S, Dall'Agnol, M, Phillips, AD and Nataro, JP (2001) Roles of Fis and YafK in biofilm formation by enteroaggregative *Escherichia coli*. *Molecular Microbiology* **41**:983–997.

Shu, M, Wong, L, Miller, JH and Sissons, CH (2000) Development of multi-species consortia biofilms of oral bacteria as an enamel and root caries model system. *Archives of Oral Biology* **45**:27–40.

Stickler, D (1999) Biofilms. *Current Opinions in Microbiology* **2**:270–275.

Stickler, DJ, Morris, NS, McLean, RJC and Fuqua, C (1998) Biofilms on indewelling urethral catheters produce quorum-sensing signal molecules *in situ* and *in vivo*. *Applied and Environmental Microbiology* **64**:3486–3490.

Stoodley, P, Boyle, JD, Dodds, I and Lappin-Scott, HM (1997) Consensus model of biofilm structure. In: Biofilms, Community Interactions and Control (Eds. Wimpenny, J, Handley, P, Gilbert, P, Lappin-Scott, H and Jones, M), BioLine, Cardiff, pp. 1–9.

Stoodley, P, Dodds, I, Boyle, JD and Lappin-Scott, HM (1999) Influence of hydrodynamics and nutrients on biofilm structure. *Journal of Applied Microbiology* **85**:19–28.

Stoodley, P, Hall-Stoodley, L, Boyle, JD, Jørgensen, F and Lappin-Scott, HM (2000) Environmental and genetic factors influencing biofilm structure. In: Community Structure and Co-operation in Biofilms (Eds. Allison, D, Gilbert, P, Lappin-Scott, H and Wilson, M), Cambridge University Press, Cambridge, pp. 53–64.

Suci, P, Mittelman, MW, Yu, FP and Geesey, GG (1994) Investigation of ciprofloxacin penetration into *Pseudomonas aeruginosa* biofilms. *Antimicrobial Agents and Chemotherapy* **38**:2125–2133.

Surette, MG and Bassler, BL (1998) Quorum sensing in *Escherichia coli* and *Salmonella typhimurium*. *Proceedings of the National Academy of Sciences* **95**:7046–7050.

Sutherland, IW (2001) The biofilm matrix—an immobilized but dynamic microbial environment. *Trends in Microbiology* **9**:222–227.

Tang, HB, DiMango, E, Bryan, R, Gambello, M, Iglewski, BH, Goldberg, JB and Prince, A (1996) Contribution of specific *Pseudomonas aeruginosa* virulence factors to pathogenesis of pneumonia in a neonatal mouse model infection. *Infection and Immunity* **64**:37–43.

Tannock, GW (1994) The acquisition of the normal microflora of the gastrointestinal tract. In: Human Health: The Contribution of Microorganisms (Ed. Gibson, SAW), Springer, London, pp. 1–16.

Van Delden, C and Iglewski, B (1998) Cell-to-cell signaling and *Pseudomonas aeruginosa* infections. *Emerging Infectious Diseases* **4**:551–560.

Van Loosdrecht, MCM, Eikelboom, D, Gjaltema, A, Mulder, A, Tijhuis, L and Heijnen, JJ (1995) Biofilm structures. *Water Science and Technology* **32**:35–43.

Vaudaux, PE, Lew, DP and Waldvogel, FA (1994) Host factors predisposing to and influencing therapy of foreign body infections. In: Infections Associated with Indwelling Medical Devices (Eds. Bisno, AL and Waldvogel, FA), American Society for Microbiology Press, Washington, DC, pp. 1–29.

Walker, JT, Mackerness, CW, Rogers, J and Keevil, CW (1995) Heterogeneous mosaic biofilm—a haven for waterborne pathogens. In: Microbial Biofilms (Eds. Lappin-Scott, HM and Costerton, JW), Cambridge University Press, Cambridge, pp. 233–250.

Xu, KD, Stewart, SP, Xia, F, Huang, CT and McFeters, GA (1998) Spatial physiological heterogeneity in *Pseudomonas aeruginosa* biofilm is determined by oxygen availability. *Applied and Environmental Microbiology* **64**:4035–4039.

Ziebuhr, W, Krimmer, V, Rachid, S, Lößner, I, Götz, F and Hacker, J (1999) A novel mechanism of phase variation of virulence in *Staphylococcus epidermidis*: evidence for control of the polysaccharide intercellular adhesin synthesis by alternating insertion and excision of the insertion sequence element IS*256*. *Molecular Microbiology* **32**:345–356.

ZoBell, CE (1943) The effect of solid surfaces upon bacterial activity. *Journal of Bacteriology* **46**:39–56.

2 Biofilms Associated with Medical Devices and Implants

2.1 Problems of Biofilms Associated with Medical Devices and Implants

RODNEY M. DONLAN
Division of Healthcare Quality Promotion,
National Center for Infectious Diseases,
Center for Disease Control, Atlanta, GA 30333, USA

INTRODUCTION

Indwelling medical devices have been shown to provide a suitable habitat for the development of microbial biofilms. Though distinctly different from biofilms encountered in aquatic and industrial water systems (e.g. different types of organism, nutrient requirement, substratum, and temperature requirement), biofilms in medical devices nonetheless have characteristics in common, such as an extracellular polymeric substance matrix, tenacious association with a surface, altered growth rates and dramatically decreased susceptibility to antimicrobial agents. In addition, these biofilms are particularly relevant because of their impact on human health.

Human infections may result from the use of any of a variety of indwelling medical devices. Organisms responsible include Gram-positive and Gram-negative bacteria and yeasts, and are commonly found as pure cultures (though polymicrobial biofilms may occur, especially in devices used for extended periods). Biofilms may also affect the device operation or integrity. The exact mechanisms involved in biofilm-associated infections are still poorly defined, though there are specific characteristics of biofilms that increase the probability of infection, such as the concentration of surface-attached organisms which may vastly exceed the number in the liquid phase. Figure 2.1.1 shows a biofilm that has developed on the surface of a urinary catheter. A discussion of biofilms on medical devices

Medical Biofilms. Edited by J. Jass, S. Surman, J. Walker

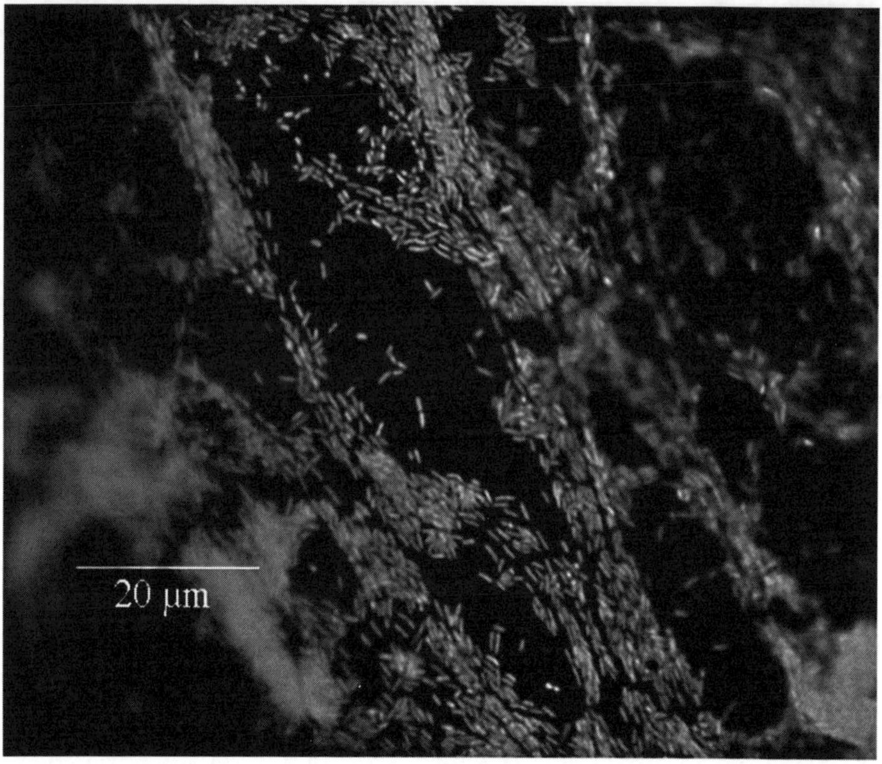

Figure 2.1.1. Laboratory-grown biofilm on the surface of a urinary catheter. Photograph shows *P. aeruginosa* biofilm stained with 4′,6-diamidino-2-phenylindole. Image by Amy Spoering.

follows, with a focus on the problems caused by this microbiological process.

INCIDENCE AND TYPES OF DEVICE-RELATED INFECTION

There is a direct association between the use of indwelling medical devices and infection. Tables 2.1.1–2.1.3 show rates of urinary-catheter-associated urinary tract infections (UTIs), central line-associated blood-stream infections, and ventilator-associated pneumonia for various intensive care facilities in the USA (Centers for Disease Control and Prevention NNIS System 2000). The pooled mean value given is the number of infections per 1000 device days. Though the infection rates vary depending upon the

Table 2.1.1. Urinary-catheter-associated UTI rate in intensive care units (ICUs)[a]

Type of ICU	No. of units	Urinary catheter-days	Pooled mean
Burn	16	38 212	10.2
Coronary	96	326 839	5.8
Medical	125	776 197	6.8
Paediatric	67	166 299	5.1
Surgical	144	963 902	5.2

[a]Data from Centers for Disease Control and Prevention NNIS System, 2000.

Table 2.1.2. Central line-asssociated blood-stream infection rate in ICUs[a]

Type of ICU	No. of units	Central line-days	Pooled mean
Burn	16	32 390	10.0
Coronary	95	203 909	4.6
Medical	126	548 124	6.1
Paediatric	70	234 100	7.7
Surgical	144	756 718	5.3

[a]Data from Centers for Disease Control and Prevention NNIS System, 2000.

Table 2.1.3. Ventilator-associated pneumonia rate in ICUs[a]

Type of ICU	No. of units	Ventilator-days	Pooled mean
Burn	16	22 591	14.9
Coronary	93	140 269	8.9
Medical	124	522 137	7.5
Paediatric	70	233 886	5.2
Surgical	144	535 349	13.6

[a]Data from Centers for Disease Control and Prevention NNIS System, 2000.

intensive care unit sampled and type of device used, it is clear that, for all categories listed, a certain percentage of individuals utilizing these devices will develop an infection. It is probable that a significant proportion of these device-related infections are due to biofilm formation on the device. These infections may pose serious consequences for the patient and place a significant financial burden on society. For example, 40–50% of patients with a prosthetic heart valve who contract endocarditis will die during initial hospitalization (Douglas and Cobbs 1992). Few of these patients can be cured by antibiotic therapy alone (Hancock 1994), as commonly, the only strategy for curing the infection is to remove the device and replace it. Infection rates for central venous catheters (CVCs) are 3–5% (Maki 1994), and may be as high as 10–50% in patients undergoing short-term urinary catheterization (Stickler 1996). As with prosthetic heart valves, often the only solution is to remove the device and treat the patient with antibiotics until the infection is resolved. One estimate of the incidence of infection following hip arthroplasty (replacement of the hip joint with a prosthesis) is approximately 1% over the lifetime of the prosthesis. This equates to more than 1000 new cases for the 100 000 total hip replacements performed each year as of 1992 (Nasser 1992).

INDWELLING MEDICAL DEVICES THAT MAY DEVELOP BIOFILMS

Table 2.1.4 lists a number of indwelling medical devices that have been shown to develop biofilms. Several of these devices are discussed, with an emphasis on how they are used in the patient and the types of infection that may result when biofilms develop.

CVCs may be inserted for administration of fluids, blood products, medications, total parenteral nutrition (TPN) solutions, or haemodynamic monitoring (Flowers *et al.* 1989). Raad *et al.* (1993) showed that biofilms were universally present on CVCs, and could be associated with either the inner lumen or outer surfaces. The organisms colonizing CVCs and forming biofilms originate either from the skin insertion site and migrate along the outer surface towards the tip or from the hub of the device, or, if as a result of manipulation by health-care workers, travel along the inner lumen (Elliott *et al.* 1997; Raad 1998). Because the device is in direct contact with the blood stream, upon insertion the surface becomes coated with platelets, plasma, and tissue proteins, including albumin, fibrinogen, and laminin, all of which may provide a conditioning film on the material surface (Raad

Table 2.1.4. Indwelling medical devices shown to develop biofilms

CVC needle-less connectors	Pacemakers
CVCs	Peritoneal dialysis catheters
Contact lenses	Prosthetic joints
Endotracheal tubes	Tympanostomy tubes
IUDs	Urinary catheters
Mechanical heart valves	Voice prostheses

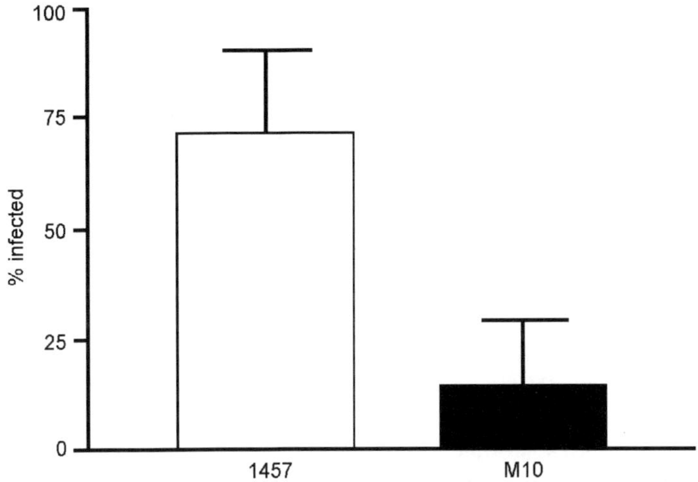

Figure 2.1.2. Rate of CVC-associated infection in animals challenged with wild-type (1457) or adhesin-deficient (M10) *S. epidermidis*. Error bars represent standard errors of the mean. Reprinted from Rupp *et al.* (1999), *Infection and Immunity* **67**:2656–2659. Reprinted by permission of American Society for Microbiology Journals Department.

1998). Several of the bacteria involved in biofilm formation on CVCs produce adhesins to specific adsorbed blood proteins, so that the nature of the conditioning film may, in part, determine the rate of initial attachment of these bacteria to the device surface. For example, *Staphylococcus aureus* adheres to fibronectin, fibrinogen, and laminin, whereas *Staphylococcus epidermidis* adheres only to fibronectin. Rupp *et al.* (1999) investigated the roles of *S. epidermidis* polysaccharide intercellular adhesin (PIA) and haemagluttin (HA) in adhesion and biofilm formation onto CVCs in a rat model system. By comparing organisms deficient in both PIA and HA with wild-type organisms, they found that the wild-type organisms produced greater biofilm levels and higher rates of infection, as shown in Figure 2.1.2. *S. epidermidis* 1457 was the wild-type strain and *S. epidermidis* M10 was the mutant strain (deficient in adhesins). Adsorbed blood proteins may also influence initial attachment by other organisms. Murga *et al.* (2001) demonstrated enhanced attachment and biofilm formation of several Gram-negative bacteria onto surfaces pre-conditioned with freshly drawn human blood.

The organisms that have been shown to colonize CVCs include coagulase-negative staphylococci (CNS), *S. aureus*, *P. aeruginosa*, *Klebsiella pneumoniae*, *Enterococcus faecalis*, and *Candida albicans*. Colonization and biofilm formation on CVCs may occur as rapidly as 3 days following catheterization (Anaissie *et al.* 1995). Catheters in place for less than 10 days tended to develop more extensive biofilm formation on the external surfaces. Biofilm formation on long-term catheters (up to 30 days) has been shown to be more extensive on the inner lumen (Raad *et al.* 1993).

Urinary catheters are tubular latex or silicone devices that are inserted through the urethra into the bladder to measure urine output, collect urine during surgery, prevent urinary retention, or control incontinence. Urinary catheters may be configured as either an open system, where the catheter drains into an open collection system, or closed, where the catheter empties into a closed collecting bag (Kaye and Hessen 1994). Foley urinary catheters have an inflatable balloon near the tip designed to hold the catheter in place in the bladder. The catheter is connected to a drainage tube and collection bag (Stickler and Hughes 1999). Patients commonly develop UTIs within 4 days when an open system configuration is used, whereas with closed systems the patients are much less susceptible to a UTI, and the urine can remain sterile for 10–14 days in approximately 50% of the patients (Kaye and Hessen 1994). Regardless of whether the system is open or closed, it has been observed that 10–50% of patients undergoing short-term catheterization (up to 7 days) develop a UTI, and all patients undergoing long-term catheterization (greater than 28 days) develop a UTI (Stickler 1996). McClean *et al.* (1995) also noted that the risk of developing a catheter-related UTI increased by approximately 10% for each day the catheter was in place.

Bacteria that attach to urinary catheters and develop biofilms may be introduced into the urethra or bladder when the catheter is inserted, entering through the sheath of exudate that surrounds the catheter, or travel intraluminally from the inside of the tubing or collection bag (Kaye and Hessen 1994). Catheters are colonized initially by single species, including *S. epidermidis, E. faecalis, Escherichia coli* or *Proteus mirabilis.* With time, the number and diversity of organisms increases, with the development of mixed communities containing *Providencia stuartii, P. aeruginosa, Proteus mirabilis, K. pneumoniae, Morganella morganii, Acinetobacter calcoaceticus,* and *Enterobacter aerogenes* (Stickler 1996; Stickler *et al.* 1993, 1998). Based upon evidence gained from scanning and transmission electron microscopy, it also appears that there may be other organisms present that do not grow on standard culture media (Nickel *et al.* 1989).

Prosthetic heart valves can be categorized as either mechanical valves or bioprostheses (tissue valves) (Braunwald 1997). Prosthetic valve endocarditis may occur in patients with both types of valve, and the rates of infection are similar for both categories. When the prosthetic valve is implanted in the patient, the resulting tissue damage leads to an accumulation of platelets and fibrin at the suture site and on the device

(Douglas and Cobbs 1992). It has been noted that prosthetic valve endocarditis is predominantly caused when microorganisms colonize the sewing cuff fabric portion of the valve (Illingworth *et al.* 1998). Organisms responsible for prosthetic valve endocarditis may be grouped into 'early' or 'late' colonizers. The predominant 'early' colonizers are CNS and probably originate from the initial contamination of the surgical site during implantation (Hancock 1994; Karchmer and Gibbons 1994). Organisms responsible for 'late' prosthetic valve endocarditis (which can be defined as that condition occurring from 12 months onward) may be streptococci, CNS, enterococci, *S. aureus*, Gram-negative coccobacilli, and fungi. In one study, *Streptococcus viridans* was the most common organism isolated during 'late' prosthetic valve endocarditis (Hancock 1994).

Contact lenses may be classified as either soft contact lenses, which are constructed of hydrogel or silicone and are designed to allow oxygen to diffuse through the lens material to the cornea, or hard contact lenses, which are constructed of polymethylmethacrylate and move with each blink, allowing oxygen-containing tears to flow underneath the lenses (Dart 1996). Bacteria adhere to both types of lens (Miller and Ahearn 1987; Stapleton *et al.* 1993; Stapleton and Dart 1995; Dart 1996). The organisms commonly isolated include *P. aeruginosa, S. aureus, S. epidermidis, Serratia* spp., *E. coli, Proteus* spp. and *Candida* (Dart 1996). Biofilms may also contain multiple species of bacteria or bacteria and fungi (McLaughlin-Borlace *et al.* 1998). These organisms may originate from contaminated lens disinfectant solutions or from the lens storage case itself (McLaughlin-Borlace *et al.* 1998). One study determined that 80% of asymptomatic contact lens users had contaminated storage cases (McLaughlin-Borlace *et al.* 1998) and that organisms isolated from the lens case were identical to organisms isolated from the corneas of infected patients in the majority of these patients. These authors also detected biofilms on the surfaces of 20 contact lens samples collected from patients diagnosed with microbial keratitis.

Intrauterine devices (IUDs) may be made of a non-absorbable material such as polyethylene impregnated with barium sulphate or designed to release a chemically active substance such as copper or a progestational agent. IUDs normally have a tail in order to locate the device for removal, and these tails are composed of plastic monofilament enclosed by a nylon sheath. IUD usage had been demonstrated to result in pelvic inflammatory disease (Chesney 1994; Lewis 1998; Wolf and Kreiger 1986), and IUDs removed from asymptomatic women have been shown to be contaminated with *S. epidermidis*, enterococci, and anaerobic lactobacilli (Wolf and Kreiger 1986). A study by Marrie and Costerton (1983) isolated *Lactobacillus plantarum, S. epidermidis, Corynebacterium* spp., Group B *Streptococcus, Micrococcus* spp., *C. albicans, S. aureus* and *Enterococcus* spp. In women with pelvic inflammatory disease, IUDs may also contain β haemolytic streptococci, *S. aureus*, *E. coli* and some anaerobic bacteria (Wolf and Kreiger 1986).

Artificial voice prostheses (AVPs) are inserted into patients following surgical treatment for extensive cancer of the larynx or hypopharynx (laryngectomy), in order to improve voice acquisition. The AVP is generally constructed of silicone and consists of a valve placed in the surgically created tracheo-oesophageal shunt. This allows air passage from the respiratory tract to the pharynx and mouth, while preventing contents from the digestive tract from passing into the respiratory tract. An AVP may be used to produce speech by closing the stoma with a finger and forcing air through the valve to the upper digestive tract, where the muscular structures at the oesophageal entrance function as an alternative sound source. Because AVPs are placed in a moist and nutrient-rich environment, they quickly become colonized by microorganisms. AVPs are replaced clinically on average every 4 months due to biofilm formation, which results in food and liquid leakage or increased air-flow resistance (Everaert *et al.* 1998a).

Organisms most commonly isolated from AVPs are *C. albicans, Candida tropicalis, Streptococcus mitis, Streptococcus sobrinus, S. salivarius, S. epidermidis, Rothia dentrocariosa, Stomatococcus mucilaginous* and other staphylococci (Busscher *et al.* 1997; Everaert *et al.* 1998a). As is the case with other indwelling medical devices, the nature of the conditioning film and the surface properties of the AVP are important in the initial adhesion of organisms to the silicone surfaces and in the susceptibility to detachment. Figure 2.1.3 compares the number of microorganisms remaining attached to treated and untreated AVPs following subjection to shear forces. Everaert *et al.* (1998a) added cultures of either *S. salivarius, S. epidermidis, C. albicans*, or *C. tropicalis* to a parallel plate flow chamber containing treated and untreated surfaces, and, after passing an air bubble through this chamber (a simulated shear force), determined the number of cells adhering. In general, for each of the four cultures examined, detachment occurred much more readily from the treated surfaces.

Prosthetic joints may become colonized by microorganisms, resulting in acute septic arthritis with joint pain and erythema, swelling, fever, and other systemic symptoms such as loosening of the joint and chronic pain (Steckelberg and Osmon 1994). Prosthetic joint infections (PJIs) can be classified based upon the elapsed time from implantation of the device to occurrence of infection. In acute infection, there is an obvious wound infection soon after surgery and PJI will result within 3 months. PJI occurring within the following 12 months can be classified as subacute PJI. In this case there may be wound inflammation or infection, fever, and an increase in erythrocyte sedimentation rate; these symptoms will usually resolve without antibiotics, but several months later PJI and joint loosening

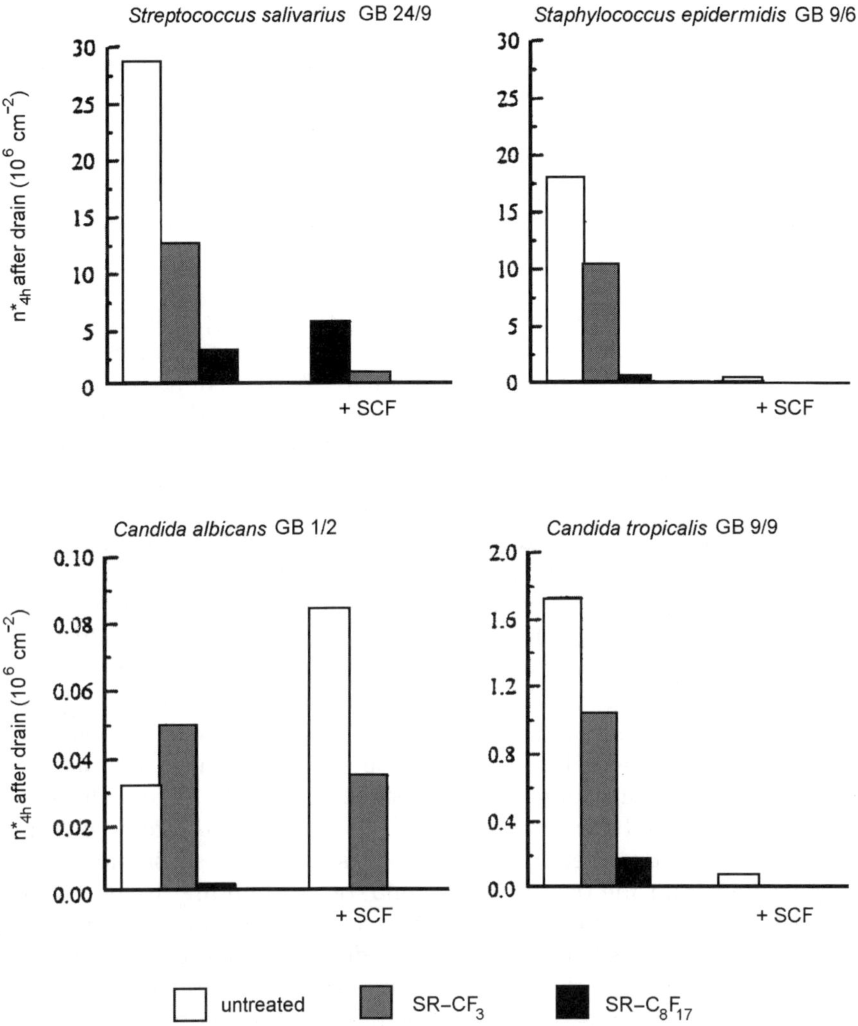

Figure 2.1.3. Numbers of microorganisms (adhesion time 4 h) able to withstand the passage of an air bubble through the flow chamber on Ar-SR-CF_3 and Ar-SR-C_8F_{17} surfaces and untreated silicone rubber in the absence and presence of a salivary conditioning film (SCF). Reprinted from *Colloids and Surfaces B: Biointerfaces* **10**, Everaert *et al.* (1998b) Adhesion of yeasts and bacteria to fluoro-alleylsiloxane layers chemisorbed on silicone rubber, pp. 179–190, with permission from Elsevier Science.

may occur. Late infections usually occur more than 12 months after the implantation (Maderazo *et al.* 1988). The criteria for determining whether a patient has a PJI are:

(a) two or more cultures from sterile joint aspirates or intra-operative cultures positive for the same organism;
(b) purulence at the time of surgical inspection;
(c) acute inflammation consistent with infection on histopathological examination of intra-capsular tissue;
(d) presence of a sinus tract that communicates with the joint space (Steckelberg and Osmon 1994).

Microorganisms causing PJI may originate during insertion of the prosthetic joint into the body, as a result of a transient bacteraemia, or through the exit site of the device from the body (Tunney *et al.* 1996). Steckelberg and Osmon (1994) presented data obtained from the Mayo Clinic for 1969–1991 showing that CNS comprised 25% of the organisms responsible for PJI. Other organisms commonly isolated included *S. aureus*, *β*-haemolytic streptococci, viridans streptococci, enterococci, *E. coli*, *P. mirabilis*, *Bacteriodes* spp. and other strict anaerobes. Polymicrobial infections were also observed. This study did not detect any significant differences in the proportion of organisms in early as opposed to late infections. Nasser (1992) found that *S. aureus* was the predominant organism in hip PJI, followed by *S. epidermidis*, *E. coli*, *Proteus* spp., anaerobes and *P. aeruginosa*. However, several investigators have suggested that routine sampling and culture procedures may fail to detect the causative organisms in PJI. This may be in part because the standard protocol for detecting organisms responsible for PJI is to culture the organisms obtained by aspiration of fluid collected from the area surrounding the prosthetic joint. In this case, only those organisms that are detached from the device surface and/or surrounding tissue surfaces would be detected. For example, Gristina and Costerton (1985) found that there were differences between organisms detected using routine culturing techniques and what was observed using scanning electron microscopy (SEM). They commonly observed polymicrobial infections using SEM, whereas culture techniques indicated the presence of a single organism. Tunney *et al.* (1999c) examined the organisms removed from prosthetic hip joints by sonication and found that 72% of culture-negative samples were positive by 16S rRNA, indicating the presence of organisms that would not (or could not) grow on culture media. The results from a study by Tunney *et al.* (1999c) are shown in Table 2.1.5.

Tympanostomy tubes are implanted through the tympanic membrane into the middle ear to alleviate pressure build-up and hearing loss in

Table 2.1.5. Comparison of the detection rates of prosthetic hip infection by different methods[a]

Method of detection	No. of samples	No. of positive samples	% of positive samples
Culture of tissue only	120	5	4
Culture of tissue and implants	120	26	22
Immunofluorescence microscopy	113	71	63
16S rRNA	118	85	72
Inflammatory cell infiltration	81	59	73

[a]From Tunney *et al.* (1999c), *Journal of Clinical Microbiology* **37**:3281–3290. Reproduced by permission of American Society for Microbiology Journals Department.

patients with otitis media. Tympanostomy tubes may become contaminated with microorganisms and biofilms can develop on their inner lumen (Saidi *et al.* 1999). The most common complication resulting from tympanostomy tube insertion is otorrhea, a release of purulent or mucopurulent fluid from the ear. This requires aggressive antibiotic therapy and, not uncommonly, removal of the tube (Beidlingmaier *et al.* 1998).

Biedlingmaier *et al.* (1998) found that *Pseudomonas* and *Staphylococcus* were the two organisms most often cultured from tympanostomy tubes removed from patients with chronically draining ears. Brook *et al.* (1998) collected middle-ear aspirates from children who were experiencing chronic otorrhea immediately following removal of the tube and recovered both aerobic and anaerobic bacteria. Aerobic organisms isolated included *P. aeruginosa, S. aureus, Proteus* spp., *Moraxella catarrhalis, K. pneumoniae* and non-typable *Haemophilus influenzae.* Anaerobes isolated included *Peptostreptococcus* sp., *Prevotella* sp., *Bacteriodes* sp. and *Fusobacterium* sp.

Beidlingmaier *et al.* (1998) investigated the *in vitro* colonization of tympanostomy tubes of different material by *P. aeruginosa, S. aureus,* or *S. epidermidis* in Trypticase Soy Broth. Materials examined included tubes constructed of Armstrong-style silicone, fluoroplastic, ionized-treated modified silicone, and silver-oxide-coated Armstrong-style silicone. After exposure for 5 days at 37°C, *P. aeruginosa* developed biofilms on all surfaces except ionized-treated modified silicone. For both *S. aureus* and *S. epidermidis,* all materials except ionized-treated modified silicone and fluoroplastic developed biofilms. Saidi *et al.* (1999) performed a similar study using an animal model system. They found that *S. aureus* inoculated into the ears of guinea pigs colonized all materials, though the ion-bombarded silicone surfaces contained significantly fewer attached cells than did silicone, silver-oxide-impregnated silicone, fluoroplastic, or silver-oxide-impregnated fluoroplastic surfaces. SEM showed that biofilms had developed after the 10-day implantation period on all tubes with the exception of the ion-bombarded silicone material.

RELATING BIOFILM FORMATION ON MEDICAL DEVICES TO DISEASE

Though it is clear from epidemiologic evidence that biofilms of indwelling medical devices may be associated with infection, the mechanisms for this process are poorly understood. Biofilms on indwelling medical devices may affect patient health by one or more of the following processes: (a) microorganisms in biofilms develop populations far in excess of numbers in the surrounding medium; (b) cells or cell aggregates may detach from biofilms; (c) biofilms may produce endotoxins; (d) biofilm-associated organisms are resistant to the host immune system; (e) biofilms provide a niche for the generation of resistant organisms (through resistance plasmid exchange).

Concentration of Organisms

Organisms within biofilms can develop populations that far exceed numbers of the same organism in the bulk fluid. For example, Costerton *et al.* (1987) showed that biofilm populations exceeded planktonic populations in aquatic systems by more than three orders of magnitude. Rogers *et al.* (1996) and Donlan *et al.* (1994) found the same relationship for urinary catheters and drinking water pipes. This concentration effect could provide a focus of infection.

Detachment

Cells and aggregates of cells will detach from biofilms in or on the medical device, and these cells could then colonize the patient's blood stream or urinary tract to cause an infection. The process of detachment has not been well characterized from medical device biofilms, but studies on *in vitro* biofilms have demonstrated that flow effects (Characklis *et al.* 1990), changes in nutrient concentration (Characklis 1990), or perhaps quorum-sensing by the biofilm-associated organisms (Davies *et al.* 1998) could result in detachment.

Production of Endotoxins

Gram-negative bacteria produce endotoxins, and biofilms containing these organisms have been associated with pyrogenic reactions in patients with indwelling medical devices (Riofol *et al.* 1999). Vincent *et al.* (1989) showed, in an *in vitro* study, that surface-associated endotoxin concentrations correlated with biofilm levels, though they did not determine a relationship between biofilm levels and endotoxin release.

Resistance to Host Immune System Clearance

Studies have shown that *E. coli* cells grown in a biofilm and then resuspended were as sensitive to phagocytosis by human polymorphonuclear leucocytes (PMNLs) as non-biofilm bacteria (Yasuda *et al.* 1994). The study also demonstrated that these same organisms were less susceptible to the killing activity of the active oxygen species in the PMNLs. Others have shown that the extracellular slime produced by *S. epidermidis* actually interfered with macrophage phagocytosis (Shiau and Wu 1998). Using a rabbit model system, Ward *et al.* (1992) showed that biofilm-associated growth rates of *P. aeruginosa* over a 42-day period in the animal were unaffected by the immune system of the animal; there was no difference between pre-immunized and non-immunized animals. In this case the pre-immunized animals had a 1000-fold higher titre of the antibody specific against *P. aeruginosa.* Anwar *et al.* (1992) compared the susceptibility of young (2-day-old) and aged (7-day-old) *P. aeruginosa* biofilms to the bactericidal action of serum. They found that the susceptibility patterns of planktonic and young biofilm cells were similar; aged biofilms were significantly less susceptible (Figure 2.1.4). These results taken together provide evidence that organisms within biofilms on medical devices can overcome the host immune system to persist in the host and ultimately cause an infection.

A Niche for Resistant Organisms

Biofilm-associated bacteria exchange plasmids by conjugation (Ehlers and Bouwer 1999; Hausner and Wuertz 1999; Roberts *et al.* 1999). Ehlers and Bouwer (1999) showed that rates of conjugation were significantly higher in *Pseudomonas* spp. biofilms than for the same organisms grown under planktonic conditions. The physical proximity of cells within microcolonies and the absence of shear forces might favour conjugation. This phenomenon takes on greater significance if it can be considered that resistance factors to a number of antibiotics are encoded on plasmids.

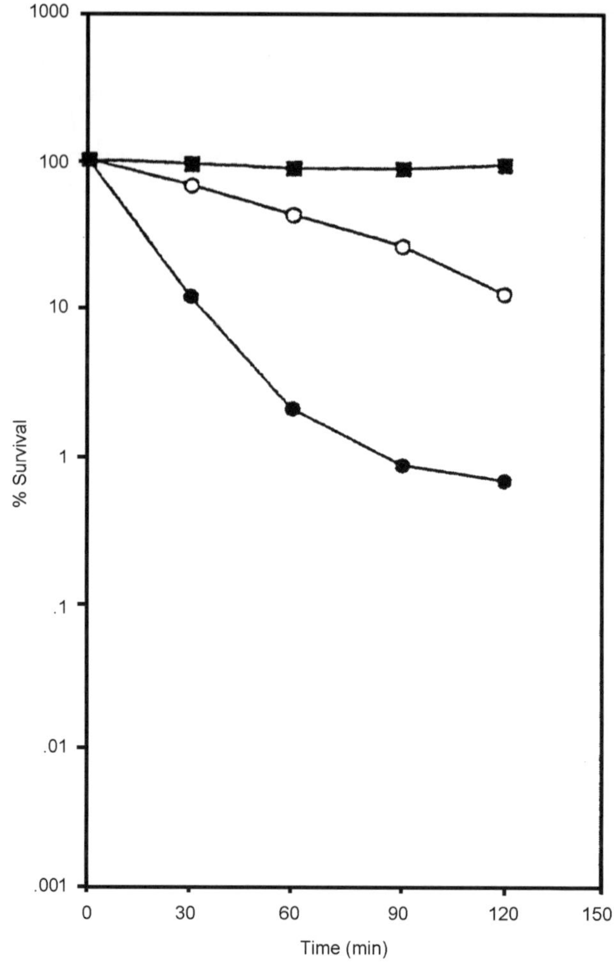

Figure 2.1.4. Sensitivity of mucoid *P. aeruginosa* to bactericidal action of serum. Symbols: (●) planktonic cells; (○) young biofilm; (■) ageing biofilm. Reprinted from *FEMS Microbiology Letters* **92**, Anwar *et al.* (1992). Susceptibility of biofilm cells of *Pseudomonas aeruginosa* to bactericidal actions of whole blood and serum, pp. 235–242, with permission from Elsevier Science.

EFFECT OF BIOFILMS ON MEDICAL DEVICE OPERATION

Biofilm formation on medical devices may also affect the operation or integrity of the device. Problems might include destruction of tissues surrounding prosthetic heart valves, resulting in leakage, material destruction of AVPs, obstruction of urinary catheters, or loosening of artificial hip and knee prostheses. Karchmer and Gibbons (1994) observed that microorganisms attach to and invade the valve annulus into which the prosthetic valve has been sewn. The ensuing infection may result in leakage when the sutures that anchor the prosthetic valve to the tissue pull away. Neu *et al.* (1993) examined explanted AVPs and showed, using SEM, that yeast cells deteriorated the silicone material and in some cases actually grew through the silicone into the inner valve region of the device. When this occurred the devices either leaked or had increased air-flow resistance. Urease-producing organisms may colonize urinary catheters. In addition to developing biofilms on these devices, they may also hydrolyse the urea in urine to form free ammonia. This free ammonia will, in turn, alter the pH at the biofilm surface and cause precipitation of minerals such as calcium phosphate and magnesium ammonium phosphate. These minerals form encrustations in catheters (Tunney *et al.* 1999b). Stickler *et al.* (1998) provided a case study where the urinary catheter was completely blocked within 4–5 days of use and the minerals in the biofilms were elevated in calcium, magnesium, and phosphorus. Tunney *et al.* (1999a) reported that a percentage of patients with culture-positive hip implants also had joint loosening. Although aseptic, mechanical loosening of implants is the most frequent complication following hip replacement surgery (Fitzgerald 1992); a percentage of cases of joint loosening may be due to infection. Tunney *et al.* (1999b) showed that the incidence of prosthetic joint infection was significantly underestimated by current culture detection techniques.

CONCLUSIONS

A wide variety of indwelling medical devices are currently used by the health-care industry. Many of these devices have been associated with patient infection, and often these infections are caused by development of biofilms on or in the device. Biofilms form when microorganisms interact with the device surface, adhere, and develop an established community of microbial cells and extracellular polymeric substances. The link between biofilm formation on an indwelling medical device and patient infection is not well understood. However, the following is known:

(a) biofilms provide a mechanism whereby cells can be concentrated on a surface by several orders of magnitude;
(b) cells within biofilms may detach and cause an infection;
(c) Gram-negative biofilm-associated cells may release endotoxins;
(d) biofilm-associated cells may resist host immune defence mechanisms; and

(e) organisms within biofilms may exchange extra-chromosomal genetic elements, such as plasmids, that encode for antibiotic resistance.

Biofilms may also be responsible for the malfunction of a number of indwelling devices by several different mechanisms. Research is needed to elucidate the mechanisms of cellular adhesion and biofilm formation on indwelling medical devices and to understand the mechanisms whereby biofilms cause infection.

ACKNOWLEDGEMENTS

The author would like to thank Tiene Bauters for providing supporting literature on artificial voice prostheses and Amy Spoering, an Emerging Infectious Disease Training Fellow at CDC, for the biofilm image used in Figure 2.1.1.

REFERENCES

Anaissie, EG, Samonis, G, Kontoyiannis, D, Costerton J, Sabharwal, U, Bodey, G and Raad, I (1995) Role of catheter colonization and infrequent hematogenous seeding in catheter-related infections. *European Journal of Clinical Microbiology and Infectious Disease* **14**:135–137.

Anwar, H, Strap, JL and Costerton, JW (1992) Susceptibility of biofilm cells of *Pseudomonas aeruginosa* to bactericidal actions of whole blood and serum. *FEMS Microbiology Letters* **92**:235–242.

Beidlingmaier, J, Samaranayke, R and Whelen, P (1998) Resistance to biofilm formation on otologic implant materials. *Otolaryngology Head and Neck Surgery* **118**:444–451.

Braunwald, E (1997) Valvular heart disease. In: Heart Disease, 5th edition, vol. 2 (Ed. Braunwald, E), W.B. Saunders Co., Philadelphia, pp. 1007–1076.

Brook, I, Yocum, P and Shah, K (1998) Aerobic and anaerobic bacteriology of otorrhea associated with tympanostomy tubes in children. *Acta Otolaryngologica (Stockholm)* **118**:206–210.

Busscher, HJ, Geertsema-Doornbusch, GI and van der Mei, HC (1997) Adhesion to silicone rubber of yeasts and bacteria isolated from voice prostheses: influence of salivary conditioning films. *Journal of Biomedical Materials Research* **34**:201–210.

Centers for Disease Control and Prevention NNIS System (2000) National Nosocomial Infections Surveillance (NNIS) system report, data summary from January 1992–April 2000, issued June 2000. *American Journal of Infection Control* **28**:429–448.

Characklis, WG (1990) Biofilm processes. In: Biofilms (Eds. Characklis, WG and Marshall, KC), John Wiley & Sons, Inc., New York, pp. 195–231.

Characklis, WG, McFeters, GA and Marshall, KC (1990) Physiological ecology in biofilm systems. In: Biofilms (Eds. Characklis, WG and Marshall, KC), John Wiley & Sons, New York, pp. 341–394.

Chesney, PJ (1994) Infections of the female genital tract. In: Infections Associated with Indwelling Medical Devices, 2nd edition (Eds. Bisno, AL and Waldovogel, FA), American Society for Microbiology, Washington, pp. 347–374.

Costerton, JW, Cheng K-J, Geesey, GG, Ladd, TI, Nickel, JC, Dasgupta, M and Marrie, TJ (1987) Bacterial biofilms in nature and disease. *Annual Reviews of Microbiology* **41**:435–464.

Dart, JKG (1996) Contact lens and prosthesis infections. In: Duane's Foundations of Clinical Ophthamology (Eds. Tasman, W and Jaeger, EA), Lippincott–Raven, Philadelphia, pp. 1–30.

Davies, DG, Parsek, MR, Pearson, JP, Iglewski, BH, Costerton, JW and Greenberg, EP (1998) The involvement of cell-to-cell signals in the development of a bacterial biofilm. *Science* **280**:295–298.

Donlan, RM, Pipes, WO and Yohe, TL (1994) Biofilm formation on cast iron substrata in water distribution systems. *Water Research* **28**:1497–1503.

Douglas, JL and Cobbs, CG (1992) Prosthetic valve endocarditis. In: Infective Endocarditis, 2nd edition (Ed. Kaye, D), Raven Press, New York, pp. 375–396.

Ehlers, LJ and Bouwer, EJ (1999) RP4 plasmid transfer among species of *Pseudomonas* in a biofilm reactor. *Water Science Technology* **7**:163–171.

Elliott, TSJ, Moss, HA, Tebbs, SE, Wilson, IC, Bonser, RS, Graham, TR, Burke, LP and Faroqui, MH (1997) Novel approach to investigate a source of microbial contamination of central venous catheters. *European Journal of Clinical Microbiology and Infectious Disease* **16**:210–213.

Everaert, EPJM, Van de belt-Gritter, B, van der Mei, HC, Busscher, HJ, Verkerke, GJ, Dijk, F, Mahieu, HF and Reitsma, A (1998a) *In vitro* and *in vivo* microbial adhesion and growth on argon plasma-treated silicone rubber voice prostheses. *Journal of Material Science: Materials in Medicine* **9**:147–157.

Everaert, EPJM, van der Mei, HC and Busscher, HJ (1998b) Adhesion of yeasts and bacteria to fluoro-alleylsiloxane layers chemisorbed on silicone rubber. *Colloids and Surfaces B: Biointerfaces* **10**: 179–190.

Fitzgerald, RH, Jr (1992) Total hip arthroplasty sepsis. *Orthopedic Clinics of North America* **23**:259–264.

Flowers, RH, Schwenzer, KJ, Kopel, RF, Fisch, MJ, Tucker, SI and Farr, BM (1989) Efficacy of an attachable subcutaneous cuff for the prevention of intravascular catheter-related infection. *Journal of the American Medical Association* **261**:878–883.

Gristina, AG and Costerton JW (1985) Bacterial adherence to biomaterials and tissue. *Journal of Bone and Joint Surgery, American Volume* **67**:264–273.

Hancock, EW (1994) Artificial valve disease. In: The Heart Arteries and Veins, 8th edition, vol. 2 (Eds. Schlant, RC, Alexander, RW, O'Rourke, RA, Roberts, R and Sonnenblick, EH), McGraw-Hill, New York, pp. 1539–1545.

Hausner, M and Wuertz, S (1999) High rates of conjugation in bacterial biofilms as determined by quantitative *in situ* analysis. *Applied and Environmental Microbiology* **65**:3710–3713.

Illingworth, BL, Twendon, K, Schroeder, RF and Cameron, JD (1998) *In vivo* efficacy of silver-coated (silzone) infection-resistant polyester fabric against a biofilm-producing bacteria, *Staphylococcus epidermidis. Journal of Heart Valve Disease* **7**:524–530.

Karchmer, AW and Gibbons, GW (1994) Infections of prosthetic heart valves and vascular grafts. In: Infections Associated with Indwelling Medical Devices, 2nd edition (Eds. Bisno, AL and Waldovogel, FA), American Society for Microbiology, Washington, pp. 213–249.

Kaye, D and Hessen, MT (1994) Infections associated with foreign bodies in the urinary tract. In: Infections Associated with Indwelling Medical Devices, 2nd edition (Eds. Bisno, AL and Waldovogel, FA), American Society for Microbiology, Washington, pp. 291–307.

Lewis, R (1998) A review of bacteriological culture of removed intrauterine contraceptive devices. *British Journal of Family Planning* **24**:95–97.

Maderazo, EG, Judson, S and Pasternak, H (1988) Late infections of total joint prostheses. *Clinical Orthopaedics Related Research* **229**:131–142.

Maki, DG (1994) Infections caused by intravascular devices used for infusion therapy: pathogenesis, prevention, and management. In: Infections Associated with Indwelling Medical Devices, 2nd edition (Eds. Bisno, AL and Waldovogel, FA), American Society for Microbiology, Washington, pp. 155–212.

Marrie, TJ, and Costerton JW (1983) A scanning and transmission electron microscopic study of the surfaces of intrauterine contraceptive devices. *American Journal of Obstetrics and Gynecology* **146**:384–394.

McLaughlin-Borlace, L, Stapleton, F, Matheson, M and Dart, JKG (1998) Bacterial biofilm on contact lenses and lens storage cases in wearers with microbial keratitis. *Journal of Applied Microbiology* **84**:827–838.

McLean, RJC, Nickel, JC and Olson, ME (1995) Biofilm associated urinary tract infections. In: Microbial Biofilms (Eds. Lappin-Scott, HM and Costerton JW), Cambridge University Press, Cambridge, pp. 261–273.

Miller, MJ and Ahearn, DG (1987) Adherence of *Pseudomonas aeruginosa* to hydrophilic contact lenses and other substrata. *Journal of Clinical Microbiology* **25**:1392–1397.

Murga, R, Miller, JM and Donlan, RM (2001) Biofilm formation by gram-negative bacteria on central venous catheter connectors: Effect of conditioning films in a laboratory model. *Journal of Clinical Microbiology* **39**: 2294–2297.

Nasser, S (1992) Prevention and treatment of sepsis in total hip replacement surgery. *Orthopedic Clinics of North America* **23**:265–277.

Neu, TR, Van der Mei, HC, Busscher, HJ, Dijk, F and Verkerke, GJ (1993) Biodeterioration of medical-grade silicone rubber used for voice prostheses: a SEM study. *Biomaterials* **14**:459–464.

Nickel, JC, Downey JA and Costerton, JW (1989) Ultrastructural study of microbiologic colonization of urinary catheters. *Urology* **34**:284–291.

Raad, I (1998) Intravascular-catheter-related infections. *Lancet* **351**:893–898.

Raad, I, Costerton, W, Sabharwal, U, Sacilowski, M, Anaissie, W and Bodey, GP (1993) Ultrastructural analysis of indwelling vascular catheters: a quantitative relationship between luminal colonization and duration of placement. *Journal of Infectious Disease* **168**:400–407.

Riofol, C, Devys, C, Meunier, G, Perraud, M and Goullet, D (1999) Quantitative determination of endotoxins released by bacterial biofilms. *Journal of Hospital Infections* **43**:203–209.

Roberts, AP, Pratten, J, Wilson, M and Mullany, P (1999) Transfer of a conjugative transposon, Tn5397 in a model oral biofilm. *FEMS Microbiology Letters* **177**:63–66.

Rogers, J, Norkett, DI, Bracegirdle, P, Dowsett, AB, Walker, JT, Brooks, T and Keevil, CW (1996) Examination of biofilm formation and risk of infection associated with the use of urinary catheters with leg bags. *Journal of Hospital Infection* **32**:105–115.

Rupp, ME, Ulphani, JS, Fey, PD and Mack, D (1999) Characterization of *Staphylococcus epidermidis* polysaccharide intercellular adhesin/hemagglutin in the pathogenesis of intravascular catheter-associated infection in a rat model. *Infection and Immunity* **67**:2656–2659.

Saidi, IS, Biedlingmaier, JF and Whelan, P (1999) *In vivo* resistance to bacterial biofilm formation on tympanostomy tubes as a function of tube material. *Otolaryngology Head and Neck Surgery* **120**:621–627.

Shiau, A-L and Wu, C-L (1998) The inhibitory effect of *Staphylococcus epidermidis* slime on the phagocytosis of murine peritoneal macrophages is interferon-independent. *Microbiology and Immunology* **42**:33–40.

Stapleton, F and Dart, J (1995) *Pseudomonas* keratitis associated with biofilm formation on a disposable soft contact lens. *British Journal of Ophthalmology* **79**:864–865.

Stapleton, F, Dart, JK, Matheson, M and Woodward, EG (1993) Bacterial adherence and glycocalyx formation on unworn hydrogel lenses. *Journal of British Contact Lens Association* **16**:113–117.

Steckelberg, JM, and Osmon, DR (1994) Prosthetic joint infections. In: Infections Associated with Indwelling Medical Devices, 2nd edition (Eds. Bisno, AL and Waldovogel, FA), American Society for Microbiology, Washington, pp. 259–290.

Stickler, DJ (1996) Bacterial biofilms and the encrustation of urethral catheters. *Biofouling* **94**:293–305.

Stickler, D and Hughes, G (1999) Ability of *Proteus mirabilis* to swarm over urethral catheters. *European Journal of Clinical Microbiology and Infectious Disease* **18**:206–208.

Stickler, D, Ganderton, L, King, J, Nettleton, J and Winters, C (1993) *Proteus mirabilis* biofilms and the encrustation of urethral catheters. *Urology Research* **21**:407–411.

Stickler, DJ, Morris, NS, McLean, RJC and Fuqua, C (1998) Biofilms on indwelling urethral catheters produce quorum-sensing signal molecules *in situ* and *in vitro*. *Applied and Environmental Microbiology* **64**:3486–3490.

Tunney, MM, Gorman, SP and Patrick, S (1996) Infection associated with medical devices. *Reviews of Medical Microbiology* **7**:195–205.

Tunney, MM, Patrick, MD, Curran, G, Ramage, G, Hanna, D, Nixon, JR, Gorman, SP, Davis, RI and Anderson, N (1999a) Detection of prosthetic hip infection at revision arthroplasty by immunofluorescence microscopy and PCR amplification of the bacterial 16S rRNA gene. *Journal of Clinical Microbiology* **37**:3281–3290.

Tunney, MM, Jones, DS and Gorman, SP (1999b) Biofilm and biofilm-related encrustations of urinary tract devices. In: Methods in Enzymology, vol. 310 (Ed. Doyle, RJ), Academic Press, San Diego, pp. 558–566.

Tunney, MM, Patrick, S, Curran, MD, Ramage, G, Anderson, N, Davis, RI, Gorman, SP and Nixon, JR (1999c) Detection of prosthetic joint biofilm infection using immunological and molecular techniques. In: Methods in Enzymology, vol. 310 (Ed. Doyle, RJ), Academic Press, San Diego, pp. 566–576.

Vincent, FC, Tibi, AR and Darbord, JC (1989) A bacterial biofilm in a hemodialysis system. Assessment of disinfection and crossing of endotoxin. *American Society for Artificial Internal Organs Transactions* **35**:310–313.

Ward, KH, Olson, ME, Lam, K and Costerton, JW (1992) Mechanism of persistent infection associated with peritoneal implants. *Journal of Medical Microbiology* **36**:406–413.

Wolf, AS and Kreiger, D (1986) Bacteriological colonization of intrauterine devices (IUDs). *Archives of Gynecology* **239**:31–37.

Yasuda, H, Akiki, Y, Aoyama, J and Yokota, T (1994) Interaction between human polymorphonuclear leucocytes and bacteria released from *in-vitro* bacterial biofilm models. *Journal of Medical Microbiology* **41**:359–367.

2.2 Pathogenesis and Detection of Biofilm Formation on Medical Implants

CHRISTOF VON EIFF and GEORG PETERS
Institute of Medical Microbiology, University of Münster Hospital and Clinics, Münster, Germany

INTRODUCTION

Indwelling or implanted foreign polymer bodies such as orthopaedic devices, prosthetic heart valves, and cardiac pacemakers, as well as intravascular catheters, renal dialysis shunts, cerebrospinal fluids shunts, or continuous ambulatory peritoneal dialysis (CAPD) catheters, have become a common and indispensable part of modern medical care. These devices are increasingly used in almost all fields of medicine for diagnostic and/or therapeutic procedures. However, the use of foreign material has led to associated complications because the insertion or implantation of medical devices is associated with a definitive risk of microbial infection. According to underlying patient characteristics, type of device and microorganism, the morbidity and mortality of device-associated infections may vary, however, polymer-associated infections contribute significantly to the increasing problem of nosocomial infections (National Nosocomial Surveillance System Report 2000).

Though a variety of bacteria have been implicated as causative organisms in polymer-associated infections, staphylococci, particularly *Staphylococcus epidermidis* and other coagulase-negative staphylococci (CNS), account for the majority of infections both of temporarily inserted and of permanently implanted material. The ability to adhere to materials and form biofilms is an important feature of the pathogenicity of these bacteria (von Eiff *et al.* 1998, 1999). Normally, the CNS live in balanced harmony on our skin, forming the major component of the cutaneous microflora. Outside the

Medical Biofilms. Edited by J. Jass, S. Surman, J. Walker

setting of a medical device, these organisms rarely cause infections. However, in the appropriate clinical setting, specifically when there is an infection of an implanted device, the CNS can cause severe disease and even death. This chapter deals with the various virulence factors involved in the pathogenesis of polymer-associated staphylococcal infection that have been defined and characterized during the last few years and the microbiological diagnosis and methods of detection of polymer-associated infections, particularly those associated with intravascular catheters.

MECHANISMS OF BIOFILM FORMATION IN THE PATHOGENESIS OF POLYMER-ASSOCIATED INFECTIONS

The ability to adhere to materials and form biofilms is an important feature in the pathogenesis of foreign-body-associated infection due to colonization of the polymer surface and forming multi-layered cell clusters, which are embedded in an amorphous extracellular material. Infection of the polymer probably occurs by inoculation with only a few bacteria from the patient's skin or mucous membranes during implantation of the medical device. The colonizing bacteria together with the extracellular material, which is mainly composed of cell wall teichoic acids, are referred to as biofilm. The presence of large adherent biofilms, including multilayered staphylococcal cell clusters on explanted intravascular catheters has been demonstrated by scanning electron microscopy (see Figures 2.2.1–2.2.3; Locci *et al.* 1981; Peters *et al.* 1981, 1982).

Biofilm formation proceeds in two stages: a rapid attachment of the bacteria to the polymer surface is followed by a more prolonged accumulation phase that involves cell proliferation and intercellular adhesion (Mack *et al.* 1996; Heilmann *et al.* 1997). For years, efforts have been made to identify the bacterial factors responsible for each of the two phases.

Adhesion of the Bacteria to Biomaterials

Microbial adherence to foreign bodies depends on the cell surface characteristics of the microorganisms and on the nature of the polymer material. Factors involved include physico-chemical forces such as polarity, London–van der Waal's forces and hydrophobic interactions (Dickinson and Bisno 1989; Herrmann and Peters 1997). Cell surface hydrophobicity and initial adherence have been attributed to bacterial surface-associated proteins. The staphylococcal surface proteins SSP-1 and SSP-2, which are

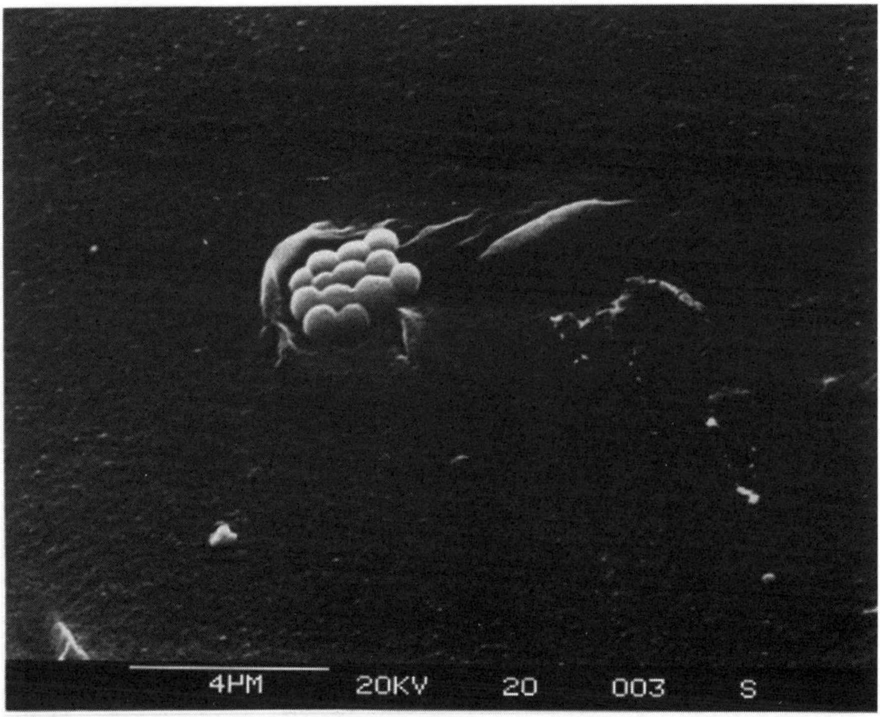

Figure 2.2.1. Electron micrograph of the inner surface of a polyethylene catheter showing adherent staphylococci after perfusion with saline containing *S. epidermidis*. If catheters were perfused *in vitro* with a buffer solution inoculated with CNS, staphylococcal cells could be seen preferentially adhering to surface defects a few minutes after starting the perfusion experiment (Locci, Peters and Pulverer 1981).

organized as fimbria-like polymers, have been described as contributing to *S. epidermidis* adherence to polystyrene (Veenstra *et al.* 1996). In addition, the 148 kDa autolysin, AtlE, of *S. epidermidis* has been identified as a surface-associated protein mediating primary attachment of bacterial cells to a polymer surface (Heilmann *et al.* 1997). Aside from proteins, a polysaccharide structure called capsular polysaccharide/adhesin has been associated with initial adherence (Muller *et al.* 1993).

While the direct interaction between microorganisms and naked foreign body surfaces plays a crucial role in the early stages of the adherence process *in vitro*, and probably also *in vivo*, additional factors may be important in the later stages of adherence *in vivo*, because implanted material rapidly becomes coated with plasma and connective tissue proteins such as fibronectin, fibrinogen, vitronectin, thrombospondin,

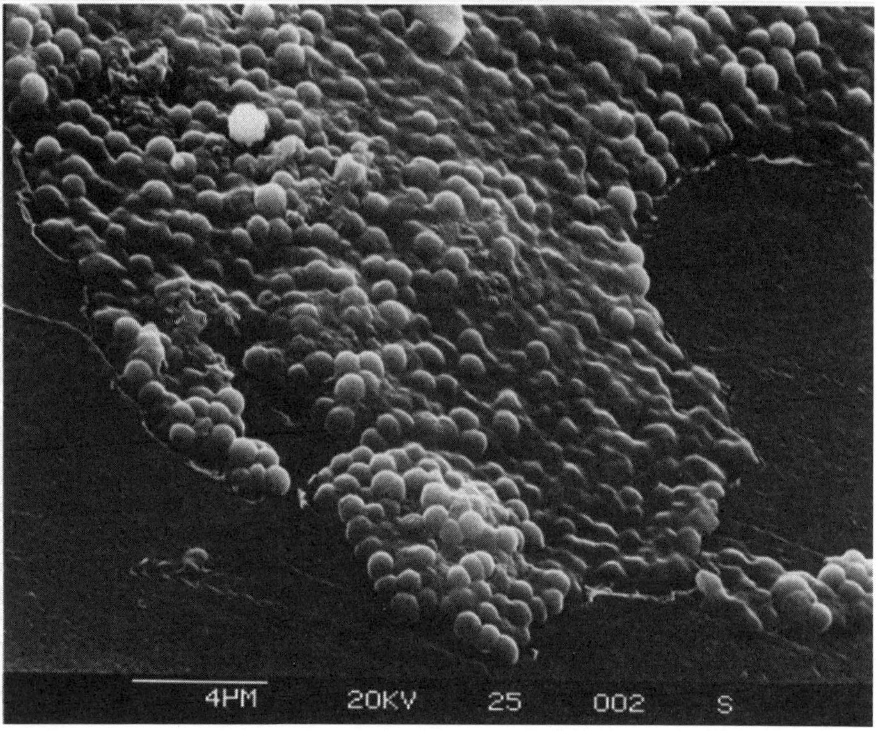

Figure 2.2.2. Extracellular slime substance produced by *S. epidermidis* on the inner surface of a polyethylene catheter. Starting with an incubation time of about 12 h, the adherent cells become covered by a thin film of material that steadily increases in thickness (Locci, Peters and Pulverer 1981).

laminin, collagen and von Willebrand factor (vWf) (Herrmann *et al.* 1988, 1997). Some of these host factors might serve as specific receptors for colonizing bacteria. In the vascular system at sites of increased flow, e.g. in the capillary system, vWf may also play an important role in adhesion of staphylococcal cells to polymer surfaces because, under high shear rates, platelets do not appreciably bind to extracellular matrix proteins other than vWF (Herrmann *et al.* 1997). Several host factor-binding proteins from *Staphylococcus aureus* have been cloned and sequenced, among them the fibrinogen receptor ClfA (clumping factor) (McDevitt *et al.* 1994) and the fibrinogen-binding protein FbpA (Cheung *et al.* 1995).

Recently, Palma *et al.* (1999) described a novel mechanism for enhancement of adherence of *S. aureus* to host components. A secreted protein, designated as Eap (extracellular adherence protein), which can form oligomeric structures, was purified from the supernatant and was found

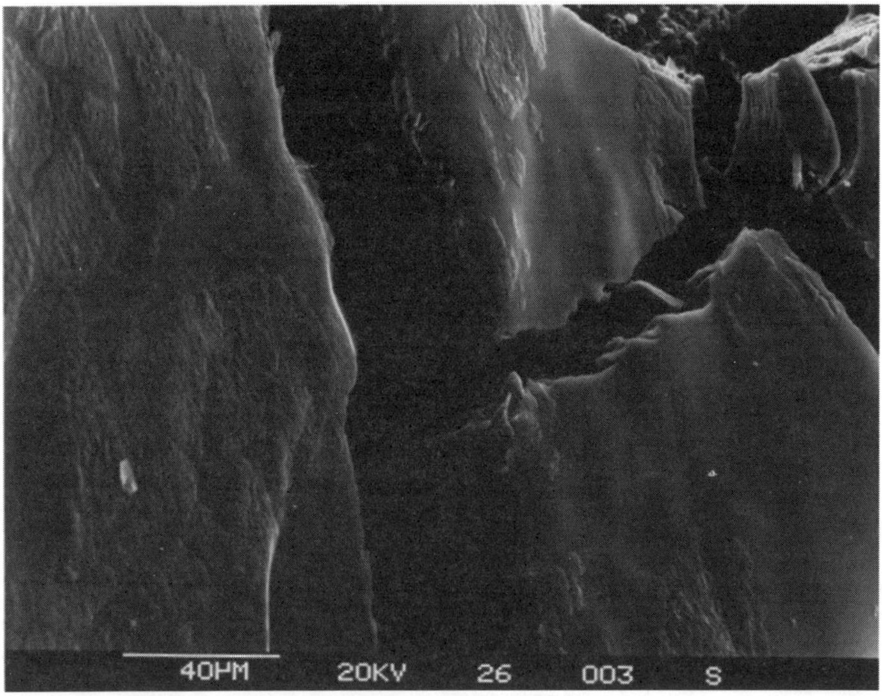

Figure 2.2.3. Thick matrix deposited on the inner surface of a catheter infected by *S. epidermidis*. The colonizing bacteria and the extracellular material, which is mainly composed of cell-wall teichoic acids, are referred to as biofilm (Locci, Peters and Pulverer 1981).

to be able to bind to at least seven plasma proteins, e.g. fibronectin, the alpha-chain of fibrinogen, and prothrombin, and to the surface of *S. aureus*. Unlike *S. aureus*, there is little data on host factor-binding proteins of CNS available. Adherence of most CNS strains is poorly promoted by fibrinogen, because these bacteria lack a clumping factor. The autolysin AtlE from *S. epidermidis* mediating primary attachment to a polystyrene surface was also found to exhibit vitronectin-binding activity, suggesting not only a function in the early stages of adherence, but also a contribution to later stages of adherence involving specific interactions with plasma proteins deposited on the polymer surface (Heilmann *et al.* 1997).

Accumulation of a Multilayered Biofilm

Once adhered to the polymer surface, bacteria proliferate and accumulate in multilayered cell clusters, which requires intercellular adhesion. A specific polysaccharide antigen, termed polysaccharide intercellular adhesin (PIA), involved in intercellular adhesion and biofilm accumulation has been detected and analysed (Mack *et al.* 1994, 2000). Transposon mutants lacking the antigen are not able to accumulate in multilayered cell clusters. Purification and structural analysis of PIA revealed that it is a linear β-1,6-linked glucosaminoglycan mainly composed of at least 130 2-deoxy-2-amino-D-glucopyranosyl residues, of which 80–85% are *N*-acetylated (Mack *et al.* 1996). The *icaADBC* operon, which mediates cell clustering and PIA synthesis in *S. epidermidis*, has been cloned and sequenced (Heilmann *et al.* 1996; Gerke *et al.* 1998). Most recently, Mack *et al.* (2000) identified three other gene loci that have a direct or indirect regulatory influence on expression of the synthetic genes for PIA and biofilm formation. Another antigenic marker of slime production and accumulation designated as slime-associated antigen (SAA), a high-molecular-weight compound rich in *N*-acetylglucosamine was formerly identified. Changes in the purification procedure have shown that the composition of the SAA differs from that originally described and that SAA consists mainly of *N*-acetylglucosamine. Hence, it has been concluded that SAA and PIA may have the same antigenic structure (Baldassarri *et al.* 1996). Recent investigations showed that PIA, at least in part, also mediates haemagglutination (HA) (Fey *et al.* 1999; Mack *et al.* 1999).

In order to assess the importance of PIA/HA-mediated biofilm production in the pathogenesis of biomaterial-based infection, a mouse infection model was used (Rupp *et al.* 1999a). A PIA/HA-positive *S. epidermidis* strain was significantly more likely to cause a subcutaneous abscess than its isogenic PIA/HA-negative mutant ($P < 0.01$) and was significantly less likely to be eradicated from the inoculation site by host defence. Furthermore, the wild-type strain was found to adhere to the implanted catheters more abundantly than the PIA/HA-negative mutant. In an additional study assessing the importance of biofilm production in a rat central venous catheter (CVC)-associated infection model, the wild-type *S. epidermidis* strain was significantly more likely to cause a CVC-associated infection (71 versus 14%, $P < 0.03$) resulting in bacteraemia and metastatic disease than its isogenic PIA/HA-negative mutant (Rupp *et al.* 1999b).

Other factors are also necessary for accumulation and biofilm formation of *S. epidermidis*, as demonstrated by Hussain *et al.* (1997). The 140 kDa extracellular protein AAP (accumulation-associated protein), which was detected only in extracellular products from bacteria grown under sessile conditions and which was missing in the accumulation-negative mutant M7, was proven to be essential for accumulative growth in certain *S. epidermidis* strains on polymer surfaces. Of 58 coagulase-negative staphylococci studied, 55% were 140 kDa antigen positive and produced significantly larger amounts of biofilm than the other strains that were 140 kDa antigen negative.

The clinical experience with polymer-associated staphylococcal infections reveals that the host defence mechanisms often seem to be unable to handle infection and, in particular, to eliminate staphylococci from the infected polymer device. In addition, antibacterial chemotherapy is frequently not able to cure these infections, despite the use of antibiotics with proven *in vitro* activity. Thus, the biofilm may protect the embedded staphylococci against host response mechanisms as well as against antibiotics (Herrmann and Peters 1997; von Eiff *et al.* 1999).

CONVENTIONAL MICROBIOLOGICAL DIAGNOSIS AND DETECTION OF BACTERIA EMBEDDED IN BIOFILMS IN POLYMER-ASSOCIATED INFECTIONS

Infections associated with foreign bodies are usually low grade and caused by commensal bacteria. The symptoms and signs can, therefore, be mild and go undetected, especially as devices such as vascular catheters are common in seriously ill patients, who may have other reasons for subtle signs of infection, such as a moderate fever and episodes of chills.

Early diagnosis is crucial, to prevent morbidity and excessive length of stay in hospital. Polymer-associated infections are difficult to treat and often require removal of the device. Furthermore, replacement of infected prostheses is costly and not without risks. Delay in diagnosis may allow the microorganisms to colonize the device and form thick layers of bacteria and extracellular material, sometimes together with a fibrin sheath, which may mean that removal of the device is essential for complete eradication of the infection. Antimicrobial treatment alone is much more likely to be successful if it is started before the organisms have become established within biofilms (Kristinsson 1997).

In the past, there have been many attempts to find simple and reliable methods to diagnose polymer-associated infections. Most require examination of the medical device itself after it has been removed and, therefore, can only provide a diagnosis in retrospect. This means that a large number of devices, particularly intravascular catheters, will be removed unnecessarily, as they have not been colonized, and some may be removed too late. The microbiological methods available to diagnose polymer-associated infections before removal of the device, for example by performing blood cultures and cultures of skin and exit sites, are not as accurate, but, when taken together with clinical symptoms, give a good indication of infection.

Among all device-related infections, those that are catheter-associated are by far the most common. Although some of the methods for laboratory diagnosis of infection are transferable to devices other than catheters, some techniques evidently have limitations with respect to differences in the construction of materials used. The objective of all the culture methods mentioned is to detect a polymer-associated infection following removal of the device by differentiating the polymers associated with infection from those that are not.

Broth Culture of Catheter Segments

The traditional and easiest method to detect bacterial growth on the surface of catheters is to culture vascular catheter segments qualitatively in liquid media. However, a drawback of this method is that it fails to distinguish heavily colonized or infected catheters from those merely contaminated with very small numbers of organisms when the catheter is removed. While the sensitivity of these cultures is excellent, they are not very specific and have a very low positive predictive value. Thus, it is fair to say that broth cultures are poor predictors of catheter colonization and catheter-related infections. However, conversely, when broth cultures are negative, the probability of a catheter-associated infection is low. In most places, however, a more semi-quantitative or quantitative culture method is used (Kristinsson 1997).

Semi-quantitative Culture of the Catheter Surface

Quantification of some sort is necessary to differentiate catheters that are related to infection from those that are not. Maki *et al.* (1977) introduced a new technique to identify intravascular catheter-related infections. Catheter segments were transferred to the surface of a blood agar plate, and rolled back and forth across the surface at least four times, with a slight downward pressure exerted by pre-flamed forceps. Catheters yielding 15 or more colonies were more frequently associated with local inflammation, whereas $\geqslant$15 CFU (colony forming units) suggested a significant 'infection'. It is generally accepted that, when using the roll plate method, the detection of $\geqslant$15 CFU in the absence of any clinical symptoms is considered as 'catheter colonization', whereas a catheter-associated infection is probable in the case of $\geqslant$15 CFU together with accompanying symptoms. Further studies have attempted to improve the sensitivity and specificity of the method by evaluating the different criteria for assessing the presence of a genuine infection. Direct comparisons of different studies are difficult, however, as there are variable numbers of patients, different patient population/risk groups, different catheter types, and varying definitions for infection in each study (Kristinsson 1997; Christensen *et al.* 2000). Furthermore, in many studies, the finding of a positive catheter culture has been a part of the definition of catheter-related bacteraemia.

This could overestimate the sensitivity of the methods if the catheter cultures themselves are falsely negative. Raad *et al.* (1993) have shown that almost all indwelling CVCs have visible adherent microorganisms on their surfaces, but only a few of these are culturable.

Although this semi-quantitative culture technique is the most labour-efficient method available at present to diagnose catheter-related infection, the method has some limitations with regard to determining the strength of adherence, inaccurate assessment of colony numbers due to problems with interference from adjacent colonies, i.e. inadequate colony separation, and in assessing the effect of the construction of the materials. However, the method has also been applied, with minor modification, to other materials, such as those used in vascular grafts, and can be easily modified to address particular experimental questions (Wengrovitz *et al.* 1991; Greenfeld *et al.* 1995).

Quantitative Culture of the Inside of the Catheter

An alternative approach to diagnosing intravascular catheter-related infections was introduced by Cleri *et al.* (1980), who studied short intravenous catheters and other intravascular inserts by flushing the catheter and performing cultures on the resulting flush fluid. This group cultured the inside of the intradermal and intravascular segments of the catheter quantitatively and found that growth of more than 1000 colonies from the inside of a segment was significantly associated with bacteraemia. The increased risk of catheter-associated bacteraemia increased from 29% with 10^3 to 10^4 colonies isolated per segment to 100% with $>10^6$ colonies per segment.

In order to determine the routes of infection in 20 cases of catheter-associated bacteraemia/fungaemia, Linares *et al.* (1985) used both the semi-quantitative method of Maki *et al.* (1977) to culture the outside and the quantitative method of Cleri *et al.* (1980) to culture the inside of intravascular and subcutaneous segments of long intravascular catheters. The intravascular tip had >15 colonies on the outside in 18 cases (90%) and >1000 colonies on the inside in 19 cases (95%), whereas the numbers for the subcutaneous segment were 13 (65%) and 15 (75%) respectively. Kristinsson *et al.* (1989) evaluated both techniques with slight modifications for 236 unselected long intravascular catheters. They found a good correlation between results from both methods, however, cultures from the outside of catheters produced more false positive results.

Quantitative Culture after Vortexing or Ultrasonication

Brun-Buisson *et al.* (1987) recognized that it may be technically difficult to roll a long segment of catheter across an agar surface in a way that dislodges most bacteria from the catheter surface onto the surface of the agar. Likewise, flushing the inside of catheters may be difficult. These investigators modified the method described by Cleri *et al.* (1980) and vortexed the catheter segment in sterile water for 1 min before diluting the solution for quantitative cultures. Most of the catheter tips growing 10^3 CFU ml^{-1} or more were associated with clinical signs of infection. Catheters yielding less than 10^3 CFU ml^{-1} were usually either contaminated or colonized, without clinical symptoms of sepsis. A 'cut-off' concentration of 10^3 CFU ml^{-1} showed 97.5% sensitivity and 88% specificity to diagnose catheter-associated sepsis.

A slightly different method was used to study the association between bacterial growth at the catheter insertion site and colonization of the catheter in patients receiving total parenteral nutrition. After placing 2 ml of sterile fluid in the tube containing the catheter segment, the tube was vigorously centrifuged for 90 seconds to elute microorganisms from the catheter, which were subsequently enumerated by standard dilution colony count methods. Five of the 74 catheters studied were associated with bloodstream infections, all of which had $>10^3$ CFU ml^{-1} cultured from the intravascular catheter segment (Bjornson *et al.* 1982).

Investigators have used a variety of means, such as vortexing or scraping, to remove organisms actively from the surface of an object. It has been suggested that some bacteria may adhere so avidly to polymer surfaces that they are not dislodged by rolling, flushing or vortexing. Sherertz *et al.* (1990) used sonication to dislodge the bacteria from catheters to determine whether this was an appropriate method to remove sessile adherent organisms embedded in the biofilm layer. Over a 3-year period they cultured 1681 catheters by placing them in 10 ml of broth, sonicating for 1 min (55 kHz, 125 W), and vortexing for 15 seconds before taking a 0.1 ml sample into either 0.9 or 9.9 ml of broth; 0.1 ml of these dilutions and 0.1 ml of the broth were surface-plated on blood agar. This technique allowed quantification of the number of CFU removed from a catheter for between 10^2 and 10^7 CFU. For catheter cultures in which $\geqslant 10^2$ CFU grew, a linear regression equation could be calculated.

The sonication method is versatile, as it can be performed on a variety of objects with complex shapes, it is quantitative, and provides information on viable organisms. However, the technique may not uniformly strip bacteria from the surface of the device, as the sonic energy may not penetrate the material uniformly (Christensen *et al.* 2000).

Comparison of the Different Culture Techniques

None of the culture methods described is clearly optimal. Ideally, both surfaces of an intravascular catheter should be cultured, but this may not be practical. The semi-quantitative culture method introduced by Maki *et al.* (1977) is simple and has been the method most frequently used in published studies to date, however, many other methods appear to be comparable. The best predictor of true infection, in a routine hospital setting, from the methods evaluated to date, is to set a threshold of 100 CFU and culture of the inside of the catheters (Kristinsson *et al.* 1989; Kristinsson 1997).

Raad *et al.* (1993) investigated the ultrastructure of indwelling vascular catheters with scanning and transmission electron microscopy (see below) and compared the results with those of the semi-quantitative roll technique and culture after ultrasonication. External colonization was predominant in the first 10 days of catheter placement, and luminal colonization became predominant after 1 month. These results suggest that the semi-quantitative roll technique would be most useful for short-term catheters, which were predominantly studied by Maki *et al.* (1977), whereas for long-term catheters the internal surface should be cultured. The sensitivity of the semi-quantitative roll-plate technique of the catheter tips was only 42 to 45%, as opposed to 65 to 72% for culture following ultrasonication of the tips. It was concluded that, in order to isolate sessile microorganisms, the more disruptive method of sonication might be necessary.

Culture after ultrasonication seems to be the most sensitive method, and it is also easy to quantify. However, ultrasonication involves the use of special equipment and may not be practical in many laboratories. Since there is no adopted standard method, the method one chooses is not critical as long as it is a semi-quantitative or quantitative technique and preferably validated for the population/catheters under study.

Staining and Microscopy of Catheter Segments

Microbiological cultures of devices such as catheters normally need at least an overnight incubation before they can be examined for growth. Several investigators have studied the possibility of using microscopy of the catheter segments as a tool for diagnosis in order to reduce the time taken to decide whether catheters are colonized or not. The catheters were either stained by Gram or acridine orange stains before examination.

In an investigation of 330 catheters in which the segments were Gram-stained the method had 100% sensitivity and 97% specificity in predicting colonization (according to the roll-plate method). In that study, the staining solution was passed through the lumen of the catheters to stain the inside of the catheters. After they had been blotted dry on filter paper, at least 200 fields were examined under an oil immersion objective. A single bacterium per 20 fields was designated as positive (Cooper and Hopkins 1985).

To overcome the problems and to speed up the investigation of a cylindrical catheter surface, another microscopy technique has been developed where the catheter segments are rolled over the surface of a glass slide in a shallow narrow streak of sterile saline solution. The slide is subsequently fixed and stained, and the entire length of the impression smear examined. The sensitivity of this method was found to be 83% and the specificity 81%, however, the positive predictive value was only 44% (Collignon *et al.* 1987).

In a study by Zufferey *et al* (1988), 710 catheter segments were stained with a fluorochrome dye, acridine orange. The catheters were studied with a fluorescence microscope, first at ×100 magnification; if there was no fluorescence detected after 3 min of examination, it was considered negative. If fluorescent material was detected, the catheter was also inspected with an oil immersion lens at ×1000 magnification to determine the morphology of the fluorescent material. This technique was again compared with the semi-quantitative culture method and showed a sensitivity of 84% and specificity of 99%.

In a study evaluating direct catheter staining in the diagnosis of intravascular catheter-associated infections, both Gram and acridine orange staining revealed poor sensitivity in predicting results of the semi-quantitative cultures: 44% and 71% respectively. Gram-staining, however, had higher specificity, with 91% as opposed to 77% with acridine orange (Coutlee *et al.* 1988).

Although the microscopy of catheter segments may reduce the time taken to decide whether a catheter is colonized or not, the main problems associated with this technique are as follows: frequent fine focusing is necessary owing to the cylindrical shape of the catheter; opaque catheters can be difficult/impossible to examine; and the technique is time consuming and seems to lack sensitivity.

DETECTION OF BACTERIAL ADHERENCE AND BIOFILM

Direct Observation of Microbial Attachment and Colonization

Under the proper circumstances, light microscopy is the quickest way to characterize the interaction between microorganisms and a surface. This method enables the microscopist to estimate the amount of microbial attachment and colonization, as well as to study the nature of the attachment. By examination of the specimen, microbes bound to the surface in individual cells, in clumps of cells, in mats, or with the expression of extracellular materials or associated with special surface structures may be observed. Being relatively inexpensive, simple to use, and readily available, the light microscope has proven to be a versatile instrument for studying microbial attachment to surfaces of devices (Christensen *et al.* 2000).

However, newer, more sophisticated microscopes, such as the transmission electron microscope, the scanning electron microscope and the scanning confocal laser microscope, have allowed investigators to study biofilms in more detail (Surman *et al.* 1996).

The use of the transmission electron microscope necessitates that the substratum has to be sectioned, but it does have the advantage that it enables the visualization and characterization of internal and external microbial adherence structures (Knutton 1995). By combining this technique with gold-labelled antibodies, specific antigenic structures can be located. The resolving power of the transmission electron microscope greatly exceeds the power of the other methods; however, like light microscopy, the procedure destroys the biofilm and the microorganisms and does not provide information on microbial viability.

The advent of the scanning electron microscope has enabled scientists to study in fine detail the attachment of microorganisms to surfaces (Peters *et al.* 1981; Knutton 1995). This method has the greatest utility for exploring the attachment of microorganisms to various polymers, such as examining where bacteria preferentially attach to an object. Furthermore, this technique furnishes us with information about the nature of attachment and the three-dimensional appearance of microbial biofilms. This procedure has the advantage that direct counts of the number of organisms attached to opaque or highly textured surfaces can be determined. Because the instrument visualizes the specimen at an angle from the side, maintaining the same field size between different fields and determining the field dimensions can be difficult for objects with a uniform surface and is nearly impossible for objects with a convoluted surface (Christensen *et al.* 2000).

Morphological investigations on various types of intravascular catheter using scanning electron microscopy and transmission electron microscopy have led to the first insight into the pathogenesis of foreign-body-associated infections (see above: Mechanisms of biofilm formation in the pathogenesis of polymer-associated infections; Figures 2.2.1–2.2.3) (Locci *et al.* 1981; Peters *et al.* 1981, 1982).

The scanning confocal laser microscope generates a high-resolution, three-dimensional image of the specimen, which includes both internal and external structures. Sanford *et al.* (1996) used this technique to examine slime layers produced by *S. epidermidis*. They determined the living architecture of the slime layer: rather than a uniform biofilm, the bacteria grew in conical multicellular structures separated by channels that presumably allowed the deepest layers of the biofilm to receive nutrients and release wastes. Like the optical microscope, the wavelength of light limits the resolving power of the confocal microscope, so that most submicrobial structures cannot be seen.

Tube Test and Plate Test for Detection of Slime-forming Bacteria

Bacterial adherence is believed to be an important mediator of infection at the site of implanted biomaterials. It is known that *S. epidermidis* becomes an important pathogen when a foreign body, such as a medical device, is implanted. Bacteria such as *S. epidermidis* adhere to materials, particularly to polymers and some strains adhere more than others. The most noteworthy technical problem, at least for some strains of CNS, is that the biofilm production is subject to phase variation (Christensen *et al.* 1987; Ziebuhr *et al.* 1997). This means that slime production is a heterogeneous phenomenon in which there is unequal expression of slime by individual daughter cells from the same strain. As a consequence, the clinical detection of slime can vary from isolate to isolate of an infecting strain of CNS, and the incidence of actual slime-producing cells in a slime-positive culture may be quite low. In the laboratory, Christensen *et al.* (1994) demonstrated that as few as one slime-producing cell per 16 000 non-slime-producing cells results in a culture that produces a gross amount of slime. Finally, since they are adhesive, slime-positive daughter cells may be difficult to recover from an infected foreign body, whereas the non-adherent slime-negative daughter cells may be easily recovered (Christensen *et al.* 1994). In conclusion, firmly attached microbial colonies are easily stained and visualized. The appearance of these deposits can form the basis for qualitative and semi-quantitative visual estimates of microbial colonization and for quantitative assays of microbial colonization by spectrophotometric determination of the optical density of the colony.

Christensen *et al.* (1982) developed the easiest method for detecting bacterial adherence. They demonstrated the presence of slime by a simple procedure known as the tube test, which consisted of emptying the contents of a culture test tube and staining the residual adherent film of bacteria using either tryptan blue or safranin. They showed that 63% of the pathogenic strains produced slime, and only 37% of the nonpathogenic strains produced slime. Since the test-tube procedure is simple, inexpensive, and expedient, it has been used by a number of investigators, primarily in the context of clinical isolates of CNS, although it has also been extended to *S. aureus*.

In later studies, Christensen *et al.* (1985) upgraded this qualitative test into a quantitative assay (known as the plate test) by substituting microtitre wells for culture tubes and by measuring the optical density of the stained adherent bacterial film with an automatic spectrophotometer. Using this quantitative assay, strains associated with intravascular catheter sepsis were shown to produce significantly thicker bacterial films than blood culture contaminants or skin strains. Whereas some studies described a link between the production of slime and persistence of infection, other investigators did not find any epidemiological association between the production of slime and the production of disease by pathogenic isolates of CNS. A major problem in interpreting these studies is that there are no standard assays for measuring slime production, yet technical factors greatly influence the observed presence of slime. Biofilm formation is dependent upon environmental factors such as the culture medium, the presence of carbohydrates or iron, CO_2 content, and oxygenation (Christensen *et al.* 1994). The tube test was shown to be variable and highly subjective and was demonstrated to produce different results than from the plate test. The demonstration of slime depends upon both the fixation method used and the age of the culture. The importance of these technicalities was shown by Deighton and coworkers (Deighton *et al.* 1988; Deighton and Balkau 1990), who, when using the tube test in an early 1988 study and in a later 1990 study, failed to find an epidemiological association between slime production and clinical isolates of CNS. When this group applied the quantitative plate test to the 1990 collection, however, they successfully found such an association.

In summary, both techniques described by Christensen *et al.* (1994), the tube test and the plate test, categorize the bacteria as adherent or non-adherent, but cannot determine adherence to nontransparent materials.

Radiolabelling Experiments for Studies of Adhesion

One of the most sensitive and versatile methods in the study of microbial adhesion to surfaces of foreign bodies is to use radioisotope labelling of the organism. There is a wide variety of radionuclides, microorganisms, and substrata used in radiolabelling experiments reported in the literature. Various techniques have been reported using bacteria radiolabelled with ^{111}In-oxine, [^{14}C]glucose, [^{3}H]thymidine and [^{35}S]methionine (Ardehali and Mohammad 1993). ^{111}In-oxine has been used to measure the attachment of *S. epidermidis* and *Pseudomonas aeruginosa* to fibrin-coated glass cover slips, [^{14}C]glucose to measure the adhesion of *Candida* to intravascular catheters, and [^{3}H]thymidine to label various Gram-positive and Gram-negative bacteria to different foreign-body materials (Vaudaux *et al.* 1993; Benson *et al.* 1996). The primary limitation in the use of radionuclides, apart from the restriction involving handling of radioisotopes, is that the ratio of counts-per-minute (cpm) to microbe is unstable and dependent on the isotope and material labelled. Microbial replication dilutes the concentration of the label, and metabolic processes may destroy the radionucleotide–bacterium link. Under experimental conditions, the investigators can follow experiments using radiolabelled microorganisms for short periods only (a few hours), so limiting the use of this technique to experiments concerning the microbial adhesion phase of colonization. Further restrictions for using such methods involve the use of special equipment, licensing and disposal (Christensen *et al.* 2000).

Dye Elution Technique for Detection of Adherence

Merritt *et al.* (1998) developed a simple modification of the techniques of Christensen *et al.* (1982, 1985) to detect bacterial adherence to non-transparent materials without using radioactive labelling. The materials, to which the bacteria had adhered, were rinsed in saline, fixed in formalin, stained with crystal violet and air dried. The dye was then eluted with ethanol and then the optical density of the solution was read with a 540 nm filter. The effectiveness of this technique was confirmed using polystyrene as the material for adherence. Polystyrene test tubes were compared using a tube with a visual film present from a known biofilm producer with a tube from a known non-producer; the bacteria could be easily distinguished. The slime-forming standard strain *S. epidermidis* RP62A showed distinctly more adherence than the biofilm-negative strain. The pre-incubation of the polymer with bovine serum albumin decreased the adherence of RP62A, whereas fibrinogen did not change the adherence of RP62A. When this technique was applied to other materials, e.g. to polyethylene, Dacron or silicone rubber, it was demonstrated that RP62A adhered more to all the materials than did the control strain. When additional strains of *S. epidermidis* were tested, a great variation in adherence capability to medical devices was noted. The relationship of this to actual infection at the site of indwelling devices remains unproven, but various studies have indicated a relationship between infection and isolation of an organism adherent to polystyrene or to the material at the infected site, or between adherence and the ability to establish infection in an animal model. Further work needs to be done on the characterization of organisms that are strongly adherent and those that are not adherent to help define the mechanisms involved in biofilm formation (Merritt *et al.* 1998).

16S rRNA **in situ** *Hybridization Technique for the Detection and Differentiation of Fastidious Microorganisms*

Recently, Krimmer *et al.* (1999) reported improved detection and identification of *S. aureus* and *S. epidermidis* by an *in situ* hybridization method with fluorescence-labelled oligonucleotide probes specific for staphylococcal 16S rRNA. The fluorescent hybridization technique with the *S. epidermidis*-specific probe SEP-1 proved to be suitable for the *in situ* detection of *S. epidermidis* cells even when they are embedded in a thick matrix of extracellular polysaccharides. In a control experiment with a biofilm-forming *S. aureus* isolate, no specific staining was obtained. These findings confirmed the specificity of SEP-1 for *S. epidermidis* and also ruled out the possibility that the positive signal was generated by non-specific binding to extracellular proteins or polysaccharides. The authors concluded that SEP-1 is an appropriate diagnostic tool for the detection and differentiation of *S. epidermidis* isolates even when they produce biofilms. In addition, it was suggested that the 16S rRNA *in situ* hybridization technique could represent a powerful diagnostic tool for the detection and differentiation of many other fastidious microorganisms.

The detection of bacterial rRNA genes as an indicator of the presence of bacteria is a reliable method that offers many advantages, particularly for the detection of bacteria adhered to biomaterial surfaces (Krimmer *et al.* 1999; Tunney *et al.* 1999). Each bacterial cell contains multiple copies of the 16S rRNA in its ribosomes. Thus, the technique is sensitive enough to detect single bacterial cells. In addition, 16S rRNA genes are highly conserved throughout bacterial evolution. They consist of regions that are common to all eubacteria and of other regions that are extremely species specific. By using appropriate gene probes, it is possible either to detect any targeted bacterial pathogen or, when highly specific probes are used, to identify single bacterial species. Also, this technique allows for the identification of microorganisms independently of bacterial growth rates and metabolic activities. This is of particular use in the detection of dormant and metabolically inactive bacteria, such as those bacteria embedded in biofilms, since the number of ribosomes is not significantly affected in such organisms.

Detection by Immunofluorescence Microscopy and PCR Amplification of the Bacterial 16S rRNA Gene

In a study by Tunney *et al.* (1999) that included 120 patients with total hip revision surgery, the detection rates of bacterial infection of hip prostheses by culture and non-culture methods were compared with each other. It was the first study to combine sampling by mild ultrasonication to dislodge bacteria growing within adherent biofilms with the use of strict anaerobic techniques. Whereas the incidence of infection by culture of material dislodged from retrieved prostheses after ultrasonication was 22%, bacteria were observed by immunofluorescence microscopy in 63% of sonicated samples with a monoclonal antibody specific for *Propionibacterium acnes* and polyclonal antiserum specific for *Staphylococcus* spp. The bacteria were present either as single cells or in aggregates of up to 300 bacterial cells, which were not seen prior to sonication to dislodge the biofilm. Discrimination between bacteria from skin-flake contamination and infecting bacteria was possible by immunofluorescence microscopy, since examination of skin scrapings did not reveal large aggregates of bacteria but did reveal skin cells. Bacteria were detected in all of the culture-positive samples and in single cases in which only one type of bacterium was identified by culture, both coccoid and coryneform bacteria were observed by immunofluorescence microscopy. Bacterial DNA was detected in 72% of sonicated samples by PCR amplification of a region of the bacterial 16S rRNA gene with universal primers. All of the culture-positive samples were also positive for bacterial DNA. Evidence of high-level infiltration, either of neutrophils or of lymphocytes or macrophages into associated tissue, was shown in 73% of patients. These results indicate that the use of non-culture methods, as opposed to culture methods, significantly increases the level of detection in infected prostheses. Furthermore, immunfluorescence microscopy allows a rapid quantitative and qualitative assessment of infected medical devices and distinguishes the bacteria from the infected prostheses from bacteria that may result from skin contamination (Tunney *et al.* 1999).

REFERENCES

Ardehali, R and Mohammad, SF (1993) ^{111}Indium labeling of microorganisms to facilitate the investigation of bacterial adhesion. *Journal of Biomedical Materials Research* **27**:269–275.

Baldassarri, L, Donnelli, G, Gelosia, A, Voglino, MC, Simpson, AW and Christensen, GD (1996) Purification and characterization of the staphylococcal slime-associated antigen and its occurrence among *Staphylococcus epidermidis* clinical isolates. *Infection and Immunity* **64**:3410–3415.

Benson, DE, Burns, GL and Mohammad, SF (1996) Effects of plasma on adhesion of biofilm forming *Pseudomonas aeruginosa* and *Staphylococcus epidermidis* to fibrin substrate. *American Society for Artificial Internal Organs Journal* **42**:M655–M660.

Bjornson, HS, Colley, R, Bower, RH, Duty, VP, Schwartz-Fulton, JT and Fischer, JE (1982) Association between microorganism growth at the catheter insertion site and colonization of the catheter in patients receiving total parenteral nutrition. *Surgery* **92**:720–727.

Brun-Buisson, C, Abrouk, F, Legrand, P, Huet, Y, Larabi, S and Rapin, M (1987) Diagnosis of central venous catheter-related sepsis. Critical level of quantitative tip cultures. *Archives of Internal Medicine* **147**:873–877.

Cheung, AI, Projan, SJ, Edelstein, RE and Fischetti, VA (1995) Cloning, expression, and nucleotide sequence of a *Staphylococcus aureus* gene (*fbpA*) encoding a fibrinogen-binding protein. *Infection and Immunity* **63**:1914–1920.

Christensen, GD, Simpson, WA, Bisno, AL and Beachey, EH (1982) Adherence of slime-producing strains of *Staphylococcus epidermidis* to smooth surfaces. *Infection and Immunity* **37**:318–326.

Christensen, GD, Simpson, WA, Younger, JJ, Baddour, LM, Barrett, FF, Melton, DM and Beachey, EH (1985) Adherence of coagulase-negative staphylococci to plastic tissue culture plates: a quantitative model for the adherence of staphylococci to medical devices. *Journal of Clinical Microbiology* **22**:996–1006.

Christensen, GD, Baddour, LM and Simpson, WA (1987) Phenotypic variation of *Staphylococcus epidermidis* slime production *in vitro* and *in vivo. Infection and Immunity* **55**:2870–2877.

Christensen, GD, Baldassarri, L and Simpson, AW (1994) Colonization of medical devices by coagulase-negative staphylococci. In: Infections Associated with Indwelling Medical Devices (Eds. Bisno, AL and Waldvogel, FA), ASM Press, Washington, pp. 45–78.

Christensen, GD, Simpson, AW, Anglen, JO and Gainor, BJ (2000) Methods for evaluating attached bacteria and biofilms. In: Handbook of Bacterial Adhesion: Principles, Methods, and Applications (Eds. An, YM and Friedman, RJ), Humana Press, Totowa, pp. 213–233.

Cleri, DJ, Corrado, ML and Seligman, SJ (1980) Quantitative culture of intravenous catheters and other intravascular inserts. *Journal of Infectious Diseases* **141**:781–786.

Collignon, P, Chan, R and Munro, R (1987) Rapid diagnosis of intravascular catheter-related sepsis. *Archives of Internal Medicine* **147**:1609–1612.

Cooper, GL and Hopkins, CC (1985) Rapid diagnosis of intravascular catheter-associated infection by direct Gram staining of catheter segments. *New England Journal of Medicine* **312**:1142–1147.

Coutlee, F, Lemieux, C and Paradis, JF (1988) Value of direct catheter staining in the diagnosis of intravascular-catheter-related infection. *Journal of Clinical Microbiology* **26**:1088–1090.

Deighton, MA and Balkau, B (1990) Adherence measured by microtiter assay as a virulence marker for *Staphylococcus epidermidis* infections. *Journal of Clinical Microbiology* **28**:2442–2447.

Deighton, MA, Franklin, JC, Spicer, WJ and Balkau, B (1988) Species identification, antibiotic sensitivity and slime production of coagulase-negative staphylococci isolated from clinical specimens. *Epidemiology and Infection* **101**:99–113.

Dickinson, GM and Bisno, AL (1989) Infections associated with indwelling devices: concepts of pathogenesis; infections associated with intravascular devices. *Antimicrobial Agents and Chemotherapy* **33**:597–601.

Fey, PD, Ulphani, JS, Götz, F, Heilmann, C, Mack, D and Rupp, ME (1999) Characterization of the relationship between polysaccharide intercellular adhesin and hemagglutination in *Staphylococcus aureus*. *Journal of Infectious Diseases* **179**: 1561–1564.

Gerke, C, Kraft, A, Süssmuth, R, Schweitzer, O, and Götz, F (1998) Characterization of the *N*-acetylglucosaminyltransferase activity involved in the biosynthesis of the *Staphylococcus epidermidis* polysaccharide intercellular adhesin. *Journal of Biological Chemistry* **273**:18 586–18 593.

Greenfeld, JI, Sampath, L, Popilskis, SJ, Brunnert, SR, Stylianos, S and Modak, S (1995) Decreased bacterial adherence and biofilm formation on chlorhexidine and silver sulfadiazine-impregnated central venous catheters implanted in swine. *Critical Care Medicine* **23**:894–900.

Heilmann, C, Schweitzer, O, Gerke, C, Vanittanakom, N, Mack, D and Götz, F (1996) Molecular basis of intercellular adhesion in the biofilm-forming *Staphylococcus epidermidis*. *Molecular Microbiology* **20**:1083–1091.

Heilmann, C, Hussain, M, Peters, G and Götz, F (1997) Evidence for autolysin-mediated primary attachment of *Staphylococcus epidermidis* to a polystyrene surface. *Molecular Microbiology* **24**:1013–1024.

Herrmann, M and Peters, G (1997) Catheter-associated infections caused by coagulase-negative staphylococci: clinical and biological aspects. In: Catheter-related Infections (Eds. Seifert, H, Jansen, B and Farr, BM), Marcel Dekker, New York, pp. 79–109.

Herrmann, M, Vaudaux, PE, Pittet, D, Auckenthaler, R, Lew, PD, Schumacher-Perdreau, F, Peters, G and Waldvogel, FA (1988) Fibronectin, fibrinogen, and laminin act as mediators of adherence of clinical staphylococcal isolates to foreign material. *Journal of Infectious Diseases* **158**:693–701.

Herrmann, M, Hartleib, J, Kehrel, B, Montgomery, RR, Sixma, JJ and Peters, G (1997) Interaction of von Willebrand factor with *Staphylococcus aureus*. *Journal of Infectious Diseases* **176**:984–991.

Hussain, M, Herrmann, M, von Eiff, C, Perdreau, RF and Peters, G (1997) A 140-kilodalton extracellular protein is essential for the accumulation of *Staphylococcus epidermidis* strains on surfaces. *Infection and Immunity* **65**:519–524.

Knutton, S (1995) Electron microscopical methods in adhesion. *Methods in Enzymology* **253**:145–158.

Krimmer, V, Merkert, H, von Eiff, C, Frosch, M, Eulert, J, Lohr, JF, Hacker, J and Ziebuhr, W (1999) Detection of *Staphylococcus aureus* and *Staphylococcus epidermidis* in clinical samples by 16S rRNA-directed *in situ* hybridization. *Journal of Clinical Microbiology* **37**:2667–2673.

Kristinsson, KG (1997) Diagnosis of catheter-related infections. In: Catheter-related Infections (Eds. Seifert, H, Jansen, B and Farr, BM), Marcel Dekker, New York, pp. 31–57.

Kristinsson, KG, Burnett, IA and Spencer, RC (1989) Evaluation of three methods for culturing long intravascular catheters. *Journal of Hospital Infection* **14**:183–191.

Linares, J, Sitges-Serra, A, Garau, J, Perez, JL and Martin, R (1985) Pathogenesis of catheter sepsis: a prospective study with quantitative and semiquantitative cultures of catheter hub and segments. *Journal of Clinical Microbiology* **21**:357–360.

Locci, R, Peters, G and Pulverer, G (1981) Microbial colonization of prosthetic devices. III. Adhesion of staphylococci to lumina of intravenous catheters perfused with bacterial suspensions. *Zentralblatt für Bakteriologie, Mikrobiologie und Hygiene B* **173**:300–307.

Mack, D, Nedelmann, M, Krokotsch, A, Schwarzkopf, A, Heesemann, J and Laufs, R (1994) Characterization of transposon mutants of biofilm-producing *Staphylococcus epidermidis* impaired in the accumulative phase of biofilm production: genetic identification of a hexosamine-containing polysaccharide intercellular adhesin. *Infection and Immunity* **62**:3244–3253.

Mack, D, Fischer, W, Krokotsch, A, Leopold, K, Hartmann, R, Egge, H and Laufs, R (1996) The intercellular adhesin involved in biofilm accumulation of *Staphylococcus epidermidis* is a linear beta-1,6-linked glucosaminoglycan: purification and structural analysis. *Journal of Bacteriology* **178**:175–183.

Mack, D, Riedewald, J, Rohde, H, Magnus, T, Feucht, HH, Elsner, HA, Laufs, R and Rupp, ME (1999) Essential functional role of the polysaccharide intercellular adhesin of *Staphylococcus epidermidis* in hemagglutination. *Infection and Immunity* **67**:1004–1008.

Mack, D, Rohde, H, Dobinsky, S, Riedewald, J, Nedelmann, M, Knobloch, JK, Elsner, HA and Feucht, HH (2000) Identification of three essential regulatory gene loci governing expression of *Staphylococcus epidermidis* polysaccharide intercellular adhesin and biofilm formation. *Infection and Immunity* **68**:3799–3807.

Maki, DG, Weise, CE and Sarafin, HW (1977) A semiquantitative culture method for identifying intravenous catheter-related infection. *New England Journal of Medicine* **296**:1305–1309.

McDevitt, D, Francois, P, Vaudaux, P and Foster, TJ (1994) Molecular characterization of the clumping factor (fibrinogen receptor) of *Staphylococcus aureus*. *Molecular Microbiology* **11**:237–248.

Merritt, K, Gaind, A and Anderson, JM (1998) Detection of bacterial adherence on biomedical polymers. *Journal of Biomedical Materials Research* **39**:415–422.

Muller, E, Takeda, S, Shiro, H, Goldmann, D and Pier, GB (1993) Occurrence of capsular polysaccharide/adhesin among clinical isolates of coagulase-negative staphylococci. *Journal of Infectious Diseases* **168**:1211–1218.

National Nosocomial Surveillance System (2000) National Nosocomial Surveillance System report. URL: http://www.cdc.gov/ncidod/hip/surveill/NNIS.htm.

Palma, M, Haggar, A and Flock, JI (1999) Adherence of *Staphylococcus aureus* is enhanced by an endogenous secreted protein with broad binding activity. *Journal of Bacteriology* **181**:2840–2845.

Peters, G, Locci, R and Pulverer, G (1981) Microbial colonization of prosthetic devices. II. Scanning electron microscopy of naturally infected intravenous catheters. *Zentralblatt für Bakteriologie, Mikrobiologie und Hygiene B* **173**:293–299.

Peters, G, Locci, R and Pulverer, G (1982) Adherence and growth of coagulase-negative staphylococci on surfaces of intravenous catheters. *Journal of Infectious Diseases* **146**:479–482.

Raad, I, Costerton, W, Sabharwal, U, Sacilowski, M, Anaissie, E and Bodey, GP (1993) Ultrastructural analysis of indwelling vascular catheters: a quantitative relationship between luminal colonization and duration of placement. *Journal of Infectious Diseases* **168**:400–407.

Rupp, ME, Ulphani, JS, Fey, PD, Bartscht, K and Mack, D (1999a) Characterization of the importance of polysaccharide intercellular adhesion/hemagglutinin of *Staphylococcus epidermidis* in the pathogenesis of biomaterial-based infection in a mouse foreign body infection model. *Infection and Immunity* **67**:2627–2632.

Rupp, ME, Ulphani, JS, Fey, PD and Mack, D (1999b) Characterization of *Staphylococcus epidermidis* polysaccharide intercellular adhesion/hemagglutinin in the pathogenesis of intravascular catheter-associated infection in a rat model. *Infection and Immunity* **67**:2656–2659.

Sanford, BA, de Feijter, AW, Wade, MH and Thomas, VL (1996) A dual fluorescence technique for visualization of *Staphylococcus epidermidis* biofilm using scanning confocal laser microscopy. *Journal of Industrial Microbiology* **16**:48–56.

Sherertz, RJ, Raad, II, Belani, A, Koo, LC, Rand, KH, Pickett, DL, Straub, SA and Fauerbach, LL (1990) Three-year experience with sonicated vascular catheter cultures in a clinical microbiology laboratory. *Journal of Clinical Microbiology* **28**: 76–82.

Surman, SB, Walker, JT, Goddard, DT, Morton, LHG, Keevil, CW, Weaver, W, Skinner, A and Kurtz, J (1996) Comparison of microscope techniques for the examination of biofilms. *Journal of Microbiological Methods* **25**:57–70.

Tunney, MM, Patrick, S, Curran, MD, Ramage, G, Hanna, D, Nixon, JR, Gorman, SP, Davis, RI and Anderson, N (1999) Detection of prosthetic hip infection at revision arthroplasty by immunofluorescence microscopy and PCR amplification of the bacterial 16S rRNA gene. *Journal of Clinical Microbiology* **37**:3281–3290.

Vaudaux, P, Pittet, D, Haeberli, A, Lerch, PG, Morgenthaler, JJ, Proctor, RA, Waldvogel, FA and Lew, DP (1993) Fibronectin is more active than fibrin or fibrinogen in promoting *Staphylococcus aureus* adherence to inserted intravascular catheters. *Journal of Infectious Diseases* **167**:633–641.

Veenstra, GJ, Cremers, FF, van Dijk, H and Fleer, A (1996) Ultrastructural organization and regulation of a biomaterial adhesin of *Staphylococcus epidermidis*. *Journal of Bacteriology* **178**:537–541.

von Eiff, C, Heilmann, C and Peters, G (1998) *Staphylococcus epidermidis*: why it is so successful? *Clinical Microbiology and Infection* **4**:297–300.

von Eiff, C, Heilmann, C and Peters, G (1999) New aspects in the molecular basis of polymer-associated infections due to staphylococci. *European Journal of Clinical Microbiology and Infectious Diseases* **18**:843–846.

Wengrovitz, M, Spangler, S and Martin, LF (1991) Sonication provides maximal recovery of *Staphylococcus epidermidis* from slime-coated vascular prosthetics. *The American Surgeon* **57**:161–164.

Ziebuhr, W, Heilmann, C, Götz, F, Meyer, P, Wilms, K, Straube, E and Hacker, J (1997) Detection of the intercellular adhesion gene cluster (*ica*) and phase variation in *Staphylococcus epidermidis* blood culture strains and mucosal isolates. *Infection and Immunity* **65**:890–896.

Zufferey, J, Rime, B, Francioli, P and Bille, J (1988) Simple method for rapid diagnosis of catheter-associated infection by direct acridine orange staining of catheter tips. *Journal of Clinical Microbiology* **26**:175–177.

2.3 Control of Biofilms Associated with Implanted Medical Devices

PETER GILBERT, ANDREW J. McBAIN, ALEXANDER H. RICKARD and SARAH R. SCHOOLING
School of Pharmacy and Pharmaceutical Sciences, University of Manchester, Oxford Road, Manchester, UK

INTRODUCTION

Microbial biofilms have become inexorably linked with man's failure to control them by treatment regimes that are effective against suspended bacteria. This is particularly notable with respect to the treatment of infections associated with the surfaces of indwelling medical devices. Failure has been related to a localized concentration of bacteria and their extracellular products (exopolymers and extracellular enzymes) that moderates the access of treatment agents and starves the more deeply placed cells. Biofilms, therefore, present gradients of physiology, and of concentration for the imposed treatment agent, where small sub-populations can sometimes survive, and where death is delayed for the least-susceptible cells. Such cells are either innately insensitive to a wide variety of treatment agents or they adopt resistant phenotypes during the sub-lethal phases of treatment. It is the diversity of action mechanisms displayed by those agents towards which biofilms are resistant that makes singular explanations of resistance phenomena difficult. However, it now seems likely that biofilms provide an environment in which the presence of small subsets of cells (persisters) is encouraged. These might represent induction, through a general stress response, of a viable non-culturable state (somnicells) that can be awakened post-treatment, or they might represent suicide-less mutants. Regardless, if such infections are to be resolved, then either these persister phenotypes must be targeted by anti-infection strategies, or the biofilm itself must be prevented from forming.

Medical Biofilms. Edited by J. Jass, S. Surman, J. Walker

In this chapter we consider the nature of biofilm resistance towards antibiotics and the current therapeutic and prophylactic options. Since these have often been spectacular in their failure to resolve such infections, we shall attempt to examine prospective developments of novel treatment and preventative strategies.

RESISTANCE OF BIOFILMS TO ANTIMICROBIAL AGENTS AND ANTIBOTICS

Much of the research interest in biofilm communities stems from our inability to control or eradicate them using antibiotics. Indeed, biofilms are reportedly some 10–1000 times less susceptible, towards a wide variety of differently acting control agents, than are the equivalent planktonic cells (Allison *et al.* 2000). The multiplicity of treatment agents towards which this resistance is shown makes singular explanations difficult to support, since exceptions can always be found.

Resistance due to Extracellular Polymeric Matrices and Physico-chemical Gradients

Resistance towards antibiotics and biocides can be mediated through reaction-diffusion limitation (Costerton *et al.* 1987; Hoyle *et al.* 1992; Huang *et al.* 1995), but this is sufficient only to explain the inability of some chemically reactive molecules, and those possessing strong positive charges, to penetrate the glycocalyx. Similarly, enzyme-mediated reaction-diffusion limitation takes account of only those molecules that provide substrate for relatively specific enzymes such as β-lactamases (Giwercman *et al.* 1991; Lambert *et al.* 1993). Regardless, for long-term resistance to result in either case, the reaction capacity of the biofilm for the agent must be sufficient to deplete the bulk treatment phase (Suci *et al.* 1994; Stewart 1996; Stewart *et al.* 1998). Alternative explanations of resistance depend on the generation, through the close proximity of cells, of nutrient gradients, and thereby a plethora of phenotypes that will include unsusceptible ones (Brown *et al.* 1990; Gilbert *et al.* 1990; Wentland *et al.* 1996). Since the least susceptible phenotype might be different for each treatment agent, yet still be represented within the community, then this explanation takes some account of their diverse physical, chemical and biochemical properties. Such phenotypes, however, depend upon the physical–chemical environment in which the cells grow. This will change as susceptible community members succumb to the treatment and no longer consume nutrients. Thus, slow-growing, unsusceptible cells may increase their rate of growth and become susceptible. Neither diffusion limitation nor physiological gradients can therefore account for the long-term survival of biofilm communities during chronic exposure to inimical agents (Gilbert and Allison 1999; Allison *et al.* 2000). Rather, such mechanisms only delay killing of a subset of cells.

For long-term survival to result from such delays in killing, it has been argued that the survivors must either adapt during the sub-lethal phase of treatment or represent variants that were already resistant before the commencement of treatment. The nature of such persister cells will undoubtedly affect the outcome of the treatment, and it is towards such physiologies that future agents and treatment regimens must be directed. It is important, therefore, to consider the potential mechanisms associated with them.

Adoption/Selection of Resistance Phenotypes

It has been suggested that the long-term survival of biofilm communities might relate to the adoption, or clonal expansion, of more resistant phenotypes during the delayed action of the treatment agents. Such phenotypes might relate to growth *per se* as a biofilm, the so-called 'biofilm phenotype', to non-specific responses towards localized high cell densities (quorum-sensing) or to the proximity of a surface. The temporary presence of sub-inhibitory concentrations of the treatment agents might also induce the expression of efflux pumps and possibly select for efflux mutants.

Attachment-specific Resistance Phenotypes

Bacteria can sense the proximity of a surface, up-regulate production of extracellular polysaccharide, and rapidly alter their susceptibility towards antibiotics (Ashby *et al.* 1994) and biocides (Das *et al.* 1998). Das *et al.* (1998) showed that the susceptibility of *Pseudomonas aeruginosa* and *Staphylococcus aureus* to a range of different biocides changed rapidly after cellular attachment and biofilm formation. In some instances, three- to five-fold decreases in susceptibility occurred immediately on attachment in the presence of biocide that exceeded the minimum inhibitory concentration (MIC) for planktonic cells. Later, we (Gilbert *et al.* 2001) showed that the bactericidal mechanisms of the same biocides were unchanged in mature biofilms. This indicated that active efflux of the agent or a decrease in penetrability of the community had caused the reduced susceptibility, rather than a change in target. In a similar study, Fujiwara *et al.* (1998) demonstrated that, after 1 h incubation, the minimum bactericidal concentrations towards adherent *P. aeruginosa*, *Serratia marcescens* and *Proteus mirabilis* were markedly elevated. The magnitude of the decrease in susceptibility observed immediately after bacterial attachment, but before biofilm formation, is generally far less than that observed in mature biofilms and is insufficient to account for the reported levels of resistance in biofilm communities. Whatever the extent or implication of such change, the possibility exists that the changes are mediated through the accumulation of homoserinelactone (HSL)-like signal substances at the occluded surface (Davies *et al.* 1998) and might be circumvented by HSL-antagonists such as the furanones (see below).

Efflux Pumps

An increasingly observed resistance mechanism is the expression and overproduction of multidrug efflux pumps (Nikaido 1998). Expression of such pumps is induced in Gram-negative bacteria, through sub-lethal exposure to a plethora of agents (George and Levy 1983; Ma *et al.* 1993). These include not only small hydrophilic antibiotics but also other xenobiotics, such as pine oil, salicylate and triclosan (Miller and Sulavick 1996; McMurry *et al.* 1998a). Mutations that increase the expression of such efflux pumps result in elevated levels of resistance. Whilst efflux pumps are operational in a wide variety of Gram-negative organisms, and may be plasmid or chromosomally encoded (Nikaido 1996), multidrug efflux pumps *qac*A–G also contribute to antibiotic resistance in *S. aureus* (Rouch *et al.* 1990). Notable amongst the multidrug-resistance operons are *mar* and efflux pumps such as *acr*AB (George and Levy 1983; Ma *et al.* 1993). Moken *et al.* (1997) and McMurry *et al.* (1998a,b) have shown that mutations causing over-expression of *mar*A or *acr*AB are associated with exposure and reduced susceptibility towards antibiotics (tetracycline, chloramphenicol), biocides (quaternary ammonium compounds, pine oil) and xenobiotics such as salicylate. The importance of *mar* would be greatly increased if it were induced by growth within a biofilm *per se,* and conferred a more resistant phenotype upon the cells prior to exposure (Maira-Litran *et al.* 2000a). This has been investigated using biofilms of a group of isogenic *Escherichia coli mar* and *acr*AB mutants, with ciprofloxacin as test agent. In *E. coli,* exposure to ciprofloxacin does not induce *mar* or *acr*AB, but its expression does confer limited protection. Results showed that *mar* and *acr*AB mutants (constitutive) had reduced susceptibility to ciprofloxacin, but that there was little or no difference in the susceptibility of biofilms prepared from wild-type and *mar*-deleted or *acr*AB-deleted strains (Maira-Litran *et al.* 2000a). Clearly, neither *mar* nor *acr*AB is specifically induced within biofilms, but their expression in response to appropriate inducer substances might be enhanced. In this respect *mar* expression is inversely related to specific growth rate (Maira-Litran *et al.* 2000b). Following exposure of biofilms to sub-lethal levels of inducer substances such as β-lactams, tetracyclines and salicylates, *mar* expression will be greatest within the depths of the biofilm where growth rates are suppressed. Treatment with antimicrobials that act as substrates but that are not themselves inducers, might lead to a clonal expansion of mutant cells that are constitutive in efflux pump expression. Similar systems, under the regulation of different inducer agents, might extend this explanation of biofilm tolerance to include other treatment agents.

Suicide-less Mutants

It is becoming increasingly recognized that many species of bacteria are capable of undergoing programmed cell death (Jensen and Gerdes 1995; Naito *et al.* 1995; Yarmolinsky 1995; Franch and Gerdes 1996; Boutibonnes 1997; Hochman 1997; Cellini *et al.* 1998; Engelberg-Kulka and Glaser 1999; Lewis 2000) and related autolytic processes. Programmed cell death in bacteria cannot be viewed as advantageous if these organisms exist primarily in a planktonic mode. Indeed, such suicide is not beneficial to the individual cell, but is often a highly evolved mechanism displayed within tissues. As such, it appears to be highly probable that programmed cell death within bacteria relates to the biofilm mode of growth. A recent and novel hypothesis for the considerable recalcitrance of biofilms relates to the potential of damaged bacterial cells to undergo apoptosis or programmed cell death. In this respect, Lewis (2000, 2001) suggested that death of cells following treatment with bactericidal agents results not from direct action of the agent, but from a programmed suicide mechanism and cellular lysis (Black *et al.* 1991; Moyed and Bertrand 1983). If this is the case, then cells subjected to different treatment agents with different mechanisms of action may well die from a common process. A singular mechanism of death allows us to speculate on singular mechanisms of resistance and survival from inimical treatments. If an entire population of cells underwent programmed cell death simultaneously, as the result of a sub-lethal exposure to antimicrobial compounds, then little benefit would be derived. It is imperative that a small proportion of the population should be able to avoid such a response and ultimately be responsible for the survival and recovery of the community. It is equally important that this trait of 'selfishness' is not retained in the resultant clones. The biocide literature of the last 50 years has been punctuated with reports of low-level, persistent survival of antimicrobial treatments (tailing) where the agent has not been quenched and where the survivors do not demonstrate resistance when re-cultured or cloned (Bigger 1944). The recent evidence suggests that such cells, rather than being resistant to the agent (Koch 1987), are actually defective in programmed cell death (Brooun *et al.* 2000). Following the removal of an inimical stress, these damaged persister cells would grow rapidly in the presence of nutrients released from their lysed community partners and the community would become restored. It is also postulated that biofilm populations are enriched in 'persister' cells, either as a biofilm-specific phenotype, or due their protection within the biofilm from immune responses or inimical agents (Lewis 2000, 2001). These cells would survive treatment phases and proliferate in the post-treatment phase, thereby engendering considerable recalcitrance upon the biofilm community. Similar mechanisms have been proposed for the hostile takeover of batch cultures by killer phenotypes during the stationary phase (Zambrano and Kolter 1995).

Quiescence and the General Stress Response

There has been much speculation over the ability of non-sporulating forms of bacteria to adopt spore-like states through the adoption of a quiescent state. Here, state-specific growth rates of the cells are approximately zero (Moyer and Morita 1989) as they undergo reductive divisions to complete existing rounds of chromosome replication under conditions of stress and starvation (Novitsky and Morita 1977a,b; Moyer and Morita 1989). Such cells have generally been associated with marine biofilms, where there is poor nutrient availability (Kjelleberg *et al.* 1982), but they have also been suggested to be the dominant form within other natural environments that experience a scarcity of nutrients (Morita 1986). Similar quiescence has recently been reported in Gram-positive bacteria (Lleo *et al.* 1998) and would appear to be a universal response to extreme nutrient stress (Marin *et al.* 1989). Such phenomena have been termed the general stress response (GSR) and lead to sub-populations that by virtue of their much reduced metabolism, and their synthesis of protective poly-phosphorylated nucleotides (Rhaese *et al.* 1975; Piggot and Coote 1976) are highly resistant to a wide range of metabolically active agents (Matin *et al.* 1989; Hengge-Aronis 1996). Various terms have been ascribed to such organisms, including quiescent (Trainor *et al.* 1999), dormant (Amy *et al.* 1983), resting (Munro *et al.* 1989), ultra-microbacteria (Novitsky and Morita 1977a,b) and somnicells (Roszak and Colwell 1987). It is highly likely that the same phenomena describe the viable non-culturable state (Barer and Harwood 1999). It is notable that the GSR regulator rpoS, and presumably quiescence and antibiotic resistance, has been shown to be highly expressed in clinical samples of biofilms taken directly from patients' sputum (Foley *et al.* 1999).

Perspectives on the Resistance of Biofilms

Resistance of microbial biofilms to a wide variety of antimicrobial agents is clearly associated with the organization of cells within an extensive exopolymer matrix. Such organization is able to moderate the concentrations of antimicrobial agents and antibiotics to which the more deeply lying members of the biofilm community are exposed. Such cells are, coincidentally, slow growing, starved and express stressed phenotypes that may include the up-regulation of efflux pumps. The expressed phenotype of the deeply seated biofilm community reduces their susceptibility to the treatment agents and exacerbates the likelihood of their being exposed sub-lethally. Ultimately, the failure of antimicrobial treatment relates to the presence of a small subset of persister cells within the biofilm community. These might reflect specific randomly generated clones or the localized expression of a GSR. Regardless, treatment strategies must either be sufficiently severe and aggressive as to target these persister phenotypes, or be specifically targeted to prevent their development.

CURRENT TREATMENT OF DEVICE-ASSOCIATED INFECTIONS

The first reports of biofilm recalcitrance towards even aggressive antibiotic therapy were coincident to the realization that implanted medical devices represented a significant risk of infection with skin microorganisms, such as staphylococci, enterococci, diphtheroids and others. In such situations, a common scenario emerged whereby bacteraemia was treated, and successfully resolved, but that these recurred shortly after treatment was stopped. Analysis of the isolated organisms showed these to be sensitive to the agents deployed. In many instances the same organisms were subsequently isolated from the implanted device, either when this was removed or at post-mortem (Marrie and Costerton 1984; Gristina and Costerton 1985; Costerton *et al.* 1987; Gristina *et al.* 1987, 1989). It became apparent that growth of otherwise non-pathogenic bacteria on the surfaces of implanted plastics and other biomaterials leads to a population of microorgansims that are almost completely intractable with conventional antibiotic treatments. The situation today is that many hospital policies and clinicians believe that antibiotic therapy without concomitant surgical intervention is of little use (Bisno and Waldvogel 2000).

Antibiotic Regimes

In spite of the poor prognosis associated with the use of antibiotics to treat device-associated infections there are a number of situations where this is the only available option. For example, in the management of prosthetic vascular graft infections, the combined systemic administration of vancomycin, aminoglycoside and a broad-spectrum cephalosporin is often considered to be appropriate in the absence of a specific susceptibility profile (Goeau-Brissonniere and Coggia 2000). After re-surgery, 4 to 6 weeks of parenteral administration followed by 6 months of oral administration is often advocated (Reilly *et al.* 1987; Bandyk 1995), but some authors have recommended lifelong administration of oral antibiotics in high-risk situations (Chan *et al.* 1989). Similarly, in the context of infections of prosthetic heart valves, treatment is commonly recommended over a 6-week period, having previously determined the most appropriate agent. Consequently, it is important that the aetiology of prosthetic valve endocarditis is not obscured by premature treatment (Karchmer 2000). Indolent endocarditis, which mandates early surgical intervention, does not require immediate antimicrobial therapy. Antibiotics should be withheld briefly pending the isolation and characterization of the causative organism.

In the treatment of infections associated with prosthetic joints, antibiotic therapy without surgical intervention is not considered to be an option. In one study of 25 patients, none had a satisfactory functional outcome after an average of 1.3 years follow up (Johnson and Bannister 1986); another study found that only 3 out of 13 prostheses were retained after a mean of 37.6 months among patients treated with chronic antimicrobial suppression (Steckelberg and Osmon 2000). Accordingly, prevention of device-associated infections is significantly more successful than a cure.

CURRENT APPROACHES OF PREVENTION OF DEVICE-ASSOCIATED INFECTIONS

When considering antimicrobial prophylaxis, it is important first to define the degrees of intimacy with which the various devices interact with the body. Bayston (1995) has categorized devices as follows. Category 1 devices are totally implanted in the tissues of the body and intended to remain in place for the life of the patient. Examples include large joint replacements, prosthetic heart valves, and hydrocephalus shunts. Category 2 devices are partially implanted, and are intended to remain *in situ* for long time periods (e.g. central venous catheters, external ventricular drains). Category 3 devices are not true implants, and include urinary catheters and voice prostheses. Since devices in categories 1 and 2 are situated in normally sterile tissues and body compartments, the risk of infection will be greatest during and immediately after implantation and will normally comprise of single species. Since category 3 devices are essentially open to the environment, challenge from successions of microorganisms throughout the period of implantation will occur. Accordingly, strategies for the successful prophylaxis and treatment of category 3 devices will differ from that of categories 1 and 2.

Antibacterial Prophylaxis

When considering category 1 implants, such as large joint replacements, once an implanted medical device becomes infected, the treatment is extremely problematic and rarely successful. When infection is detected, therapy with powerful antibiotics (e.g. aminoglycosides) is frequently attempted, although the inherent toxicity of these drugs makes therapy difficult. For example, the notorious resistance of biofilm infections makes therapy based on planktonic MICs of little use. Whereas a conventional broth MIC gives a value of $1\,mg\,l^{-1}$, the minimum biofilm eradication concentration (MBEC) may need to exceed $100\,mg\,ml^{-1}$ where the agent may be toxic at levels above $50\,mg\,l^{-1}$ (Bayston 1995; Ceri *et al.* 1999). Symptoms may be reduced for the duration of the antibiotic administration, only to relapse after cessation of treatment. In most cases, removal and replacement of the prosthesis are usually required to eradicate the infection, with associated patient trauma and increased cost (Dreghorn and Hamblen 1989). Every effort, therefore, should be made to prevent the initial colonization of the implant by making use of critically clean surgical procedures, clean-air technology (laminar-flow theatres) and by irrigating with antimicrobial agents (Petty *et al.* 1988).

Prophylactic intravenous antibiotic administration has become the accepted practice for procedures involving the implantation of prosthetic devices of categories 1 and 2, and is widely endorsed by most authorities (Guglielmo *et al.* 1983; Beam 1985). This practice is effective for large joint replacement. The general strategy is to administer an intravenous antibiotic, just before the initial skin incision (Hanssen and Osmon 1999).

Antimicrobial prophylaxis is regarded as the single most significant factor in the prevention of deep wound infection for prosthetic hip surgery (Ericson *et al.* 1973). Infections of a prosthetic hip are of three types: acute contiguous, chronic contiguous, and haematogenous. Acute contiguous infections result from contamination during the time of surgery and clinical manifestations of infection become apparent within 6 months. Chronic contiguous infections are diagnosed 6–24 months postoperatively and are usually caused by intra-operative contamination. Haematogenous seeding of prosthetic joints accounts for infections that develop up to 2 years or more post-surgery (Norden *et al.* 1992).

Penicillins, cephalosporins, vancomycin, and aminoglyocides are used in various combinations. Since pathogen spectrum is so variable, no single therapeutic combination is ideal for all circumstances (Kaiser 1986). For example, where methicillin-resistant *S. aureus* (MRSA) are common, clinicians frequently substitute vancomycin for cephalosporins. Cefazolin was commonly used, since it exhibited good activity against methicillin-susceptible and Gram-negative rods, although increases in MRSA have since seen this agent fall from favour (Kaiser 1986). The extended use of antibiotics beyond the post-discharge period is discouraged because of dubious efficacy and the risk of selecting for antibiotic resistance.

The benefit of antibiotic prophylaxis for other types of implant is less clear. Variable success has been observed where vascular implants are used, although some studies demonstrated significantly reduced wound infections (Hasselgren *et al.* 1984; Worning *et al.* 1986). Importantly, however, in these studies, infections directly related to the implant material were not significantly reduced. Prophylaxis has also not been unambiguously beneficial when used with hydrocephalus shunts (Bayston 1995). Antibiotic use for the long risk period associated with category 2 devices has proved impracticable, as has use for those that fall into category 3. The lack of universal success of prophylactic antibiotic administration has led to strategies that attempt to make the implant intrinsically more colonization resistant (Bayston 1995).

Antimicrobial Polymers and Surface Coatings

Incorporation of antimicrobial compounds into various substrata has been attempted in the hope of developing materials that are intrinsically colonization resistant. With respect to category 1 implants, Darouiche *et al.* (1998) used an animal model to determine the effect of dipping stainless steel intramedullary nails in a mixture of chlorhexidine and chloroxylenol in a randomized controlled trial. Tibial fractures were created, the nails used for repair, and a bacterial inoculum of 10^6 CFU of *S. aureus* injected into the intramedullary canal. Coated nails were associated with significantly lower rates of device-related osteomyelitis ($P = 0.0003$).

Malaisrie *et al.* (1998) used an *in vivo* model to evaluate facial plastics and reconstruction materials, including titanium, silicone, ion-bombarded e-PTFE (Gore-Tex), e-PTFE with and without antimicrobial coatings (silver/chlorhexidine), for their resistance properties towards *S. aureus*. The authors suggested that the antiseptic agents impregnated into the biomaterials formed a protective coat of silver, chlorhexidine, and inflammatory cells that inhibited initial bacterial adhesion to the biomaterial surface.

Bayston (1995) studied the colonization resistance of silicone shunts, impregnated with a range of antibiotics, against single doses of *ca* 10^7 antibiotic-sensitive staphylococci and coryneforms. The shunts were perfused with liquid culture medium for 14 days before being examined for colonization. Whereas trimethoprim-, clindamycin-, spiromycin-, and sodium-fusidate-impregnated catheters did not resist colonization over this time period, those treated with rifampicin or combinations of rifampicin with either trimethoprim or clindamycin did. Following subsequent re-challenge, shunts impregnated with clindamycin and rifampicin remained uncolonized for up to 28 days and were able to resist colonization by a third challenge. This degree of protection was suggested to be sufficient to eliminate nosocomial infections associated with the implantation (Stanton and Bayston 1999).

The performance of urethral catheters coated with silver has also been ambiguous. Riley *et al.* (1995) studied a silver-coated indwelling device and failed to demonstrate efficacy in preventing urinary tract infection, and vascular catheters impregnated with silver sulphadiazine and chlorhexidine completely lost their antibacterial activity after 10 days of use (Schmitt *et al.* 1995). The evidence that these catheters resist bacterial colonization is therefore suspect (Stickler and Winters 1994; Stickler 1995). Such catheters are challenged with microorganisms throughout their implantation; indwelling medical devices, on the other hand, are only at risk from microorganisms during and immediately after implantation. Therefore, there will only be a minimal opportunity for early colonizers to deplete the antibacterial agents, and the subsequent diffusion and loss of the agents will be inconsequential. It is conceivable, therefore, that impregnation with the correct antibiotics may prove efficacious.

Such approaches are unlikely to be appropriate for all but short-term use of indwelling urinary catheters, where the likely contaminants will be Gram-negative bacteria. In this respect, other workers have reported that silicone catheters treated with ciprofloxacin failed to resist colonization by sensitive strains of *Proteus mirabilis*, *P. aeruginosa*, *E. coli* and *Providencia stuartii* over 48 h exposure periods (Stickler and Winters 1994). It seems likely that, in clinical situations, the efficacy of surface-coated devices may be compromised by antibiotic-resistant bacteria, together with the barrier effect of conditioning films that rapidly coat the biomaterials *in vivo* (Stickler 1995).

Probiotic Approaches

Fuller (1989) defined probiotics as 'live microbial feed supplements, which beneficially affects the host animal by improving its intestinal balance'. However, in reality, any therapeutic application of a live microbial preparation may arguably be considered to be a probiotic. The area of medical implants has recently seen successful probiotic applications.

Silicone rubber voice prostheses are used as a shunt-valve between the digestive tract and trachea to enable patients to speak following laryngectomy. They commonly become colonized with thick biofilms, consisting of various bacterial and yeast strains (Mahieu *et al.* 1986), which impairs the function of the device. Anecdotal evidence among patients has suggested that consumption of buttermilk (containing *Lactococcus lactis*) or large quantities of Turkish yogurt (containing *Streptococcus thermophilus*) prolongs the useful life of the device by inhibiting the attachment of yeasts. This may be analogous to the inhibition of uropathogen attachment to the epithelial cells in the gut (Reid *et al.* 1990). Busscher *et al.* (1997) have shown that *Streptococcus thermophilus* B synthesizes a biosurfactant, which significantly reduced initial yeast colonization densities, regardless of the presence of conditioning film.

FUTURE DEVELOPMENTS TO IMPROVE ANTIBIOTIC TREATMENT OF DEVICE-ASSOCIATED INFECTIONS

The use of antibiotics, *per se*, for the treatment of device-associated biofilm infections is seldom successful. As attempts to prevent the occurrence of such infection through prophylactic measures cannot be guaranteed, there have thus been a number of approaches that have attempted to increase the effectiveness of curative antibiotic treatments. Some of these approaches utilize cationic permeabilizers, such as protamine, to aid penetration into an extant biofilm, whereas others use physical agencies, such as ultrasound energy and bioelectric currents, to act in synergy with antibiotics.

Chemical Synergists of Antibiotic Action

Protamine sulphate is a mixture of basic peptide sulphates prepared from the testes of fish. It forms an inactive complex with heparin and, therefore, is used therapeutically to neutralize the haemorrhaging associated with heparin overdose. Early work with protamine has suggested that it reduces the attachment of bacteria to the bladder epithelium (Parsons *et al.* 1981). Further work showed protamine disrupted the glycosaminoglycans secreted by, and bound to, epithelial cells. Teichman *et al.* (1993) showed that protamine was bactericidal towards coagulase-negative staphylococci in its own right and hypothesized that such effects were brought about by a direct effect upon the proteoglycans of the organism's glycocalyx. Such effects have been capitalized upon in attempts to use protamine to disrupt the glycocalyx associated with biofilm infections. In this respect, Teichman *et al.* (1994) showed synergism between protamine and vancomycin when directed against biomaterial-associated infections. Soboh *et al.* (1995) also demonstrated synergy for this agent, but with ciprofloxacin treatment of *P. aeruginosa* biofilms.

Bioacoustic Enhancement of Antibiotic Action

The treatment of biofilms associated with 'long-term' medical implants is rarely successful by traditional antibiotic therapies (Khoury *et al.* 1992; Darouiche *et al.* 1994), and new methods to eradicate recalcitrant biofilms *in vivo* are being sought. One novel method currently being developed is the application of ultrasound in conjunction with exposure of the biofilm to antibiotics (Qian *et al.* 1999). Initial work by Pitt *et al.* (1994) has shown that planktonic *P. aeruginosa, E. coli, S. epidermidis* and *S. aureus* are very susceptible to the dual treatment of ultrasound and gentamicin, but not to either treatment alone. In the few preliminary investigations that have been carried out, it has been demonstrated that low-frequency ultrasound in combination with aminoglycoside treatments removes most, if not all, viable bacteria within biofilms on polymeric biomaterials *in vitro* (Johnson *et al.* 1998) and *in vivo* (Rediske *et al.* 2000). However, the application of ultrasound without the addition of antibiotic has no unambiguous effect on bacterial viability or on the structure of biofilms. Although adequate evidence has yet to be presented, it is likely that ultrasound perturbs bacterial cell membranes by cavitation, which stimulates the active and/or passive uptake of antibiotics (Qian *et al.* 1997, 1999). As a direct consequence of cavitation of regions within the biofilm, micro-streaming will also exacerbate the transfer of antibiotic into the biofilm (Qian *et al.* 1999) and may alter the biofilm structure.

Trials are currently under way to examine the validity of ultrasound to enhance biofilm attack in medical scenarios (Qian *et al.* 1999), and although the use of ultrasound is only successful in concert with antibiotics, it offers an attractive method for specifically targeting and eradicating bacterial cells within, and located near to, biofilms.

Bioelectric Enhancement of Antibiotic Action

The bioelectric effect, in which electric fields are used to enhance the efficacy of charged biocides and antibiotics in killing biofilm bacteria, has been shown *in vitro* to reduce bacterial populations within biofilms as successfully as the same population in a planktonic state (Costerton *et al.* 1994). The *in vitro* efficacy of killing of biofilm bacteria on conductive surfaces with a current of $< 100\,\mu A\,cm^{-2}$ being generated and an exposure of a charged antibiotic, such as tobramycin, is approximately eight logarithmic orders of magnitude greater than treatment with the same level of antibiotic alone (Costerton *et al.* 1994; Jass and Lappin-Scott 1996;

Wellman *et al.* 1996). Current alone does not have any significant effect on bacteria within biofilms (Wellman *et al.* 1996).

The mechanism of the bioelectric effect is still not completely understood, but it has been postulated that it is a form of electrophoresis in which charged antibiotics are driven into the biofilm by the electric current (Jass *et al.* 1995). Stoodley *et al.* (1997) have demonstrated that biofilms under electric-field-stress contract and become reduced in thickness, which may also facilitate additional permeation of antibiotic. Although no *in vivo* studies have been conducted, the treatment of biofilms by the application of bioelectric fields, with an accompanying charged antibiotic exposure, may well offer a practical alternative for the *in situ* treatment of microbially contaminated implants within patients. However, further studies are needed to determine the validity of bioelectric techniques in the treatment of established multi-species biofilms.

DEVELOPMENT OF NOVEL ANTI-BIOFILM AGENTS

Quorum-Sensing and Intercellular Signalling

It is now widely postulated that quorum-sensing of localized high cell densities is a primary modulator not only of the biofilm phenotype but also of the expression of many virulence factors associated with the manifestations of infection. It is not surprising, therefore, that interference with such signalling processes has been suggested as presenting a potential approach to the prevention and treatment of biofilm-related infection (Finch *et al.* 1998; Hardman and Wise 1998).

Quorum-sensing transcriptional regulation is a concerted, population-specific event of genetic repression and/or activation. The regulation of the process is achieved through signalling molecules that are produced constitutively by the bacterial cell and released to the environment. The signal re-enters the cell and is detected through its corresponding receptor. When a threshold concentration is reached, transcription/repression of the target genes occurs. This process is referred to as quorum-sensing. Quorum-sensing mechanisms have been shown to operate in both Gram-positive (Kleerebezem *et al.* 1997) and Gram-negative organisms (Salmond *et al.* 1995; Fuqua *et al.* 1996; Eberl 1999). The effector molecules vary (Shapiro 1998; Eberl 1999; Withers *et al.* 2001); Gram-positives utilize short peptide chains, and Gram-negatives produce *N*-acylated-L-homoserine lactones (AHLs). Even though Gram-positive organisms cause the majority of implant-associated infections, and there are different effector molecules being utilized, the mechanism with Gram-negative organisms is in essence similar.

The quorum-sensing cell–cell signalling system has been shown to coordinate the expression of virulence factors in a number of organisms (Kleerebezem *et al.* 1997; Williams *et al.* 2000). In a recent review, Rumbaugh *et al.* (2000) detailed extensive studies of the opportunistic pathogen *P. aeruginosa* and indicated that quorum-sensing plays a role in the expression of virulence factors in situations ranging from within the lung of the cystic fibrosis patient, to corneal and burn wound infections. Additionally, they also detailed the modulation of host response during infection through autoinducer molecules. Furthermore, quorum-sensing has been implicated in the formation and development of biofilms (Allison *et al.* 1998; Davies *et al.* 1998; Glessner *et al.* 1999). It has been proposed that bacterial virulence and pathogenicity may be controlled through external regulation of the quorum-sensing mechanism.

Anti-signals and Signal Analogues

Owing to the role of quorum-sensing mechanisms in the regulation of virulence factors and biofilm processes, therapeutic agents that act as blocking agents for the signals have been suggested. Rather than aiming to eradicate the pathogen, these would seek to prevent morbidity, either through attenuating the pathogenicity of an organism or by affecting the organism's ability to form a biofilm, a much more resilient and persistent mode than the planktonic state. Studies incorporating such approaches are already under way.

Natural compounds that interfere with the system have been identified (Manefield *et al.* 1999; Rasmussen *et al.* 2000). The marine alga *Delisea pulchra* produces halogenated furanones, which are structurally similar to HSLs. These interfere with the signalling pathway by acting as antagonists of the signal by competing for the binding site. In a mouse skin abscess model, virulence gene expression has been altered in *S. aureus* through blocking of the cyclic thiolactone (Mayville *et al.* 1999). A synthetic blocking agent that interferes with the signal has also been developed (Kline *et al.* 1999).

At present, studies of quorum-sensing are still in their infancy, with many questions concerning the regulation of this system. Furthermore, the response is not always predictable: quorum-sensing blocking in *S. aureus* led to enhanced biofilm formation (Vuong *et al.* 2000). Further detailed studies are required, and amongst the difficulties encountered will be accounting for the effect of the host microbial communities and potential host interactions as well. It is also worth noting that such novel methods of regulation are at present limited—most of the research has been done in Gram-negative organisms, with particular reference to *P. aeruginosa*. Additionally, even if virulence is attenuated or biofilm formation circumvented, the patient will still need a functioning immune system to mount an appropriate response to eradicate the potential pathogen.

CONCLUSIONS

Biofilm infections associated with indwelling medical devices and implants are almost impossible to resolve with conventional antibiotic therapies. If these are to stand any chance of success, then either they must be administered over an extended period of time, in combinations, and at high dose, or they must be accompanied with some form of surgical intervention, usually involving the temporary removal of the device and removal of associated tissue. Prevention is a much better alternative to cure. In this respect, the prophylactic administration of antibiotics orally or parenterally prior to and immediately after surgery is appropriate for category 1 and category 2 devices. Similarly, significant advantage has been demonstrated when the antibiotics are applied directly to the implantation site during the surgical procedure, or where they are coated onto the device and released locally.

Developments to perturb the colonization process, by modification of the biomaterials, the co-application of bioacoustic sound or low-voltage electric currents with antibiotics and the development of specific anti-biofilm agents, are still in their infancy, and very much the preserve of the laboratory rather than the clinic.

REFERENCES

Allison, DG, Ruiz, B, SanJose, C, Jaspe, A and Gilbert, P (1998) Extracellular products as mediators of the formation and detachment of *Pseudomonas fluorescens*. *FEMS Microbiology Letters* **167**:179–184.

Allison, DG, McBain, AJ and Gilbert, P (2000) Microbial biofilms: problems of control. In: Community Structure and Cooperation in Biofilms (Eds. Allison, DG, Gilbert, P, Lappin-Scott, H and Wilson, M), Society for General Microbiology Press, Reading, UK, pp. 309–327.

Amy, PS, Pauling, C and Morita, RY (1983) Recovery from nutrient starvation by a marine *Vibrio* sp. *Applied and Environmental Microbiology* **45**:1685–1690.

Ashby, MJ, Neale, JE, Knott, SJ and Critchley, IA (1994) Effect of antibiotics on non-growing cells and biofilms of *Escherichia coli*. *Journal of Antimicrobial Chemotherapy* **33**:443–452.

Bandyk, DF (1995) Surgical management of vascular graft infections. In: Perspectives in Vascular Surgery, 4th edition (Ed. Goldstone, J), Quality Medical Publishing, St Louis, MO, pp. 1–13.

Barer, MR and Harwood, CR (1999) Bacterial viability and culturability. *Advances in Microbial Physiology* **41**:93–137.

Bayston, R (1995) A process for prevention of device related infection: results of challenge *in vitro* and duration of antimicrobial activity. In: The Life and Death of Biofilm (Eds. Wimpenny, J, Handley, P, Gilbert, P and Lappin-Scott, H), BioLine, Cardiff, pp. 149–152.

Beam, TR, Jr (1985) Perioperative prevention of infection in cardiac surgery. *Antibiotic Chemotherapy* **33**:114–139.

Bigger, JW (1944) Treatment of staphylococcal infection with penicillin. *Lancet* **ii**: 497–500.

Bisno, AL and Waldvogel, FA (2000) In: Infections Associated with Indwelling Medical Devices (Eds. Waldvogel, FA and Bisno, AL), 3rd edition, ASM Press, Washington, DC.

Black, DS, Kelly, AJ, Mardis, MJ and Moyed, HS (1991) Structure and organisation of *htp*, an operon that affects lethality due to inhibition of peptidoglycan or DNA synthesis. *Journal of Bacteriology* **173**:5732–5739.

Boutibonnes, P (1997) La mort des bacteries: provoquee ou program-mee? Subie ou voulue? *Medicine Sciences* **13**:73–80.

Brooun, A, Liu, S and Lewis, K (2000) A dose–response study of antibiotic resistance in *Pseudomonas aeruginosa* biofilms. *Antimicrobial Agents and Chemotherapy* **44**:540–546.

Brown, MRW, Collier, PJ and Gilbert, P (1990) Influence of growth-rate on the susceptibility to antimicrobial agents: modification of the cell envelope and batch and continuous culture. *Antimicrobial Agents and Chemotherapy* **34**:1623–1628.

Busscher, HJ, van Hoogmoed, CG, Geertsema-Doornbursch, GI, van der Kuijl-Booij, M and van der Mei, HC (1997) *Streptococcus thermophilus* and its biosurfactants inhibit adhesion by *Candida* spp. of silicone rubber. *Applied and Environmental Microbiology* **63**:3810–3817.

Cellini, L, Robuffo, I, Di Campli, E, Di Bartolomeo, S, Taraborelli, T and Dainelli, B (1998) Recovery of *Helicobacter pylori* ATCC43504 from a viable but not culturable state: regrowth or resuscitation? *Acta Pathologica, Microbiologica, et Immunologica Scandinavica* **106**:571–579.

Ceri, H, Olson, ME, Stremick, C, Read, RR, Morck, D and Buret, A (1999) The Calgary biofilm device: new technology for rapid determination of antibiotic susceptibilities of bacterial biofilms. *Journal of Clinical Microbiology* **37**:1771–1776.

Chan, FY, Crawford, ES, Coselli, H, Safti, J and Williams, TJ (1989) *In situ* prosthetic graft replacement for mycotic aneurysm of the aorta. *Annals of Thoracic Surgery* **47**:193–303.

Costerton, JW, Cheng, KJ, Geesey, GG, Ladd, TI, Nickel, JC, Dasgupta, M and Marrie, TJ (1987) Bacterial biofilms in nature and disease. *Annual Reviews in Microbiology* **41**:435–464.

Costerton JW, Ellis, B, Lam, K, Johnson, F and Khoury, AE (1994) Mechanism of electrical enhancement of efficacy of antibiotics in killing biofilm bacteria. *Antimicrobial Agents and Chemotherapy* **38**:2803–2809.

Darouiche, RO, Dhir, A, Miller, AJ, Landon, GC, Raad, II and Musher, DM (1994) Vancomycin penetration into biofilm covering infected prostheses and effect on bacteria. *Journal of Infectious Disease* **170**:720–723.

Darouiche, RO, Farmer, J, James, J, Chaput, C, Mansouri, M, Salah, G and London, GC (1998) Anti-infective efficacy of antiseptic-coated intramedullary nails. *Journal of Bone and Joint Surgery, American Volume* **80**:1336–1340.

Das, JR, Bhakoo, M, Jones, MV and Gilbert, P (1998) Changes in the biocide susceptibility of *Staphylococcus epidermidis* and *Escherichia coli* cells associated with rapid attachment to plastic surfaces. *Journal of Applied Microbiology* **84**:852–858.

Davies, DG, Parsek, MR, Pearson, JP, Iglewski, BH, Costerton, JW and Greenberg, EP (1998) The involvement of cell-to-cell signals in the development of a bacterial biofilm. *Science* **280**:295–298.

Dreghorn, CR and Hamblen, DL (1989) Revision arthroplasty: a high price to pay. *British Medical Journal* **298**:648–649.

Eberl, L (1999) *N*-Acyl homoserine lactone-mediated gene regulation in Gram-negative bacteria. *Systematic and Applied Microbiology* **22**:493–506.

Engelberg-Kulka, H and Glaser, G (1999) Addiction modules and programmed cell death and anti-death in bacterial cultures. *Annual Review of Microbiology* **53**: 43–70.

Ericson, C, Lidgren, L and Lidberg, L (1973) Cloxacillin in the prophylaxis of postoperative infections of the hip. *Journal of Bone and Joint Surgery, American Volume* **55**:808–843.

Finch, RG, Pritchard, DI, Bycroft, BW, Williams, P and Stewart, GSAB (1998) Quorum sensing: a novel target for anti-infective therapy. *Journal of Antimicrobial Chemotherapy* **42**:569–571.

Foley, I, Marsh, P, Wellington, EM, Smith, AW and Brown, MR (1999) General stress response master regulator *rpoS* is expressed in human infection: a possible role in chronicity. *Journal of Antimicrobial Chemotherapy* **43**:164–165.

Franch, T and Gerdes, K (1996) Programmed cell death in bacteria: translational repression by mRNA end-pairing. *Molecular Microbiology* **21**:1049–1060.

Fujiwara, S, Miyake, Y, Usui, T and Suginaka, H (1998) Effect of adherence on antimicrobial susceptibility of *Pseudomonas aeruginosa, Serratia marcescens* and *Proteus mirabilis. Hiroshima Journal of Medicine* **47**:1–5.

Fuller, R (1989) Probiotics in man and animals. *Journal of Applied Bacteriology* **66**:365–378.

Fuqua, WC, Winans, SC and Greenberg, EP (1996) Census and consensus in bacterial ecosystems: the *LuxR–LuxI* family of quorum-sensing transcriptional regulators. *Annual Review of Microbiology* **50**:727–751.

George, AM and Levy, SB (1983) Amplifiable resistance to tetracycline, chloramphenicol, and other antibiotics in *Escherichia coli*: involvement of a non-plasmid-determined efflux of tetracycline. *Journal of Bacteriology* **155**:531–540.

Gilbert, P and Allison, DG (1999) Biofilms and their resistance towards antimicrobial agents. In: Dental Plaque Revisited: Oral Biofilms in Nature and Disease (Eds. Newman, HN and Wilson, M), BioLine, Cardiff, pp. 125–143.

Gilbert, P, Collier, PJ and Brown, MRW (1990) Influence of growth rate on susceptibility to antimicrobial agents: biofilms, cell cycle and dormancy. *Antimicrobial Agents and Chemotherapy* **34**:1865–1868.

Gilbert, P, Das, JR, Jones, M and Allison, DG (2001) Assessment of the biocide activity upon various bacteria following their attachment to and growth on surfaces. *Journal of Applied Microbiology* **91**:248–255.

Giwercman, B, Jensen, ET, Hoiby, N, Kharazmi, A and Costerton, JW (1991) Induction of *β*-lactamase production in *Pseudomonas aeruginosa* biofilms. *Antimicrobial Agents and Chemotherapy* **35**:1008–1010.

Glessner, A, Smith, RS, Iglewski, BH and Robinson, JB (1999) Roles of *Pseudomonas aeruginosa las* and *rhl* quorum-sensing systems in control of twitching motility. *Journal of Bacteriology* **181**:1623–1629.

Goeau-Brissonniere, OA and Coggia, M (2000) Arterial prosthetic infections. In: Infections Associated with Indwelling Medical Devices (Eds. Waldvogel, FA and Bisno, AL), 3rd edition, ASM Press, Washington, pp. 127–144.

Gristina, AG and Costerton, JW (1985) Bacterial adherence to biomaterials and tissue. The significance of its role in clinical sepsis. *American Journal of Joint and Bone Surgery* **67**:264–273.

Gristina, AG, Hobgood, CD, Webb, LX and Myrvik, QN (1987) Adhesive colonisation of biomaterials and antibiotic resistance. *Biomaterials* **8**:423–426.

Gristina, AG, Jennings, RA, Naylor, PT, Myrvik, QN and Webb, LX (1989) Comparative *in vitro* antibiotic-resistance of surface-colonizing coagulase-negative staphylococci. *Antimicrobial Agents and Chemotherapy* **33**:813–816.

Guglielmo, BJ, Hohn, DC, Koo, PJ, Hunt, TK, Sweet, RL and Conte, JE, Jr (1983) Antibiotic prophylaxis in surgical procedures: a critical review of the literature. *Archives of Surgery* **118**:943–955.

Hanssen, AD and Osmon, DR (1999) The use of prophylactic antimicrobial agents during and after hip arthroplasty. *Clinical Orthopaedics* **369**:124–138.

Hardman, G and Wise, R (1998) Quorum sensing: potential means of treating Gram-negative infections? *Lancet* **351**:848–849.

Hasselgren, P, Ivarsson, L, Risberg, B and Seeman, T (1984) Effects of prophylactic antibiotics in vascular surgery. *Annals of Surgery* **200**:86–92.

Hengge-Aronis, R (1996) Back to log phase: sigma S as a global regulator in the osmotic control of gene expression in *Escherichia coli*. *Molecular Microbiology* **21**:887–893.

Hochman, A (1997) Programmed cell death in prokaryotes. *Critical Reviews in Microbiology* **23**:207–214.

Hoyle, BD, Wong, CK and Costerton, JW (1992) Disparate efficacy of tobramycin on Ca(2+)-, Mg(2+)-, and HEPES-treated *Pseudomonas aeruginosa* biofilms. *Canadian Journal of Microbiology* **38**:1214–1218.

Huang, CT, Yu, FP, McFeters, GA and Stewart, PS (1995) Non-uniform spatial patterns of respiratory activity within biofilms during disinfection. *Applied and Environmental Microbiology* **61**:2252–2256.

Jass, J and Lappin-Scott, HM (1996) The efficacy of antibiotics enhanced by electrical currents against *Pseudomonas aeruginosa* biofilms. *Journal of Antimicrobial Chemotherapy* **38**:987–1000.

Jass, J, Costerton, JW and Lappin-Scott, HM (1995) The effect of electrical currents and tobramycin against *Pseudomonas aeruginosa* biofilms. *Journal of Industrial Microbiology* **15**:234–242.

Jensen, RB and Gerdes, K (1995) Programmed cell death in bacteria: proteic plasmid stabilised systems. *Molecular Microbiology* **17**:205–210.

Johnson, DP and Bannister, GC (1986) The outcome of infected arthroplasty of the knee. *Journal of Bone and Joint Surgery, British Volume* **68**:289–291.

Johnson, LL, Peterson, RV and Pitt, WG (1998) Treatment of bacterial biofilms on polymeric biomaterials using antibiotics and ultrasound. *Journal of Biomaterial Science, Polymer Edition* **9**:1177–1185.

Kaiser, AB (1986) Antimicrobial prophylaxis in surgery. *New England Journal of Medicine* **315**:1129–1138.

Karchmer, AW (2000) Infections of prosthetic heart valves. In: Infections Associated with Indwelling Medical Devices (Eds. Waldvogel, FA and Bisno, AL), 3rd edition, ASM Press, Washington, pp. 145–172.

Khoury, AE, Lam, K, Ellis, B and Costerton, JW (1992) Prevention and control of bacterial infections associated with medical devices. *American Society for Artificial Internal Organs Transplantation* **38**:M174–M178.

Kjelleberg, S, Humphrey, BA and Marshall, SC (1982) Effects of interfaces on small, starved marine bacteria. *Applied and Environmental Microbiology* **43**:1166–1172.

Kleerebezem, M, Quadri, LEN, Kuipers, OP and de Vos, WM (1997) Quorum sensing by peptide pheromones and two-component signal-transduction systems in Gram-positive bacteria. *Molecular Microbiology* **24**:895–904.

Kline, T, Bowman, J, Iglewski, BH, de Kievit, T, Kakai, Y and Passador, L (1999) Novel synthetic analogues of the *Pseudomonas* autoinducer. *Bioorganic and Medicinal Chemistry Letters* **9**:3447–3552.

Koch, AL (1987) Similarities and differences of individual bacteria within a clone. In: *Escherichia coli* and *Salmonella*: Cellular and Molecular Biology (Eds. Neidhardt FC, Curtiss, RI, Ingraham, JL, Lin, CCI, Low, KB, Magasanik, B, Reznikoff, WS, Riley, M, Schaechter, M and Umbarger, HE), ASM Press, Washington, DC, pp. 1640–1651.

Lambert, PA, Giwercman, B and Høiby, N (1993) Chemotherapy of *Pseudomonas aeruginosa* in cystic fibrosis. In: Bacterial Biofilms and their Control in Medicine and Industry (Eds. Wimpenny, JT, Nichols, WW, Stickler, D and Lappin-Scott, HM), BioLine, Cardiff, pp. 151–153.

Lewis, K (2000) Programmed cell death in bacteria. *Microbiology and Molecular Biology Reviews* **64**:503–514.

Lewis, K (2001) Riddle of biofilm resistance. *Antimicrobial Agents and Chemotherapy* **45**:999–1007

Lleo, MD, Tafi, MC and Canepari, P (1998) Nonculturable *Enterococcus faecalis* cells are metabolically active and capable of resuming active growth. *Systematic and Applied Microbiology* **21**:333–339.

Ma, D, Cook, DN, Alberti, M, Pong, NG, Nikaido, H and Hearst, JE (1993) Molecular cloning and characterization of *acrAB* and *acrE* genes of *Escherichia coli*. *Journal of Bacteriology* **175**:6299–6313.

Mahieu, HF, van Saene, JJM, den Besten, J and van Saene, KF (1986) Oropharynx decontamination preventing *Candida* vegetation on voice prostheses. *Archives of Otolaryngology, Head and Neck Surgery* **112**:1090–1092.

Maira-Litran, T, Allison, DG and Gilbert, P (2000a) An evaluation of the potential role of the multiple antibiotic resistance operon (*mar*) and the multi-drug efflux pump *acr*AB in the resistance of *E. coli* biofilms towards ciprofloxacin. *Journal of Antimicrobial Chemotherapy* **45**:789–795.

Maira-Litran, T, Allison, DG and Gilbert, P (2000b) Expression of the multiple resistance operon (*mar*) during growth of *Escherichia coli* as a biofilm. *Journal of Applied Microbiology* **88**:243–247.

Malaisrie, SC, Malekzadeh, S and Biedlingmeier, JF (1998) *In vivo* analysis of bacterial biofilm formation on facial plastic bioimplants. *Laryngeoscope* **108**:1733–1738.

Manefield, M, de Nys, R, Kumar, N, Read, R, Givskov, M, Steinberg, P and Kjelleberg, S (1999) Evidence that halogenated furanones from *Delisea pulchra* inhibit acylated homoserine lactone (AHL)-mediated gene expression by displacing the AHL signal from its receptor protein. *Microbiology* **145**:283–291.

Marin, A, Auger, A, Blum, PH and Schultz, JE (1989) Genetic basis of starvation survival in non-differentiating bacteria. *Annual Reviews of Microbiology* **43**: 293–316.

Marrie, TJ and Costerton, JW (1984) Morphology of bacterial attachment to cardiac pacemaker leads and power packs. *Journal of Clinical Microbiology* **19**: 911–914.

Matin, A, Auger, EA, Blum, PH and Schultz, JE (1989) Genetic basis of starvation survival in non-differentiating bacteria. *Annual Reviews of Microbiology* **43**:293–316.

Mayville, P, Ji, G, Beavis, R, Yang, G, Goger, M, Novick, RP and Weir, TW (1999) Structure–activity analysis of synthetic autoinducing thiolactone peptides from *Staphylococcus aureus* responsible for virulence. *Proceedings of the National Academy of Sciences of the United States of America* **96**:1218–1223.

McMurry, LM, Oethinger, M and Levy, SB (1998a) Triclosan targets lipid synthesis. *Nature* **394**:531–532.

McMurry, LM, Oethinger, M and Levy, SB (1998b) Over expression of *mar*A, *sox*S, or *acr*AB produces resistance to Triclosan in laboratory and clinical strains of *Escherichia coli. FEMS Microbiology Letters* **166**:305–309.

Miller, PF and Sulavick, MC (1996) Overlaps and parallels in the regulation of intrinsic multiple antibiotic resistance in *Escherichia coli. Molecular Microbiology* **21**:441–448.

Moken, MC, McMurry, LM and Levy, SB (1997) Selection of multiple antibiotic resistant (*mar*) mutants of *Escherichia coli* by using the disinfectant pine-oil: roles of the *mar* and *acrAB* loci. *Antimicrobial Agents and Chemotherapy* **41**:2770–2772.

Morita, RY (1986) Starvation survival: the normal mode of most bacteria in the ocean. In: Proceedings 4th International Symposium of Microbial Ecology, Slovene Society for Microbiology, pp. 242–248.

Moyed, HS and Bertrand, KP (1983) *hlp*A, a newly recognised gene of *E. coli* that affects frequency of persistence after inhibition of murein synthesis. *Journal of Bacteriology* **155**:768–775.

Moyer, CL and Morita, RY (1989) Effect of growth rate and starvation-survival on the viability and stability of a psychrophilic marine bacterium. *Applied and Environmental Microbiology* **55**:1122–1127.

Munro, PM, Gauthier, MJ, Breittmayer, VA and Bongiovanni, J (1989) Influence of osmoregulation processes on starvation survival of *Escherichia coli* in seawater. *Applied and Environmental Microbiology* **55**:2017–2024.

Naito, T, Kusano, K and Kobayashi, I (1995) Selfish behaviour of restriction modification systems. *Science* **267**:897–899.

Nikaido, H (1996) Multidrug efflux pumps of Gram-negative bacteria. *Journal of Bacteriology* **178**:5853–5859.

Nikaido, H (1998) Antibiotic resistance caused by Gram-negative multidrug efflux pumps. *Clinical Infectious Disease* **27**:S32–S41.

Norden, C, Nelson, JD, Mader, JT and Calandra, C (1992) Evaluation of new anti-infective drugs for the treatment of infections of prosthetic hip joints. *Clinical Infectious Diseases* **15**(Suppl. 1):S177–S181.

Novitsky, JA and Morita, RY (1977a) Morphological characterisation of small cells resulting from nutrient starvation of a psychrophilic marine vibrio. *Applied and Environmental Microbiology* **32**:617–622.

Novitsky, JA and Morita, RY (1977b) Survival of a psychrophilic marine vibrio under long-term nutrient starvation. *Applied and Environmental Microbiology* **33**: 635–641.

Parsons, CL, Stauffer, C and Schmidt, JD (1981) Impairment of antibacterial effect of bladder surface mucin by protamine sulphate. *Journal of Infectious Disease* **144**:180–184.

Petty, W, Spanier, S and Shuster, JJ (1988) Prevention of infection after total joint replacement. *American Journal of Bone and Joint Surgery* **70**:536–539.

Piggot, PJ and Coote, JG (1976) Genetic aspects of bacteria enospore formation. *Bacteriological Reviews* **40**:908–962.

Pitt, WG, McBride, MO, Lunceford, JK, Roper, RJ and Sagers, RD (1994) Ultrasonic enhancement of antibiotic action on Gram-negative bacteria. *Antimicrobial Agents and Chemotherapy* **38**:2577–2582.

Qian, Z, Sagers, RD and Pitt, WG (1997) The effect of ultrasonic frequency upon enhanced killing of *Pseudomonas aeruginosa* biofilms. *Annals of Biomedical Engineering* **25**:69–76.

Qian, Z, Sagers, RD and Pitt, WG (1999) Investigation of the mechanism of the bioacoustic effect. *Journal of Biomedical Materials Research* **44**:198–205.

Rasmussen, TB, Manefield, M, Andersen, JB, Eberl, L, Anthoni, U, Christophersen, C, Steinberg, P, Kjelleberg, S and Givskov, M (2000) How *Delisea pulchra* furanones affect quorum sensing and swarming motility in *Serratia liquefaciens* MG1. *Microbiology* **146**:3237–3244.

Rediske, AM, Roeder, BL, Nelson, JL, Robison, RL, Schaalje, GB, Robison, RA and Pitt, WG (2000) Pulsed ultrasound enhances the killing of *Escherichia coli* biofilms by aminoglycoside antibiotics *in vivo*. *Antimicrobial Agents and Chemotherapy* **44**:771–772.

Reid, G, Bruce, AW, McGroaty, JA, Cheng, K-J and Costeron, JW (1990) Is there a role for lactobacilli in prevention of urogenital and intestinal infections? *Clinical Microbiology Reviews* **3**:335–344.

Reilly, LM, Stoney, RJ, Gioldstone, J and Ehrenfeld, WK (1987) Improved management of aortic graft infection: the influence of operation sequence and staging. *Journal of Vascular Surgery* **5**:421–431.

Rhaese, H, Dichtelmüller, R, Grade, R and Groscurth, R (1975) High phosphorylated nucleotides involved in regulation of sporulation in *Bacillus subtilis*. In: Spores, VI (Eds. Gerhardt, P, Costliow, RN and Sadoff, HL), American Society for Microbiology, Washington, DC, pp. 335–340.

Riley, DK, Classen, DC, Stevens, LE and Burke, JP (1995) A large randomised clinical trial of a silver-impregnated urinary catheter: lack of efficacy and staphylococcal superinfection. *American Journal of Medicine* **98**:349–356.

Roszak, DB and Colwell, RR (1987) Metabolic activity of bacterial cells enumerated by direct viable count. *Applied and Environmental Microbiology* **53**:2889–2893.

Rouch, DA, Cram, DS, Dibernadino, D, Littlejohn, TG and Skurray, RA (1990) Efflux-mediated antiseptic gene *qacA* from *Staphylococcus aureus*: common ancestry with tetracycline and sugar transport proteins. *Molecular Microbiology* **4**:2051–2062.

Rumbaugh, KP, Griswold, JA and Hamood, AN (2000) The role of quorum sensing in the *in vivo* virulence of *Pseudomonas aeruginosa*. *Microbes and Infection* **2**:1721–1731.

Salmond, GPC, Bycroft, BW, Stewart, GSAB and Williams, P (1995) The bacterial 'enigma': cracking the code of cell–cell communication. *Molecular Microbiology* **16**:615–624.

Schmitt, SK, Knapp, C, Hall, GS, Longworth, DL, McMahon, JT and Washington, JA (1995) Impact of chlorhexidine–silver sulfadiazine-impregnated central venous catheters on *in vitro* quantitation of catheter-associated bacteria. *Journal of Clinical Microbiology* **34**:508–511.

Shapiro, JA (1998) Thinking about bacterial populations as multicellular organisms. *Annual Review of Microbiology* **52**:81–104.

Soboh, F, Khoury, AE, Zamboni, AC, Davidson, D and Mittelman, MW (1995) Effects of ciprofloxacin and protamine sulfate combinations against catheter associated *Pseudomonas aeruginosa* biofilms. *Antimicrobial Agents and Chemotherapy* **39**:1281–1286.

Stanton, C and Bayston, R (1999) Use of 'antimicrobial' polymers to prevent device-related infections In: Biofilms: The Good, the Bad and the Ugly (Eds. Wimpenny, J, Gilbert, P, Walker, J, Brading, M and Bayston, R), BioLine, Cardiff, pp. 53–63.

Steckelberg, JM and Osmon, DR (2000) Prosthetic joint infections. In: Infections Associated with Indwelling Medical Devices (Eds. Waldvogel, FA and Bisno, AL), 3rd edition, ASM Press, Washington, DC, pp. 173–209.

Stewart, PS (1996) Theoretical aspects of antibiotic diffusion into microbial biofilms. *Antimicrobial Agents and Chemotherapy* **40**:2517–2522.

Stewart, PS, Grab, L and Diemer, JA (1998) Analysis of biocide transport limitation in an artificial biofilm system. *Journal of Applied Microbiology* **85**:495–500.

Stickler, D (1995) The encrustation of urethral catheters. In: The Life and Death of Biofilm (Eds. Wimpenny, J, Handley, P, Gilbert, P and Lappin-Scott, H), BioLine, Cardiff, pp. 119–125.

Stickler, DJ and Winters, C (1994) Biofilms and urethral catheters. In: Bacterial Biofilms and their Control in Medicine and Industry (Eds. Wimpenny, J, Nichols, W, Stickler, D and Lappin-Scott, H), BioLine, Cardiff, pp. 97–104.

Stoodley, P, deBeer, D and Lappin-Scott, HM (1997) Influence of electric fields and pH on biofilm structure as related to the bioelectric effect. *Antimicrobial Agents and Chemotherapy* **41**:1876–1879.

Suci, PA, Mittelman, MW, Yu, FU and Geesey, GG (1994) Investigation of ciprofloxacin penetration into *Pseudomonas aeruginosa* biofilms. *Antimicrobial Agents and Chemotherapy* **38**:2125-2133.

Teichman, JM, Stein, PC and Parsons, L (1993) Quaternary amine (protamine sulfate) is bactericidal to *Staphylococcus epidermidis*. *Journal of Infectious Disease* **157**:1500–1501.

Teichman, JM, Abraham, VE, Stein, PC and Parsons, L (1994) Protamine sulfate and vancomycin are synergistic towards *Staphylococcus epidermidis* prostheses infection *in vivo*. *Journal of Urology* **152**:213–216.

Trainor, VC, Udy, RK, Bremer, PJ and Cook, GM (1999) Survival of *Streptococcus pyogenes* under stress and starvation. *FEMS Microbiology Letters* **176**:421–428.

Vuong, C, Saenz, HL and Götz, F (2000) Impact of the *agr* quorum-sensing system on adherence to polystyrene in *Staphylococcus aureus*. *The Journal of Infectious Diseases* **182**:1688–1693.

Wellman, N, Fortun, SM and McLeod, BR (1996) Bacterial biofilms and the bioelectric effect. *Antimicrobial Agents and Chemotherapy* **40**:2012–2014.

Wentland, EJ, Stewart, PS, Huang, CT and McFeters, GA (1996) Spatial variations in growth rate within *Klebsiella pneumoniae* colonies and biofilm. *Biotechnology Progress* **12**:316–321.

Williams, P, Camara, M, Hardman, A, Swift, S, Milton, D, Hope, VJ, Winzer, K, Middleton, B, Pritchard, DI and Bycroft, BW (2000) Quorum sensing and the population-dependent control of virulence. *Philosophical Transactions of the Royal Society of London, Series B* **355**:667–680.

Withers, H, Swift, S and Williams, P (2001) Quorum sensing as an integral component of gene regulatory networks in Gram-negative bacteria. *Current Opinion In Microbiology* **4**:186–193.

Worning, AM, Frimodt-Moller, N, Ostri, P, Nilsson, T, Hojholdt, K and Frimodt-Moler, C (1986) Antibiotic prophylaxis in vascular reconstructive surgery: a double blind placebo controlled study. *Journal of Antimicrobial Chemotherapy* **17**:105–113.

Yarmolinsky, MB (1995) Programmed cell death in bacterial populations. *Science* **267**:836–837.

Zambrano, MM and Kolter, R (1995) Changes in bacterial cell properties on going from exponential growth to stationary phase. In: Microbial Quality Assurance: A Guide Towards Relevance and Reproducibility (Eds. Brown, MRW and Gilbert, P), CRC Press, Boca Raton, pp. 21–30.

3 Microbial Adhesion and Biofilm Formation on Tissue Surfaces

3.1 Biofilm-related Infections on Tissue Surfaces

SUN NYUNT WAI[1], YOSHIMITSU MIZUNOE[2] and JANA JASS[3]
[1]*Department of Molecular Biology, Umeå University, Sweden*
[2]*Department of Bacteriology, Faculty of Medical Sciences, Kyushu University, Fukuoka, Japan*
[3]*Department of Microbiology and Immunology, University of Western Ontario and The Lawson Health Research Institute, London, ON, Canada*

INTRODUCTION

In the natural state, the bacteria exist in a sessile form as biofilms or in the freely motile planktonic form (Costerton *et al*. 1995). At specific locations within the mammalian body (e.g. mouth, gut and genital tract; Figure 3.1.1) there are natural reservoirs of bacteria attached to tissue surfaces (Macfarlane and Macfarlane 1995). It is generally accepted that these naturally occurring biofilms, which do not cause disease, may, in fact, act as a barrier to potentially pathogenic microorganisms and thus help prevent infection. Infections related to these biofilms, however, can occur if there is an alteration in the commensal population, such as may occur during antibiotic therapy or following tissue/organ damage that allows the invasion of competing bacteria or their transfer into normally sterile tissues. It is rare that detrimental infectious biofilms will form on healthy tissue surfaces, owing to the rigorous defence system, including these naturally occurring microbial populations. It is more often that infectious biofilms form in a host that is in a compromised state due to immune deficiency, drug treatment, trauma, tissue damage (burns and surgery) or has an underlying physiological disease such as cystic fibrosis (CF) or diabetes (Costerton *et al*. 1999).

The presence of a foreign body, such as an implant, often results in infections of adjacent tissues or tissues at other sites within the body, in

Medical Biofilms. Edited by J. Jass, S. Surman, J. Walker

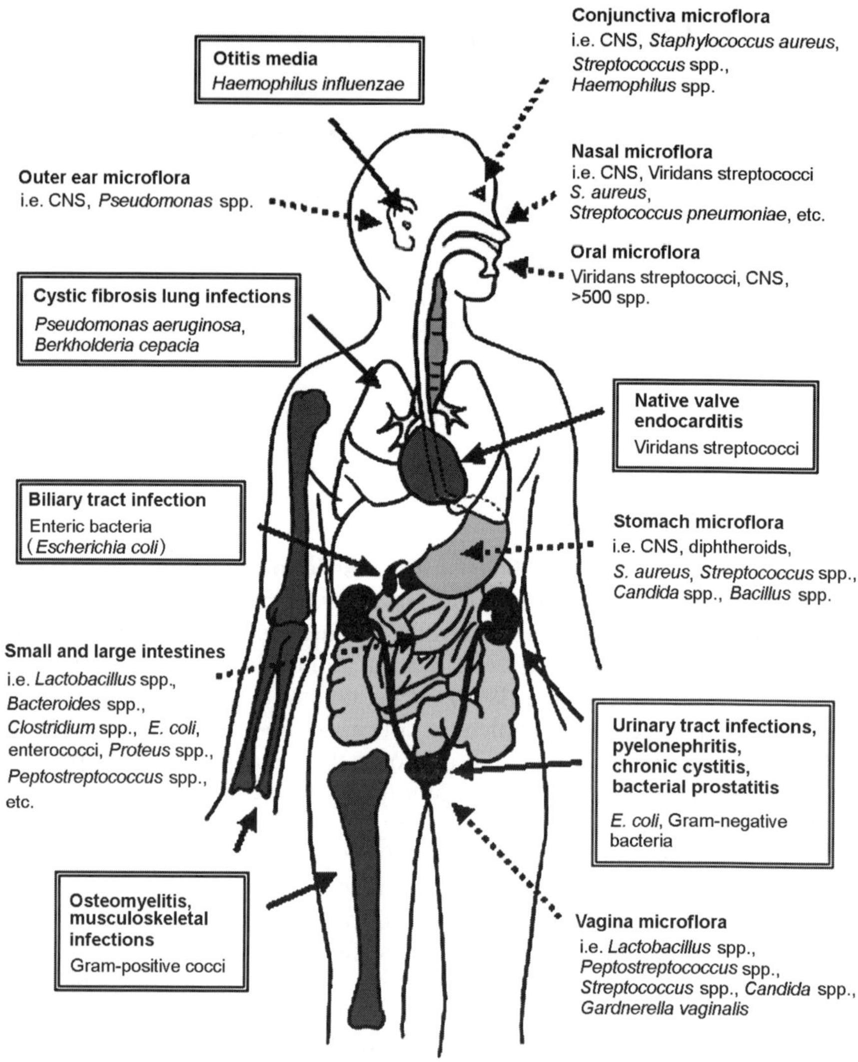

Figure 3.1.1. Biofilm-associated microbial populations colonizing different regions of the human body (CNS: coagulase-negative staphylococcus). Many locations within the body are colonized by commensal microbial populations (dashed arrow), and regions that are normally sterile may be affected by tissue-associated biofilm infections (box and solid arrows). Compiled from information provided in Costerton *et al.* (1999) and Madigan *et al.* (2000).

addition to the colonization of the biomaterial surface (see Chapter 2.1). Biofilm-related infections are primarily caused by opportunistic pathogens that, once having access to a host, are able to evade the immune system by surrounding themselves in an exopolymer matrix or a glycocalyx (Costerton *et al.* 1992). The exopolysaccharides that encompass biofilm microorganisms have been found to be less immunogenic, thus hiding the more immunogenic proteins and lipopolysaccharides on the bacterial surface (Jensen *et al.* 1990; Høiby *et al.* 1995). Microorganisms may be periodically shed into the surrounding environment or may release antigenic compounds, such as exotoxins and proteases, from the biofilms producing a strong immune response as characterized by recurrent chronic infections (Pedersen *et al.* 1990; Jensen *et al.* 1993). In some chronic infections, such as those observed in CF patients, the formation of immune complexes stimulates the release of proteolytic enzymes that destroy adjacent tissue, causing more damage than the invading organism (Høiby and Koch 1990; Kronborg *et al.* 1992, 1993).

Microbial infection in the absence of a foreign device is not generally considered as a biofilm, however, it is accepted that bacteria can persist within a host attached to tissue surfaces and may cause life-threatening infections, such as bacterial endocarditis. Such large aggregates of microorganisms and host proteins may have very similar characteristics to biofilms, allowing them to withstand the host's defence system and antibiotic treatment and to persist within the body when immobilized on a tissue surface (Dall *et al.* 1990; Costerton *et al.* 1999). These microbial communities may be described as biofilms since they are attached to a surface, in this case living cells, and, like true biofilms, are also enclosed in an exopolysaccharide matrix, which protects the cells from the immune system and increases their resistance to antimicrobial treatments (Hoyle *et al.* 1990). This is further substantiated by similarities in protein expression between *in vivo* cultures and biofilm cells. Studies of the outer membrane protein (OMP) of *in vitro Pseudomonas aeruginosa* biofilms and isolates from *in vivo* infections show similar profiles that are distinctly different from the OMP profiles of planktonic cells (Costerton and Stewart 2000). The availability of new techniques in molecular biology, proteomics, genomics and *in situ* reporter gene technology has led to many research groups comparing biofilm and planktonic cell phenotypes and *in vivo* and *in vitro* cultures.

Only recently has it been realized that the majority of human bacterial infections are biofilm related (Morris and Stickler 1999; Costerton 2000). The sites within the body associated with biofilm-like infections can be categorized into normally sterile or non-sterile environments. Here, we focus on problems associated with infections of normally sterile regions, since the biofilms in non-sterile regions are often considered beneficial.

Primary examples of biofilm-associated infections of normally sterile regions of the human body are illustrated in Figure 3.1.1. Infections can occur through the contamination of tissue during surgery or cross-infection from a non-sterile region either directly or via the circulation system. Biofilm-associated cross-infection via the circulation system occur when organisms are transported from other regions, such as the teeth, that may be colonized with heterogeneous biofilms, including potential pathogens. These have been shown to be responsible for conditions such as osteomyelitis and endocarditis, which are due to colonization of damaged regions of the bone and pericardium, respectively. Cross-infection of tissues may also occur through the transport of microorganisms from adjacent non-sterile sites. For example, the gastrointestinal and genital tracts, which are both colonized with commensal organisms, may be the source of bacteria that can cause infections of the normally sterile urinary tract and kidneys. Under some circumstances, where the body's defence system breaks down or through poor hygiene, the translocation of organisms from these sites may create an infection in the urinary tract, which, in extreme cases, can lead to severe kidney infection. This chapter concentrates on the problems caused by some of the major biofilm-related chronic infections of normally sterile tissues that affect modern society.

RESPIRATORY TRACT

Otitis Media

Otitis media is the most common disease diagnosed among children (Schappert 1992). There are four different clinical categories of otitis media: acute otitis media (AOM), otitis media with effusion (OME), persistent AOM and mucoid otitis media (MOM). AOM is defined by the presence of fever, irritability, pulling at the ears and tympanic membrane changes (Harabuchi *et al.* 1994). OME is diagnosed when fluid is visible in the middle ear in the absence of symptoms and in the absence of recent episodes of AOM (Harabuchi *et al.* 1994). Persistent AOM is defined as AOM lasting longer than 3 weeks despite one or several courses of antibiotic therapy, with the persistence of clinical and otoscopic signs of AOM. MOM is characterized by the accumulation of viscous fluid in the middle-ear cavity. OME can lead to significant hearing loss in children. Bacterial DNA has been found in a significant percentage of the effusions that were sterile when cultured, casting some doubt as to whether the DNA represents viable non-culturable or non-viable organisms, which is a common dilemma amongst molecular microbiologists. Rayner *et al.* (1998) established the presence of viable, metabolically active, intact organisms in some culture-negative OME by an RT-PCR-based assay system. Because of the increasing number of resistant middle-ear pathogens reported from different centres worldwide, an active surveillance programme of the microbiology and antimicrobial susceptibility patterns of middle-ear pathogens is required to produce an appropriate regime for antimicrobial treatment. The most prevalent pathogens causing AOM are *Streptococcus pneumoniae, Haemophilus influenzae, Moraxella catarrhalis, P. aeruginosa* and *Staphylococcus aureus* (Sih 2001).

Topical biofilm formation was detected in patients with chronic sinusitis, chronic purulent otitis media or habitual tonsillitis. Myringotomy and insertion of ventilation tubes, as a form of treatment of serous OME, has become one of the most frequently performed operations in otolaryngology. Bacterial biofilm formation has been implicated in persistent post-tympanostomy otorrhea and irreversible tube contamination. Middle-ear ventilation-tube materials are the most common predisposing factors for biofilm contamination in the middle ear. Otorrhea occurs after the insertion of tympanostomy tubes in as many as 50% of the cases. However, the majority of children had bacteriologically culture-negative middle-ear fluid. Resistant isolates from children initially treated with ampicillin or amoxicillin were sensitive to trimethoprim-sulphamethoxazole or to erythromycin and sulphisoxazole, and *vice versa*. New materials and coatings are being developed to resist permanent bacterial contamination of implanted medical devices of the ear (Biedlingmaier *et al.* 1998). Berry *et al.* (2000) demonstrated the effect of a surface treatment for fluoroplastic material used to inhibit biofilm formation of both *S. aureus* and *P. aeruginosa*. This reinforces the importance of selecting materials with appropriate surface-adherence properties, such as charge or smoothness or a combination of the two, as being more important than antibacterial treatments in preventing biofilm contamination.

Lower Respiratory Tract Infections in People with CF

CF is an inherited, autosomal recessive disease affecting exocrine gland function and usually manifests itself at birth or early childhood as a gastrointestinal or pulmonary disorder (Kuzemko 1983). It is the most prevalent of the fatal inherited diseases in the Caucasian population and is characterized by a triad of chronic pulmonary disease, pancreatic insufficiency and increased concentration of electrolytes in sweat. Patients with CF have various recurrent and chronic lung infections from early life, and most will eventually acquire chronic *P. aeruginosa* lung infection (Høiby 2000). Patients with advanced CF typically have chronic bacterial infections of the upper and lower respiratory tracts, but rarely develop extra-pulmonary infections. The lung tissue damage is due to an immune-complex-mediated chronic inflammation dominated by polymorphonuclear leucocytes releasing proteases and oxygen radicals. Altemeier *et al.* (1999) reported a case of purulent pericarditis due to *P. aeruginosa* in a patient with CF where there were no other risk factors for pericarditis, which is a previously unreported complication in CF.

Pathogenesis

Many species of bacteria produce extracellular polymers that may facilitate non-specific adhesion to surfaces and provide the framework for biofilm formation (Costerton *et al.* 1995). Whether colonization of the upper respiratory tract precedes establishment of bronchial infection with *P. aeruginosa* is unknown. *P. aeruginosa* is frequently grown from the sputum of adults with CF-related bronchiectasis, and in animal studies it adheres to buccal, nasal turbinate, tracheobronchial epithelial cells and mucus (Baker and Svanborg-Eden 1989; Pedersen 1992). The major reason why *P. aeruginosa* persists in the lungs appears to be due to its ability to produce alginate-containing microcolonies (Baker and Svanborg-Eden 1989). These bacterial populations adapt to the highly compartmentalized and anatomically deteriorating lung environment of CF patients, as well as to challenges by the body's immune defences and antibiotic therapies. Alginate has been shown to inhibit phagocytosis (Pier *et al.* 2001), and biofilm bacteria appear to stimulate a lesser oxidative burst response by neutrophils than free-swimming bacteria (Jensen *et al.* 1990). In addition, alginate, together with the host's mucus, may reduce the action of some antibiotics (Costerton 2000) and provide increased protection from toxic oxygen radicals (Simpson *et al.* 1989). All these factors contribute to the development of persistent lung infections in the CF patient (Costerton 2000). During chronic infections in CF, persistence of *P. aeruginosa* is associated with phenotypic changes that produce mucoid colony morphology and decreased systemic virulence (Deretic *et al.* 1995).

Progress of Disease and Management of Chronic Infection

Chronic *P. aeruginosa* infections, unfortunately, cannot be eradicated. In the pulmonary environment of CF patients, *P. aeruginosa* is characteristically found as a slow-growing, iron-dependent biofilm population (Brown *et al.* 1984). Chronic *P. aeruginosa* infections can be considered as mature biofilms where the population density exists in a continuous dynamic balance. The CF lung often encounters a number of different pathogens (*S. aureus*, *H. influenzae* and *Streptococcus pneumoniae*); however, *P. aeruginosa* and *Burkholderia cepacia* colonization leads to fast decline in health (Gilligan 1991; Al-Bakri *et al.* 1999). Chronic *P. aeruginosa* lung infections in CF patients can, therefore, be regarded as a model for other chronic infections caused by biofilm-forming bacteria. It can be considered the prototype of how biofilm-producing bacteria survive for decades despite a strong immune-mediated inflammatory response and intensive chemotherapy (Koch and Høiby 1993). The consequence of the chronic inflammation for the CF patient is a gradual destruction of the lung tissue, the production of copious amounts of viscous sputum and eventually respiratory failure and death. The increased knowledge of the pathogenesis of such biofilm infections will hopefully facilitate improved rationale for prophylaxis and therapy in the future.

Once acquired, *P. aeruginosa* is almost never eradicated; some patients are colonized for many years while remaining relatively well, yet others rapidly decline within months of infection by the organism (Høiby and Koch 1990). It is not known at which point in the infection that *P. aeruginosa* becomes a significant contributor to the disease state or what host or bacterial factors (e.g. toxin production) are important in the decline of patient status. Fegan *et al.* (1990) suggested that the transition of the colonizing *P. aeruginosa* strain from an initial non-mucoid, serum-resistant phenotype exhibiting a large range of exoenzymes to a mucoid, serum-sensitive organism producing a more restricted range of exoenzymes indicates the patient's deteriorating status. Analysis of the regulatory genes for alginate synthesis suggests that high osmolarity of the CF patient sputum could activate the *alg*D gene, which encodes for one of the enzymes involved in alginate biosynthesis (Berry *et al.* 1989). The alginate produced by mucoid *P. aeruginosa* in CF lung infections directly contributes to obstructive disease by forming viscous solutions or gels in the infected airways (Hentzer *et al.* 2001). It is important, therefore, to identify the early stages of infection and the appropriate time to start intensive antimicrobial therapy.

It has been suggested that alginate together with host mucus is a barrier to antibiotic diffusion (Anwar *et al.* 1990). There is overwhelming evidence of the ability of mucoid *P. aeruginosa* to survive aggressive antibiotic therapy in a CF lung. Alginate has also been implicated in non-diffusion-related resistance to aminoglycosides (Hatch and Schiller 1998) and β-lactamases (Bagge *et al.* 2000). Other studies have also described an increased exopolysaccharide-mediated antibiotic resistance of sessile *P. aeruginosa* in biofilms compared with the corresponding planktonic cells (Costerton 2000). Combination therapy of antibiotics and mucolytics or slime dispersants is reported to provide symptomatic relief by decreasing mucus viscosity, thereby facilitating its clearance from the lung (Gordon *et al.* 1991). As a consequence of the CF gene defect, mucocillary clearance of the bacteria from the lung is impaired by the viscous, dehydrated nature of the airway secretion. Furthermore, mucoid variants of *P. aeruginosa* thwart the phagocytic cells by production of alginate, a gel-like substance, within which bacteria are protected from phagocytes and antibiotics. CF patients develop progressive cytokine-mediated inflammatory lung disease, with abundant production of thick, tenacious, protease- and oxidant-rich purulent airway secretions that are difficult to clear even with physiotherapy. In the search for a potential treatment, Ghio *et al.* (1996) have tested tyloxapol, an alkylaryl polyether alcohol polymer detergent previously used as a mucolytic agent in adult chronic bronchitis. Vasconcellos *et al.* (1994) suggested that human plasma gelsolin, a protein that severs actin filaments, rapidly decreased the viscosity of CF sputum samples *in vitro*. Their results suggested that gelsolin might have therapeutic potential as a mucolytic agent in CF patients.

Thus, the ability of bacteria to colonize the lungs of CF patients is the outcome of a biological jigsaw puzzle comprising the effects of the CF gene defect, the severity of underlying lung disease and the ability of individual bacterial species to overcome the normally highly effective lung defences. Recently, Massengale *et al.* (2000) observed that neither exogenous nor endogenous alginate affects the ability of *P. aeruginosa* to invade CF/T43 respiratory epithelial cells. Most patients with CF experience recurrent and chronic endobronchial *P. aeruginosa* infections. Høiby (2000) suggested that it may be possible to prevent or delay the onset of these chronic *P. aeruginosa* infections in most patients with CF by eliminating cross-infection and by early aggressive antibiotic treatment of the first positive sputum culture and of any subsequent intermittent colonization. Pier (1997) suggested that immunoglobulin G preparations with opsonic antibodies against mucoid exopolysaccharide could provide CF patients with antibodies that they normally do not produce during chronic lung infection and may improve their clinical outcome.

GASTROINTESTINAL TRACT

Biliary Tract

The biliary tract of a healthy individual does not harbour any microorganisms. Bacteria can invade the biliary system either ascending via the sphincter of Oddi (Sung *et al.* 1992a) and/or by the haematogenous route from the portal venous system (Sung *et al.* 1991). Although the route of entry of bacteria into the biliary tract is still controversial, current observations suggest that the potential source of biliary pathogens originates from the gastrointestinal tract and infection descends via the haematogenous route in the portal circulation to invade the biliary tract (Sung *et al.* 1991).

Endoscopic biliary stenting has become a standard palliative treatment for obstructive jaundice due to inoperable malignancies of the pancreas and the hepatobiliary system, but infections and blockage of these stents by biliary sludge and bacterial biofilms may present major complications (Sung and Chung 1995). Stent infections are mainly caused by members of the Enterobacteriaceae, *P. aeruginosa*, Enterococci species and *Candida albicans*. The efficacy of endoscopic stenting is often limited by stent blockage, which develops 3 to 6 months after implantation, resulting in recurrence of jaundice, cholangitis and septicaemia (Leung *et al.* 1988).

Bacterial Biofilms Associated with Biliary Stents

In the clinical situation, biliary proteins may be adsorbed onto the surface of the stents immediately after their insertion, and some biliary proteins may be used by bacterial cells as receptors for adhesion to the surface (Yu *et al.* 1996). As a result of bacteria adhering to a surface of the biomaterial and forming adherent biofilms, blockage of indwelling biliary stents is caused by high-protein-containing bile, which may ultimately lead to stent occlusion. Three common biliary pathogens are often isolated from the bile of the patients suffering from acute suppurative cholangitis (Sung *et al.* 1993): *Escherichia coli* (O21:H25), which have pili and produce capsular polysaccharide; *E. coli* (O101:H9), which is non-piliated but with capsular polysaccharide; and *Enterococcus faecalis*, without either pili or a capsular polysaccharide. Stent blockage causes symptoms of chills and fever, and leads to deteriorating liver function and requires removal and replacement of the stent.

The pathogenesis of stent blockage is initiated by bacterial attachment to the stent surface to form a biofilm (Costerton *et al.* 1999). With time, the growth of biofilm and progressive agglomeration of bile sediments form a biliary sludge that finally results in occlusion of the lumen. The microorganisms in bacterial biofilms, covered by the glycocalyx matrix, are resistant to antimicrobial agents in normal dosages (Anwar and Costerton 1990). Biliary sludge occurring in the plastic stents promotes blockage (Leung *et al.* 1998); bacteriological cultures of this sludge have revealed a mixed infection with Gram-positive and Gram-negative bacteria. Gram-negative *E. coli* were more adherent than Gram-positive Enterococcus. Precolonization of *E. coli* facilitates subsequent attachment of Enterococcus (Leung *et al.* 1998). They concluded that there may be a synergistic effect between the different species of bacteria in adherence and biofilm formation and these are important factors in the blockage of biliary stents (Leung *et al.* 1998).

Problems in Management of Biliary Sludge

The most toxic bile salts have no effect on biofilm bacteria. Recently, Leung *et al.* (2000a,b) observed that ciprofloxacin prophylaxis eliminates

Gram-negative bacterial infection in bile and minimizes sludge formation and, therefore, may have a potential benefit in delaying stent blockage. The hydrophobic bile salts reduce bacterial adhesion on biomaterial, suggesting that incorporation of such bile salts might prevent the formation of bacterial biofilm. Sung *et al.* (1994) have also observed that the hydrophobic bile salt, taurodeoxycholate, reduces the adherence of *E. coli* in their *in vitro* studies. McAllister *et al.* (1993) suggested that the ultrasmooth surface of Vivathane does not allow bacterial adherence and biofilm deposition. They observed that this new polymer is more suitable for use in biliary stents in long-term applications. Recently, Tsang *et al.* (1999) discovered that silicone-covered self-expanding metal stents are likely to extend patency rates in malignant obstructive jaundice by providing a larger lumen for bile flow and allowing cyclical antibiotics to prevent bacterial biofilm formation. The adherence of bacteria on a plastic surface involves hydrophobic interaction of the plastic polymer and the bacterial cell wall. Clinical studies with oral antibiotic prophylaxis to prevent stent blockage have produced conflicting results. Leung *et al.* (2000b) observed that prolonged ciprofloxacin perfusion reduced *E. coli* adherence *in vitro*. They have also proposed that timely treatment with appropriate antibiotics reduces bacterial adherence *in vitro* and may be potentially beneficial in the prevention of stent blockage.

Brown Pigment Stone Formation

Acute cholangitis has been well recognized since 1877, when Charcot described the classical triad of pain, fever and jaundice in these patients. It has a high morbidity and mortality, is the third most common abdominal emergency and is an important cause of septicaemia in Southeast Asia (French *et al.* 1990). Cholangitis is commonly associated with stones obstructing the bile duct, as well as with benign and malignant structures of the biliary tree. The pathogenesis of pigment gallstones, as with the blockage of the biliary stents, is closely related to the formation of biofilms as described above (Sung *et al.* 1991), with the two major factors of mucous glycoprotein and calcification of the structure contributing to the protective environment in the biofilm (Sung *et al.* 1993).

Recurrent jaundice and cholangitis due to stent occlusion by biliary sludge is a major complication of endoscopic stenting for malignant obstructive jaundice. A scanning electron microscopy study of blocked stents revealed that the collection of amorphous materials forming a dense concretion on the surface of the stent was associated with microcolonies of live bacteria (Leung *et al.* 1989). Thus, bacterial biofilms play a role in the formation of brown pigment stones. The bacterial glycocalyx, like mucin, promotes the agglomeration of bile sediments and bacterial microcolonies in the formation of bacterial biofilms. With the trapping of more bacteria, further deconjugation and the precipitation of calcium bilirubinate, the biofilms consolidate to form a pigment stone (Stewart *et al.* 1987; Leung *et al.* 1989). Brown pigment stones can be viewed as a thick calcified bacterial biofilm that predisposes the biliary system to obstruction and recurrent infection. As viable bacterial cells are often present within the biofilm, any treatment that involves crushing the stone *in vivo* by endoscopic manoeuvre has the risk of releasing these bacteria and reactivating the infection. Such patients should receive prophylactic antibiotics prior to lithotripsy (Sung *et al.* 1992b).

URINARY AND GENITAL TRACT

Urinary tract infections (UTIs) are usually divided into two categories: the uncomplicated (or simple) UTI and the complicated UTI. The uncomplicated UTI is used to describe an afebrile infection in a patient with a structurally and functionally normal urinary tract. The majority of these patients are women with isolated or recurrent bacterial cystitis, and the primary infecting pathogens are *E. coli* that are often susceptible to and eradicated by a short course of inexpensive oral antimicrobial therapy. The complicated UTI describes an infection in a patient with pyelonephritis and/or a urinary tract with a structural or functional abnormality that would reduce the efficacy of antimicrobial therapy. These infections are also frequently caused by *E. coli*; however, in hospitals and nursing homes they are more commonly caused by nosocomial pathogens (e.g. *Pseudomonas, Klebsiella, Proteus* and *Serratia* spp.) that are resistant to antimicrobial treatment (Mizunoe *et al.* 1991; Schaeffer 1992). Most microbial infections (more than 50%) have now been associated with biofilms (Costerton *et al.* 1999), and biofilms are considered to contribute to a large proportion of complicated UTIs. In the urinary tract, biofilm-associated infections include struvite urolithiasis, chronic cystitis, epididymitis, prostatitis and device-associated infections such as catheter and stent infections (McLean *et al.* 1992; Nickel *et al.* 1994). Indeed, catheter-associated biofilm infections are one of the leading causes of nosocomial infections (Nickel and McLean 1997) where the causative organisms are protected from host defences and antimicrobial therapy (Morris *et al.* 1997; Costerton *et al.* 1999).

Struvite Urolithiasis

Approximately 0.1–0.4% of the population is thought to have renal stones every year in the USA and Europe, and 2–5% of the population in Asia. Furthermore, approximately 8–15% in Europe and North America and 20% in Saudi Arabia develop renal stones in their lifetime (Robertson 1993).

Renal stones have a tendency to recur with a rate of approximately 75% over 20 years. In most developed countries, about 80% of stones consist of calcium salts (Daudon *et al.* 1995) and 20% of stones are composed of other components, such as uric acid, struvite or carbonate apatite and cystine. Struvite or carbonate apatite renal stones are called infection-related stones because they are usually associated with UTIs caused by urease-producing bacteria. In infections of the urinary tract caused by bacteria such as *Proteus* spp., *Haemophilus* spp., *Klebsiella* spp. or *Ureaplasma urealyticum*, the hydrolysis of urea yields ammonium and hydroxyl ions. The consequent alkaline pH of the urine increases the dissociation of phosphate to form trivalent phosphate, and subsequently causes precipitation of struvite (magnesium–ammonium phosphate) and carbonate apatite (Nickel *et al.* 1994; Pak 1998). Nickel *et al.* (1994) have shown that the pH rises even more precipitously in the microenvironment of the bacterial biofilm. Infection stones, though becoming less common with the introduction of antimicrobial treatment (Pak 1998), represent a significant health problem and perhaps a greater danger to the integrity of the urinary tract than do conventional metabolic stones (Nickel *et al.* 1994). Infection stones, once formed, cannot be eliminated by antimicrobial therapy and need to be physically removed, since they harbour bacteria within their interstices (Pak 1998). If left untreated, infection-related calculi can cause failure to thrive, anaemia, chronic renal insufficiency, renal failure and death (Schwartz and Stoller 1999). Even after surgical removal, many of these stones recur with subsequent morbidity and occasional mortality (Nickel *et al.* 1994). Acetohydroxamic acid, a urease inhibitor, may lower urinary saturation of struvite by preventing the formation of ammonium and hydroxyl ions (Pak 1998).

Cystitis

The bacterial biofilms associated with cystitis seem to be more easily eradicated by antimicrobial therapy in comparison with the biofilms associated with catheters or stones. There appeared to be a synergic effect between antibiotics and host defences against these bacterial biofilms (Nickel *et al.* 1994). An even smaller dose of antibiotics has been shown to prevent bacterial adherence and subsequent biofilm formation.

Chronic Bacterial Prostatitis

In chronic bacterial prostatitis, the bacteria enter the prostate by ascending from the urethra, probably assisted by turbulent urethral flow patterns and/or intraprostatic ducral reflux (Nickel and Costerton 1992). Once the bacteria enter the prostatic ducts and acini, they undergo a rapid multiplication and induce an acute inflammation in the prostate gland. It is relatively easy to eradicate the bacteria in this planktonic state with appropriate antibiotic therapy. If untreated at this point, the bacteria can form sporadic bacterial microcolonies or biofilms adherent to the epithelium of the ductal system (Nickel *et al.* 1994), with the production of exopolysaccharide slime. It seems that bacteria persisting in the prostate gland within biofilms lead to persistent immunological stimulation and subsequent chronic inflammation with clinical symptoms including pain (Nickel and Costerton 1992). The traditional and expensive diagnostic approach employing quantitative bacterial cultures of various urine segments and expressed prostatic secretion remains the key to diagnosis (Nickel *et al.* 1994). It appears that the diagnosis of chronic bacterial prostatitis is often difficult because antimicrobial therapy prescribed before obtaining the proper specimens eradicates the planktonic bacteria but not the adherent bacteria in the biofilms, which do not appear to shed planktonic bacteria easily. These findings have been subsequently confirmed by biopsy and culture of the prostate gland of patients with chronic prostatitis and negative cultures from expressed prostatic secretion (Nickel and Costerton 1993). It is not easy to eradicate bacterial chronic prostatitis because the bacteria within show the same relative resistance to antibiotics as biofilms. Treatment regimens are being developed that deliver much higher antibiotic concentrations to the biofilm within the prostatic duct, which should improve treatment success rates (Nickel *et al.* 1994).

Complications of Urinary Catheter and Stent Infections

Indwelling urethral catheters constitute a conventional means of managing bladder dysfunction; however, they also provide access for bacteria from a contaminated, external environment into a vulnerable body cavity (Morris *et al.* 1997). Devices such as catheters and stents implanted into the urinary tract are particularly vulnerable to colonization by biofilms (Stickler and McLean 1995; Nickel *et al.* 1989). Scanning electron microscopy has demonstrated that most catheters removed from patients after 7 days are colonized by a bacterial biofilm, particularly on the inner luminal surface (Ohkawa *et al.* 1990; Nickel and Mclean 1997). However, the risk of infection increases with each day of catheterization, and bacteriuria develops in almost all of chronically catheterized patients. There is evidence with urethral catheters that colonizing bacteria within the protective biofilm matrix survive the urinary concentrations of antibiotics generated by conventional treatment regimes, even though they register as antibiotic sensitive in laboratory tests (Ganderton *et al.* 1992; Nickel *et al.* 1994). The majority of patients do not generally exhibit the clinical symptoms of UTI, and it is now common practice not to intervene with antibiotics unless symptoms of pyelonephritis or septicaemia become apparent (Warren 1994).

Catheter encrustation is a major complication in the care of the many patients enduring long-term bladder catheterization. The crystalline deposits can cause trauma to the bladder mucosa and the urethra. They can also occlude the catheter lumen and cause urinary retention or leakage of urine. Catheter blockage can precipitate episodes of pyelonephritis, septicaemia and shock. Getliffe and Mulhall (1991) reported that about half of the patients in their study experienced the problem of catheter encrustation. During the acute febrile phase of biofilm infection, antibiotic therapy is essential and may be effective because planktonic cells, not biofilm cells, account for the febrile episode (Kumon 2000). During the chronic indolent phase, however, the efficacy of antimicrobial chemotherapy is questionable, although several approaches have been reported to eradicate bacteria within biofilm effectively *in vitro* and *in vivo* (Anwar and Costerton 1990; Kumon *et al.* 1995). The increase in drug resistance and the pathogenesis of bacterial biofilms can also cause problems with the management of UTIs (Kumon 2000). It has been demonstrated that genetic information on drug resistance can be transferred among bacterial strains and species within monomicrobial and polymicrobial biofilms; this factor is of particular relevance in the management of nosocomial bacterial infections (Hausner and Wuertz 1999; Kumon 2000). Although the biofilm theory has helped to explain many of the treatment difficulties due to infections associated with indwelling catheters, at present the most effective way to reduce the incidence of such infections is to avoid indwelling catheters whenever possible, or at least to reduce the duration of catheterization.

BIOFILMS OF THE LOCOMOTIVE SYSTEM—OSTEOMYELITIS

Osteomyelitis is an inflammation of the bone marrow and adjacent tissue and is often caused by bacterial infections (Shirtliff and Mader 2000). Normal bone is fairly resistant to infection; however, a high bacterial inoculum combined with biomaterial implants and traumatized tissue and bone are susceptible to immediate and delayed infections because microbes preferentially adhere to 'inert biomaterials' or to damaged tissue surfaces (Nurnberger *et al.* 1998; Shirtliff and Mader 2000). Diabetes and other immune deficiency diseases can be a predisposing factor for soft-tissue infections to lead to osteomyelitis and the demineralization and necrosis of underlying bone (Lipsky 1997). Osteomyelitis can be diagnosed by clinical, radiographic and bacteriologic evaluation of the patient. An aggressive medical approach utilizing appropriate bactericidal antibiotics is required, along with surgical intervention in chronic cases. The increased use of implanted medical devices, such as intramedullary rods, screws, plates and artificial joints, has provided a physiological niche for pathogenic organisms that cause osteomyelitis. Prosthetic implants not only provide a substrate for bacterial adherence but also limit the ability of the host to deal adequately with the infection. Adhesion-mediated infections are extremely resistant to both antibiotics and host defences, and frequently persist until the prosthetic implants are removed. The pathogenesis of adhesive infections is related, in part, to preferential colonization of 'inert' substrates whose surfaces are not integrated with healthy tissues composed of living cells and intact extracellular polymers. Tissue integration is an interesting parallel to microbial adhesion and is a desired phenomenon for the biocompatibility of certain implants and biomaterials (Gristina *et al.* 1985). Tissue integration requires a form of eukaryocytic adhesion or compatibility with possible chemical integration to an implant surface.

Bone and joint infections are difficult to cure and will often become persistent chronic infections that cause progressive bone deterioration. Recently, studies of bone infections have primarily focused on common pathogens, mechanisms of infection, methods of preventing bacterial adherence to biomaterial surfaces and clinical preventive strategies for prosthetic infections. Although infection associated with arthroplasty is a relatively rare event, when it does occur it is of major consequence for the patient. Many organisms can cause these infections, but most are the result of Gram-positive bacteria, with the *Staphylococcus* spp. accounting for about half of the cases, predominantly *S. aureus*. Streptococcus, aerobic Gram-negative bacilli, anaerobic organisms and *Candida* spp. are responsible for another significant percentage. The ability of the organism to produce an outer slime layer or exopolysaccharide matrix seems to be a contributing virulence factor for prosthesis-associated infections. Once at the surface, bacteria (i.e. *Staphylococcus* spp.) synthesize a matrix primarily of extra-cellular polysaccharides that encompasses the cells to form a biofilm, allowing the bacteria to escape the effects of antimicrobial therapy and host clearance (Brause 1986). A colonized orthopaedic implant may rapidly result in chronic osteomyelitis, and the only successful treatment available is removal. The pathogen usually grows in coherent microcolonies in the adherent biofilm, which is often so extensive that the underlying infected bone at implant surface is obscured. Here, the exopolysaccharide matrix is mainly composed of teichoic acids (80%) and staphylococcal and host proteins (Hussain *et al.* 1993). This layer has been shown to protect the embedded pathogens from the action of antimicrobial agents and the host immune system by forming a dense physical barrier (Evans *et al.* 1998).

Local immune deficiency often occurs, since the normal phagocytic processes are directed to the removal of exopolysaccharide matrix and any foreign material. Furthermore, the glycocalyx has been shown to inhibit migration of polymorphonuclear leucocytes and thus prevent phagocytosis of the bacteria (Ferguson *et al.* 1992).

In chronic forms of osteomyelitis, which are often caused by bacterial biofilm infections within the bone, localized delivery of a suitable antibiotic is desirable. Gristina *et al.* (1985) investigated the pathogenesis in osteomyelitis by scanning electron microscopy of material obtained during surgical debridement of osteomyelitic bone and showed microorganisms growing in coherent microcolonies in adherent biofilms that were so extensive they obscured the infected bone surfaces. Furthermore, transmission electron microscopy showed this type of biofilm to contain some host cells and many bacteria surrounded by a dense fibrous matrix (Gristina *et al.* 1985; Evans *et al.* 1998). Within biofilms, bacteria express different morphologies and each bacterium is closely surrounded by fibrous exopolysaccharide polymers, which are known to mediate the adhesion of bacteria to surfaces and formation of microcolonies both in natural ecosystems and biomaterial-related infections. This indicates the similarities between biofilms of bone and tissue surfaces with those of implant-related biofilms. This adherent mode of growth is known to reduce microbial susceptibility to host clearance mechanisms and antibiotic therapy, and thus may be a fundamental factor in acute and chronic osteomyelitis. The dominating presence of an exopolysaccharide matrix cementing together host and bacterial protein materials visualized in a naturally hydrated state gives credence to its role as a physical barrier to host defences and antibiotics and identifies it as a significant factor in bacterial virulence.

Problems in Management of Osteomyelitis

Prosthetic infections following total joint replacement can have catastrophic results, both physically and psychologically for patients, leading to complete failure of the arthroplasty, possible amputation, prolonged hospitalization and even death. This type of infection is resistant to antibiotic therapy and most often requires removal of both the prosthesis and infected tissue (Gristina *et al.* 1994). For example, *S. aureus* forms a fibrin-rich biofilm in the presence of plasma that is highly resistant to attack by the human immune system and to antimicrobial therapy. Habib *et al.* (1999) suggested that ofloxacin microspheres in biodegradable polymers should be used for continuous treatment of chronic infections of bone in which bacterial biofilms can occur. The sustained release of ofloxacin from the microspheres was biphasic, with an initial burst release followed by a slow-release phase. Itokazu *et al.* (1998) have proposed the use of hydroxyapatite blocks to administer antibiotics or anticancer drugs because their porous structure allows the gradual administration of the pharmacological agents. Doyle (2000) suggested that a large number of bacterial and fungal pathogens depend on hydrophobic interactions for successful colonization of a host and that effective anti-adhesins may be based on the inhibition of hydrophobic interactions between the host and the pathogen and thus may be beneficial for management of adhesin-mediated infections. Vacheethasanee and Marchant (2000) describe a series of surfactant polymers designed as surface-modifying agents to shield a biomaterial from adhesive bacterial interactions. Both altering the surface of biomaterials to attempt the prevention of colonization and increasing the antimicrobial concentrations to eradicate established biofilms have proven ineffective to date in controlling osteomyelitis infections.

INFECTIVE ENDOCARDITIS: BIOFILM OF THE CARDIOVASCULAR SYSTEM

Infective endocarditis describes a family of persistent microbial infections of the heart valves. Typically targeting previously damaged or diseased valves, infecting microbes colonize within a thrombus-like mass of platelets and fibrin. The clinical presentation of infective endocarditis usually includes non-specific symptoms of malaise, fever, sweating, myalgia, weight loss and sustained elevations of C-reactive protein, erythrocyte sedimentation rate and other inflammatory markers (Bayer *et al.* 1998). Infective endocarditis is associated with *Streptococcus viridans* in about 50% of cases, *S. aureus* in about 30% of cases and approximately 20% of cases are culture-negative (Kurland *et al.* 1999). Prosthetic valve endocarditis (PVE) is an important cause of the morbidity and mortality associated with heart valve replacement surgery and occurs in 4% of prosthetic valve carriers.

Infective Vegetations as Biofilms

Oral and dental infections appear to be the most common portals of entry for infecting bacteria, with *S. viridans* the most frequent aetiological agent, with *Streptococcus* spp. associated twice as frequently as *Staphylococcus* spp. (Wells *et al.* 1990). *Streptococcus sanguis*, which is a primary colonizer of the tooth surface, can form the foundation for the complex multiple-species biofilm known as dental plaque. In addition, these bacteria can colonize native and prosthetic heart valves and are a common cause of endocarditis (Froeliger and Fives-Taylor 2001). Primarily, virulent microbes may infect or colonize abnormal heart valves briefly to cause vegetations or may colonize and persist to cause subclinical disease. After infection, microbial colonies on heart valves adapt to communal life and behave differently than when observed in pure culture *in vitro* (Costerton *et al.* 1999). The colonies are often polymicrobial, reflecting in part the predisposing bacteraemia. The challenging environment and the need to coexist with microbial partners and competitors result in communal organization, with species synthesizing an exopolymeric superstructure. The hydralase superstructure permits diffusion of water-soluble ions but separates colonies from one another and protects them from elements of the host's defence system and therapeutic agents, such as antibiotics. The persistence phenotypes of the microbial community may differ widely from that expected in pure culture.

Adherence of bacteria to fibrin–platelet deposits on the endocardium of cardiac valves is the important initial event in the pathogenesis of infective endocarditis. Despite major advances in cardiovascular surgical techniques and routine use of prophylactic antimicrobial agents, PVE continues to complicate the course of a small percentage of patients after cardiac valve replacement. The majority of cases require surgical removal and replacement of the infected prosthesis (Hyde *et al.* 1998). Antimicrobial coatings of prosthetic heart valve sewing cuffs have been used as a potentially effective method for preventing PVE. Illingworth *et al.* (2001) suggested that the use of silver-coated polyester fabric in sewing cuff fabrication is intended to inhibit the colonization and attachment to the sewing cuff by microorganisms commonly associated with PVE and to reduce the incidence of PVE. The capsular polysaccharide/adhesin antigen of *Staphylococcus epidermidis* was required to produce endocarditis in a rabbit model in which infection resulted from haematogenous spread of bacteria from a contaminated catheter in the jugular vein (Shiro *et al.* 1995). Fey *et al.* (1999) identified that a strong association existed between biofilm formation and strains of *S. epidermidis* that produce polysaccharide intercellular adhesin and haemagglutination. Pulliam *et al.* (1985) revealed that failure to eradicate streptococci from vegetations positively correlates with exopolysaccharide production. Animals infected with *S. sanguis* II and *Streptococcus morbillorum*, both vigorous exopolysaccharide matrix producers, continued to have infected cardiac vegetations after 5 days of antibiotic therapy, whereas animals infected with *Streptococcus salivarius* and a different *S. sanguis* II, both deficient in exopolysaccharide production, had sterile vegetations after 2 days of therapy. This suggests that large dense vegetations formed from a matrix of bacterial exopolysaccharide and blood-clotting proteins may cause serious obstruction in the heart valves. To a lesser extent, bacteria in the inner core of the microcolony may be difficult to sterilize because of the physical barrier. These factors are partly responsible for the prolonged antibiotic therapy necessary for the treatment of endocarditis and may contribute to occasional antibiotic failure. Antibiotics that penetrate the exopolysaccharide matrix or decrease its production may be more efficient in the treatment of this infection. Enzymatic digestion of the bacterial polysaccharides by dextranase has augmented the effect of penicillin by decreasing the physical barrier to penicillin penetration (Dall *et al.* 1987).

SUMMARY

In recent years it has become apparent that microbial biofilms are not only associated with medical implants and foreign devices, but also with tissue surfaces. Medical implants may predispose the person to infections both on the implant and adjacent tissue surface. There is an increasing awareness that a large number of infections, especially those that become chronic, have more properties similar to biofilms than with the planktonic phase of growth (Erickson *et al.* 2002). Characteristics are an increased resistance to antibiotic treatment, persistence, evasion of host immune systems (thus exhibiting an altered immune response), expression of different proteins and of quorum-sensing molecules. The presence of biofilms also explains the nature of chronic infections that keep recurring after antibiotic treatment ceases. Here, we have summarized a number of infections associated with different regions of the body that can be regarded as tissue-associated biofilms. With increasing understanding of the pathogenesis of different infections it becomes clear that many have similarities to biofilm and must be treated as such. Although there may be more tissue-associated biofilm infections than discussed here, it is clear that bacteria adapt to their environment for best survival success and the human body is a good example of this fact.

ACKNOWLEDGEMENTS

Financial support from the Swedish Medical Research Council, the STINT (the Swedish Foundation for International Cooperation in Research and Higher Education), the Wenner–Gren Foundations and the Faculty of Medicine and Odontology of Umeå University is gratefully acknowledged.

REFERENCES

Al-Bakri, AG, Gilbert, P and Allison, D (1999) Mixed species biofilms of *Burkholderia cepacia* and *Pseudomonas aeruginosa*. In: Biofilms: The Good, the Bad and the Ugly (Eds. Wimpenny, J, Gilbert, P, Walker, J, Brading, M and Bayston, R), BioLine, Cardiff, pp. 327–337.

Altemeier, WA, Tonelli, MR and Aitken, ML (1999) *Pseudomonal pericarditis* complicating cystic fibrosis. *Pediatric Pulmonology* **27**:62–64.

Anwar, H and Costerton, JW (1990) Enhanced activity of combination of tobramycin and piperacillin for eradication of sessile biofilm cells of *Pseudomonas aeruginosa*. *Antimicrobial Agents and Chemotherapy* **34**:1666–1671.

Anwar, H, Dasgupta, MK and Costerton, JW (1990) Testing the susceptibility of bacteria in biofilms to antibacterial agents. *Antimicrobial Agents and Chemotherapy* **34**:2043–2046.

Bagge, N, Ciofu, O, Skovgaard, LT and Høiby, N (2000) Rapid development *in vitro* and *in vivo* of resistance to ceftazidime in biofilm-growing *Pseudomonas aeruginosa* due to chromosomal beta-lactamase. *Acta Pathologica, Microbiologica, et Immunologica Scandinavica* **108**:589–600.

Baker, NR and Svanborg-Eden, C (1989) Role of alginate in the adherence of *Pseudomonas aeruginosa*. *Antimicrobial Chemotherapy* **42**:72–79.

Bayer, AS, Bolger, AF, Taubert, KA, Wilson, W, Steckelberg, J, Karchmer, AW, Levison, M, Chambers, HF, Dajani, AS, Gewitz, MH, Newburger, JW, Gerber, MA, Shulman, ST, Pallasch, TJ, Gage, TW and Ferrieri, P (1998) Diagnosis and management of infective endocarditis and its complications. *Circulation* **98**:2936–2948.

Berry, A, DeVault, JD and Chakrabarty, AM (1989) High osmolarity as a signal for enhanced *alg*D transcription in mucoid and nonmucoid *Pseudomonas aeruginosa*. *Journal of Bacteriology* **171**:2312–2317.

Berry, JA, Biedlingmaier, JF and Whelan, PJ (2000) *In vitro* resistance to bacterial biofilm formation on coated fluoroplastic tympanostomy tubes. *Otolaryngology, Head and Neck Surgery* **123**:246–251.

Biedlingmaier, JF, Samaranayake, R and Whelan, P (1998) Resistance to biofilm formation on otologic implant materials. *Otolaryngology, Head and Neck Surgery* **118**:444–451.

Brause, BD (1986) Infections associated with prosthetic joints. *Clinic in Rheumatic Diseases* **12**:523–536.

Brown, MRW, Anwar, H and Lambert, PA (1984) Evidence that mucoid *Pseudomonas aeruginosa* in cystic fibrosis lung grows under iron-restricted conditions. *FEMS Microbiology Letters* **21**:113–117.

Costerton, JW (2000) Cystic fibrosis pathogenesis and the role of biofilms in persistent infection. *Trends in Microbiology* **9**:50–52.

Costerton, JW and Stewart, PS (2000) Biofilms and device-related infections. In: Persistent Bacterial Infections (Eds. Nataro, JP, Blaser, MJ and Cunningham-Rundles, S), ASM Press, Washington, DC, pp. 423–439.

Costerton, JW, Lappin-Scott, HM and Cheng, K-J (1992) Glycocalyx, bacterial. In: Encyclopaedia of Microbiology, vol. 2 (Eds. Leaderberg, J, Alexander, M, Hopwood, DA, Iglewski, BH and Laskin, AI), Academic Press, pp. 311–317.

Costerton, JW, Lewandowski, Z, Cadwell, DE, Korber, DR and Lappin-Scott, HM (1995) Microbial biofilms. *Annual Reviews of Microbiology* **49**:711–745.

Costerton, JW, Stewart, PS and Greenberg, EP (1999) Bacterial biofilms: a common cause of persistent infections. *Science* **284**:1318–1322.

Dall, L, Barnes, WG, Lane, JW and Mills, J (1987) Enzymatic modification of glycocalyx in the treatment of experimental endocarditis due to viridans streptococci. *Journal of Infectious Disease* **156**:736–740.

Dall, L, Keihofner, M, Herndon, B, Barnes, W and Lane, J (1990) Clindamycin effect on glycocalyx production in experimental viridans streptococci endocarditis. *Journal of Infectious Diseases* **161**:1221–1224.

Daudon, M, Donsimoni, R, Hennequin, C, Fellahi, S, Le Moel, G, Paris, M, Troupel, S and Lacour, B (1995) Sex- and age-related composition of 10 617 calculi analyzed by infrared spectroscopy. *Urology Research* **23**:319–326.

Deretic, V, Schurr, MJ and Yu, H (1995) *Pseudomonas aeruginosa*, mucoidy and the chronic infection phenotype in cystic fibrosis. *Trends in Microbiology* **3**:351–356.

Doyle, RJ (2000) Contribution of the hydrophobic effect to microbial infection. *Microbes and Infection* **2**:391–400.

Erickson, DL, Endersby, R, Kirkham, A, Stuber, K, Vollman, DD, Rabin, HR, Mitchell, I and Storey, DG (2002) *Pseudomonas aeruginosa* Quorum-sensing systems may control virulence factor expression in the lungs of patients with cystic fibrosis. *Infection and Immunity* **70**:1783–1790.

Evans, RP, Nelson, CL, Bowen, WR, Kleve, MG and Hickmon, SG (1998) Visualization of bacterial glycocalyx with a scanning electron microscope. *Clinical Orthopaedics and Related Research* **347**:243–249.

Fegan, M, Francis, P, Hayward, AC, Davis, GH and Fuerst, JA (1990) Phenotypic conversion of *Pseudomonas aeruginosa* in cystic fibrosis. *Journal of Clinical Microbiology* **28**:1143–1146.

Ferguson, DA Jr, Veringa, EM, Mayberry, WR, Overbeek, BP, Lambe, DW Jr and Verhoe, J (1992) *Bacteroides* and *Staphylococcus* glycocalyx: chemical analysis, and the effects on chemiluminescence and chemotaxis of human polymorphonuclear leucocytes. *Microbios* **69**:53–65.

Fey, PD, Ulphani, JS, Gotz, F, Heilmann, C, Mack, D and Rupp, ME (1999) Characterization of the relationship between polysaccharide intercellular adhesin and hemagglutination in *Staphylococcus epidermidis*. *Journal of Infectious Diseases* **179**:1561–1664.

French, GL, Cheng, AF, Duthie, R and Cockram, CS (1990) Septicaemia in Hong Kong. *Journal of Antimicrobial Chemotherapy* **25**(Suppl. C):115–125.

Froeliger, EH and Fives-Taylor, P (2001) *Streptococcus parasanguis* fimbria-associated adhesin fap1 is required for biofilm formation. *Infection and Immunity* **69**:2512–2519.

Ganderton, L, Chawla, J, Winters, C, Wimpenny, J and Stickler, D (1992) Scanning electron microscopy of bacterial biofilms on indwelling bladder catheters. *European Journal Clinical Microbiology and Infectious Disease* **11**:789–796.

Getliffe, KA and Mulhall, AB (1991) The encrustation of indwelling catheters. *British Journal of Urology* **67**:337–341.

Ghio, AJ, Marshall, BC, Diaz, JL, Hasegawa, T, Samuelson, W, Povia, D, Kennedy, TP and Piantodosi, CA (1996) Tyloxapol inhibits NF-kappa B and cytokine release, scavenges HOCI, and reduces viscosity of cystic fibrosis sputum. *American Journal of Respiratory and Critical Care Medicine* **154**:783–788.

Gilligan, PH (1991) Microbiology of airway disease in patients with cystic fibrosis. *Clinical Microbiological Reviews* **4**:35–51.

Gordon, CA, Hodges, NA and Marriott, C (1991) Use of slime dispersants to promote antibiotic penetration through the extracellular polysaccharide of mucoid *Pseudomonas aeruginosa*. *Antimicrobial Agents and Chemotherapy* **35**:1258–1260.

Gristina, AG, Oga, M, Webb, LX and Hobgood, CD (1985) Adherent bacterial colonization in the pathogenesis of osteomyelitis. *Science* **228**:990–993.

Gristina, AG, Shibata, Y, Giridhar, G, Kreger, A and Myrvik, QN (1994) The glycocalyx, biofilm, microbes, and resistant infection. *Seminars in Arthroplasty* **5**:160–170.

Habib, M, Onyilofor, S, Ebube, NK and Owusu-Ababio, G (1999) Preparation and characterization of ofloxacin microspheres for the eradication of bone associated bacterial biofilm. *Journal of Microencapsulation* **16**:27–37.

Harabuchi, Y, Faden, H and Yamanaka, N (1994) Nasopharyngeal colonization with nontypable *Haemophilus influenzae* and recurrent otitis media. *Journal of Infectious Diseases* **170**:862–866.

Hatch, RA and Schiller, NL (1998) Alginate lyase promotes diffusion of aminoglycosides through the extracellular polysaccharide of mucoid *Pseudomonas aeruginosa*. *Antimicrobial Agents and Chemotherapy* **142**:974–977.

Hausner, M and Wuertz, S (1999) High rates of conjugation in bacterial biofilms as determined by quantitative *in situ* analysis. *Applied and Environmental Microbiology* **65**:3710–3713.

Hentzer, M, Teitzel, GM, Balzer, GJ, Heydorn, A, Molin, S, Givskov, M and Parsek, MR (2001) Alginate overproduction affects *Pseudomonas aeruginosa* biofilm structure and function. *Journal of Bacteriology* **183**:5395–5401.

Høiby, N (2000) Prospects for the prevention and control of pseudomonal infection in children with cystic fibrosis. *Paediatric Drugs* **2**:451–463.

Høiby, N and Koch, C (1990) *Pseudomonas aeruginosa* infections in cystic fibrosis and its management. *Thorax* **45**:881–884.

Høiby, N, Fomsgaard, A, Jensen, ET, Johansen, HK, Kronborg, G, Pedersen, SS, Pressler, T and Kharazmi, A (1995) The immune response to bacterial biofilms. In: Microbial Biofilms (Eds. Lappin-Scott, HM and Costerton, JW), Cambridge University Press, Cambridge, pp. 233–250.

Hoyle, BD, Jass, J and Costerton, JW (1990) The biofilm glycocalyx as a resistance factor. *Journal of Antimicrobial Chemotherapy* **26**:1–6.

Hussain, M, Wilcox, MH and White, PJ (1993) The slime of coagulase-negative staphylococci: biochemistry and relation to adherence. *FEMS Microbiology Review* **10**:191–207.

Hyde, JA, Darouiche, RO and Costerton, JW (1998) Strategies for prophylaxis against prosthetic valve endocarditis: a review article. *Journal of Heart Valve Disease* **7**: 316–326.

Illingworth, B, Bianco, RW and Weisberg, S (2001) *In vivo* efficacy of silver-coated fabric against *Staphylococcus epidermidis*. *Journal of Heart Valve Disease* **9**:135–141.

Itokazu, M, Aoki, T, Nonomura, H, Nishimoto, Y and Itoh, Y (1998) Antibiotic-loaded porous hydroxyapatite blocks for the treatment of osteomyelitis and postoperative infection. A preliminary report. *Bulletin of the Hospital for Joint Diseases (New York)* **57**:125–129.

Jensen, ET, Kharazmi, A, Lam, K, Costerton, JW and Høiby, N (1990) Human polymorphonuclear leukocyte response to *Pseudomonas aeruginosa* grown in biofilm. *Infection and Immunity* **58**:2383–2385.

Jensen, ET, Kharazmi, A, Garred, P, Kronborg, G, Formsgaard, A, Mollness, TE and Høiby, N (1993) Complement activation of *Pseudomonas aeruginosa* biofilms. *Microbial Pathogenesis* **15**:377–388.

Koch, C and Høiby, N (1993) Pathogenesis of cystic fibrosis. *Lancet* **341**:1065–1069.

Kronborg, G, Shand, GH, Formsgaard, A and Høiby, N (1992) Lipopolysaccharide is present in immune complexes isolated from sputum in patients with cystic fibrosis and chronic *Pseudomonas aeruginosa* lung infection. *Acta Pathologica, Microbiologica, et Immunologica Scandinavica* **100**:75–180.

Kronborg, G, Hansen, M, Svenson, M, Formsgaard, A, Høiby, N and Bendtzen, K (1993) Cytokines in sputum and serum from patients with cystic fibrosis and chronic *Pseudomonas aeruginosa* infections as markers of destructive inflammation in the lungs. *Pediatric Pulmonology* **15**:292–297.

Kumon, H (2000) Management of biofilm infections in the urinary tract. *World Journal of Surgery* **24**:1193–1196.

Kumon, H, Ono, N, Iida, M and Nickel, JC (1995) Combination effect of fosfomycin and ofloxacin against *Pseudomonas aeruginosa* growing in a biofilm. *Antimicrobial Agents and Chemotherapy* **39**:1038–1044.

Kurland, S, Enghoff, E, Landelius, J, Nystrom, SO, Hambraeus, A and Friman, G (1999) A 10-year retrospective study of infective endocarditis at a university hospital with special regard to the timing of surgical evaluation in *S. viridans* endocarditis. *Scandinavian Journal of Infectious Disease* **31**:87–91.

Kuzemko, JA (1983) Evolution of lung disease in cystic fibrosis. *Lancet* **261**:448–449.

Leung, JW, Ling, TK, Kung, JL and Vallance-Owen, J (1988) The role of bacteria in the blockage of biliary stents. *Gastrointestinal Endoscopy* **134**:19–22.

Leung, JW, Sung, JY and Costerton, JW (1989) Bacteriological and electron microscopy examination of brown pigment stones. *Journal of Clinical Microbiology* **127**:915–921.

Leung, JW, Liu, YL, Desta, T, Libby, E, Inciardi, JF and Lam, K (1998) Is there a synergistic effect between mixed bacterial infection in biofilm formation on biliary stents? *Gastrointestinal End*oscopy **48**:250–257.

Leung, JW, Liu, YL, Desta, TD, Libby, ED, Inciardi, JF and Lam, K (2000a) *In vitro* evaluation of antibiotic prophylaxis in the prevention of biliary stent blockage. *Gastrointestinal Endoscopy* **51**:296–303.

Leung, JW, Libby, ED, Morck, DW, McKay, SG, Liu, YL, Lam, K and Olson, ME (2000b) Is prophylactic ciprofloxacin effective in delaying biliary stent blockage? *Gastrointestinal Endoscopy* **52**:175–182.

Lipsky, BA (1997) Osteomyelitis of the foot in diabetic patients. *Clinical Infectious Diseases* **25**:1318–1326.

Macfarlane, GT and Macfarlane, S (1995) Human intestinal 'biofilm' communities. In: The Life and Death of Biofilm (Eds. Wimpenny, J, Handley, P, Gilbert, P and Lappin-Scott, HM), BioLine, Cardiff, pp. 83–87.

Madigan, MT, Martinko, JM and Parker, J (2000) Brock: Biology of Microorganisms, Prentice Hall, London, pp. 773–800.

Massengale, AR, Quinn, FJ, Williams, A, Gallagher, S and Aronoff, SC (2000) The effect of alginate on the invasion of cystic fibrosis respiratory epithelial cells by clinical isolates of *Pseudomonas aeruginosa*. *Experimental Lung Research* **26**:163–178.

McAllister, EW, Carey, LC, Brady, PG, Heller, R and Kovacs, SG (1993) The role of polymeric surface smoothness of biliary stents in bacterial adherence, biofilm deposition, and stent occlusion. *Gastrointestinal Endoscopy* **39**:422–425.

McLean, RJC, Downey, JA, Lablans, AL, Clark, JM, Dumanski, AJ and Nickel, JC (1992) Modelling biofilm-associated urinary tract infections. *International Biodeterioration and Biodegradation* **30**:201–216.

Mizunoe, Y, Matsumoto, T, Nakamura, M and Kumazawa, J (1991) Chronological changes in bacteria isolated from urinary tract infections. *Nishinihon Journal of Urology* **53**:1408–1417.

Morris, NS and Stickler, DJ (1999) The development of bacterial biofilms on indwelling urethral catheters. *World Journal of Urology* **17**:345–350.

Morris, NS, Stickler, DJ and Winters, C (1997) Which indwelling urethral catheters resist encrustation by *Proteus mirabilis* biofilms? *British Journal of Urology* **80**:58–63.

Nickel, JC and Costerton, JW (1992) Coagulase-negative staphylococcus in chronic prostatitis. *Journal of Urology* **147**:398–400.

Nickel, JC and Costerton, JW (1993) Bacterial localization in antibiotic-refractory chronic bacterial prostatitis. *Prostate* **23**:107–114.

Nickel, JC and McLean, RJC (1997) Urologic perspectives on bacterial biofilms. *Infectious Urology* **(November/December)**:169–175.

Nickel, JC, Downey, JA and Costerton, JW (1989) Ultrastructural study of microbiologic colonization of urinary catheters. *Urology* **34**:284–291.

Nickel, JC, Costerton, JW, McLean, RJ and Olson, M (1994) Bacterial biofilms: influence on the pathogenesis, diagnosis and treatment of urinary tract infections. *Journal of Antimicrobial Chemotherapy* **33**(Suppl. A): 31–41.

Nurnberger, M, Treadwell, T, Lin, B and Weintraub, A (1998) Pacemaker lead infection and vertebral osteomyelitis presumed due to *Cardiobacterium hominis*. *Clinical Infectious Diseases* **27**:890–891.

Ohkawa, M, Sugata, T, Sawaki, M, Nakashima, T, Fuse, H and Hisazumi, H (1990) Bacterial and crystal adherence to the surfaces of indwelling urethral catheters. *Journal of Urology* **143**:717–721.

Pak, CY (1998) Kidney stones. *Lancet* **351**:1797–1801.

Pedersen, SS (1992) Lung infection with alginate-producing mucoid *Pseudomonas aeruginosa* in cystic fibrosis patients. *Acta Pathologica et Microbiologica Scandinavica* **100**(Suppl. 28):5–79.

Pedersen, SS, Kharazmi, A, Espersen, F and Høiby, N (1990) *Pseudomonas aeruginosa* alginate in cystic fibrosis sputum and the inflammatory response. *Infection and Immunity* **50**:3363–3368.

Pier, GB (1997) Rationale for development of immunotherapies that target mucoid *Pseudomonas aeruginosa* infection in cystic fibrosis patients. *Behring Insitutet Mitteilungen* **98**:350–360.

Pier, GB, Coleman, F, Grout, M, Franklin, M and Ohman, DE (2001) Role of alginate O acetylation in resistance of mucoid *Pseudomonas aeruginosa* to opsonic phagocytosis. *Infection and Immunity* **69**:895–901.

Pulliam, L, Dall, L, Inokuchi, S, Wilson, W, Hadley, WK and Mills, J (1985) Effects of exopolysaccharide production by viridans streptococci on penicillin therapy of experimental endocarditis. *Journal of Infectious Disease* **151**:153–156.

Rayner, MG, Zhang, Y, Gorry, MC, Chen, Y, Post, JC and Ehrlich, GD (1998) Evidence of bacterial metabolic activity in culture-negative otitis media with effusion. *The Journal of the American Medical Association* **279**:296–299.

Robertson, WG (1993) Urinary tract calculi. In: Metabolic Bone and Stone Disease (Eds. Nordin, BEC, Need, AG and Morris, HA), Churchill Livingstone, New York, pp. 249–311.

Schaeffer, AJ (1992) Infections of the urinary tract. In: Campbell's Urology Sixth Edition (Eds. Walsh, PC, Retik, AB, Stamey, TA and Vaughan, JED), W. B. Saunders Company, Philadelphia, pp. 731–806.

Schappert, SM (1992) Office visits for otitis media: United States, 1975–90. Atlanta: Centers for Disease Control and Prevention. Advance data no. 214.

Schwartz, BF and Stoller, ML (1999) Nonsurgical management of infection-related renal calculi. *The Urologic Clinics of North America* **26**:765–778.

Shiro, H, Meluleni, G, Groll, A, Muller, E, Tosteson, TD, Goldmann, DA and Pier, GB (1995) The pathogenic role of *Staphylococcus epidermidis* capsular polysaccharide/ adhesin in a low-inoculum rabbit model of prosthetic valve endocarditis. *Circulation 1* **92**:2715–2722.

Shirtliff, ME and Mader, JT (2000) Osteomyelitis. In: Persistent Bacterial Infections (Eds. Nataro, JP, Blaser, MJ and Cunningham-Rundles, S), ASM Press, Washington, DC, pp. 375–396.

Sih, TM (2001) Acute otitis media in Brazilian children: analysis of microbiology and antimicrobial susceptibility. *The Annals of Otology, Rhinology & Laryngology* **110**: 662–666.

Simpson, JA, Smith, SE and Dean, RT (1989) Scavenging by alginate of free radicals released by macrophages. *Free Radical Biology & Medicine* **6**:347–353.

Stewart, L, Smith, AL, Pellegrini, CA, Motson, RW and Way, LW (1987) Pigment gallstones form as a composite of bacterial microcolonies and pigment solids. *Annals of Surgery* **206**:242–250.

Stickler, DJ and McLean, RJC (1995) Biomaterials associated infections: the scale of the problem. *Cells and Materials* **5**:167–182.

Sung, JJ and Chung, SC (1995) Endoscopic stenting for palliation of malignant biliary obstruction. A review of progress in the last 15 years. *Digestive Diseases and Sciences* **40**:1167–1173.

Sung, JY, Shaffer, EA, Olson, ME, Leung, JW, Lam, K and Costerton, JW (1991) Bacterial invasion of the biliary system by way of the portal-venous system. *Hepatology* **14**:313–317.

Sung, JY, Leung, JW, Shaffer, EA, Lam, K, Olson, ME and Costerton, JW (1992a) Ascending infection of the biliary tract after surgical sphincterotomy and biliary stenting. *Journal of Gasteroenterology and Hepatology* **17**:240–245.

Sung, JY, Costerton, JW and Shaffer, EA (1992b) Defense system in the biliary tract against bacterial infection. *Digestive Diseases and Sciences* **37**:689–696.

Sung, JY, Leung, JW, Shaffer, EA, Lam, K and Costerton, JW (1993) Bacterial biofilm, brown pigment stone and blockage of biliary stents. *Journal of Gastroenterology and Hepatology* **8**:28–34.

Sung, JY, Shaffer, EA, Lam, K, Rususka, I and Costerton, JW (1994) Hydrophobic bile salt inhibits bacterial adhesion on biliary stent material. *Digestive Diseases Sciences* **39**:999–1006.

Tsang, TK, Pollack, J and Chodash, HB (1999) Silicone-covered metal stents: an *in vitro* evaluation for biofilm formation and patency. *Digestive Diseases and Sciences* **44**:1780–1785.

Vacheethasanee, K and Marchant, RE (2000) Surfactant polymers designed to suppress bacterial (*Staphylococcus epidermidis*) adhesion on biomaterials. *Journal of Biomedical Materials Research* **50**:302–312.

Vasconcellos, CA, Allen, PG, Wohl, ME, Drazen, JM, Janmey, PA and Stossel, TP (1994) Reduction in viscosity of cystic fibrosis sputum *in vitro* by gelsolin. *Science* **263**:969–971.

Warren, JW (1994) Catheter-associated bacteriuria in long-term care facilities. *Infection Control and Hospital Epidemiology* **15**:557–562.

Wells, AU, Fowler, CC, Ellis-Pegler, RB, Luke, R, Hannan, S and Sharpe, DN (1990) Endocarditis in the 80s in a general hospital in Auckland, New Zealand. *The Quarterly Journal of Medicine* **76**:753–762.

Yu, JL, Andersson, R and Ljungh, A (1996) Protein adsorption and bacterial adhesion to biliary stent materials. *Journal of Surgical Research* **62**:69–73.

3.2 Interaction of Biofilms with Tissues

MERLE E. OLSON, HOWARD CERI and DOUGLAS W. MORCK
Biofilm Research Group, Department of Microbiology and Infectious Diseases, Department of Biological Sciences, University of Calgary, Calgary, Alberta, Canada

INTRODUCTION

Microbial biofilms are recognized to play a major role in many industrial processes and in the environment. The field of medicine has been primarily concerned with biofilm formation on implanted medical devices, such as catheters, shunts, stents, pacemakers and heart valves, as colonization of implants has led to device failures, infection and death (Habash and Reid 1999). Microbial biofilms on tissue surfaces have been identified from biopsies and at the time of post-mortem examination, but they have not been well studied (Costerton *et al.* 1999). In order to investigate the role of microbial biofilms on tissue surfaces, tissue/organ cultures or animal models are necessary. Microbial biofilms within tissue often elicit a significant host inflammatory response in spite of the presence of low numbers of microorganisms. Indeed, greater than 99% of the biofilms within tissues are composed of host products (Buret *et al.* 1991).

Microbial biofilms within tissue usually result in chronic infections that frequently fail to respond to antibiotics and elimination by host clearance mechanisms. The medical community probably does not yet appreciate the importance that microbial biofilms play in their day-to-day dealings with infections that do not respond to classical therapeutic protocols. It has become clear to microbiologists working in industry investigating issues such as biofouling and corrosion, that biofilms must be addressed. Certainly, the future of medical microbiology lies in approaching many tissue infections as biofilms. Recent *in vitro* techniques and certain animal models described in this chapter provide methods that begin to investigate

Medical Biofilms. Edited by J. Jass, S. Surman, J. Walker

the pathogenesis and control of tissue biofilms. We believe that such research is necessary to control difficult-to-treat microbial infections within tissue.

BIOFILM FORMATION ON TISSUE SURFACES

Mammalian tissue is composed of cells with surfaces containing polysaccharides and plasma proteins overlying their membrane. The cells are surrounded by extracellular matrix molecules (collagen), enzymes (e.g. collagenase), surfactants and antibodies. Under normal circumstances, tissues are designed to resist invasion of microorganisms, but perturbations in their structure can permit pathogens and normal microflora to invade, multiply and cause disease. Microbial biofilms have been associated with a wide range of tissues under normal conditions and with disease (Costerton *et al.* 1999). Almost all tissue surfaces exposed to the environment are colonized by a natural autochthonous flora, which are non-pathogenic and usually serve to protect that tissue surface from pathogens (Rosee *et al.* 1982; Costerton *et al.* 1995). This includes the skin and most mucosal surfaces (gastrointestinal tract, urogenital tract, respiratory tract) of animals, but it also applies to plant leaves and roots. Certain bacteria, such as *Lactobacillus* spp., on the epithelial surface of the gastrointestinal and urinary tracts have a probiotic function, and pathogenic bacteria can also colonize and translocate across the mucosal barrier (Reid and Bruce 1995). The initial adhesion of bacteria to a tissue surface can be mediated by the presence of specific ligands on the bacteria and receptor sites on the tissue. Once the microorganisms attach to the tissue they can divide on its surface, forming a biofilm, or invade to form microcolonies within the tissue (Figure 3.2.1). Biofilms that initially form on tissue surfaces are vulnerable to the many defence mechanisms (e.g. mucus flow, ciliary clearance, phagocytes) of the host. However, once the biofilm becomes established on the tissue surface, the structure is resistant to elimination by certain host defence cells. Organisms that colonize epithelial surfaces usually secrete toxins that can act systemically or locally to cause tissue destruction and inflammation. *Clostridium difficile* pseudomembranous colitis and corneal oedema (caused by bacterial lipopolysaccharide (LPS) on the corneal surface) are examples of such a biofilm–tissue disease. Microorganisms may also invade through the epithelial barrier, through either a paracellular or intracellular route protected within endosomes (Figure 3.2.1). On skin, the microorganism must translocate the multicellular-thick epidermis, which provides a strong protective barrier. Once the organisms have overcome these barriers they form biofilm microcolonies that are resistant to host cell elimination and antibiotics. The tissue microcolonies stimulate a localized inflammatory

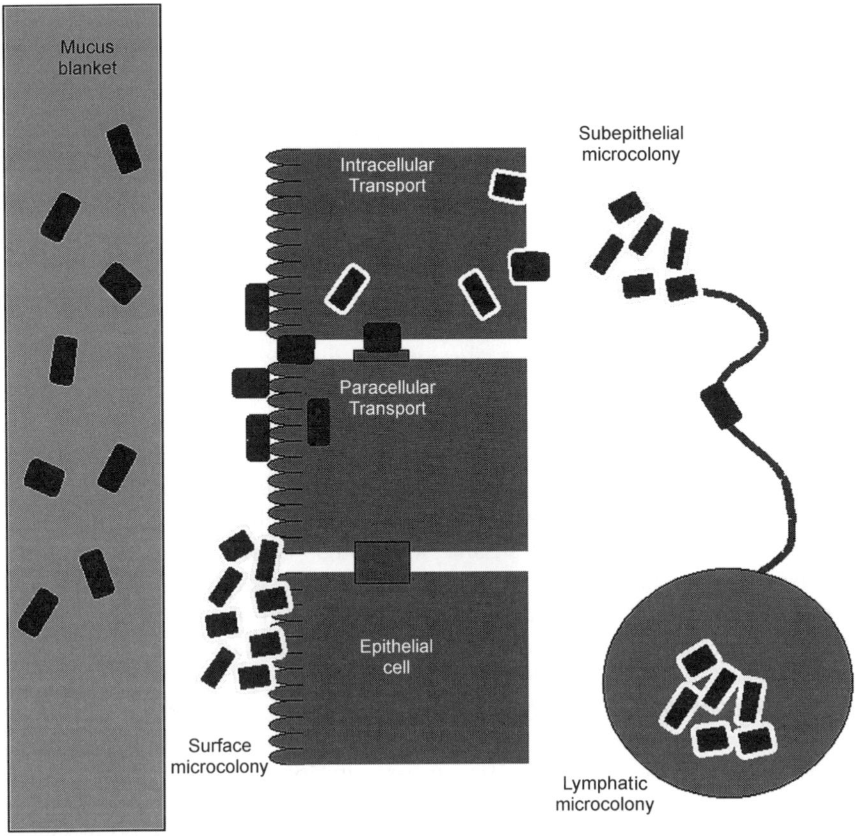

Figure 3.2.1. The process of attachment, invasion and biofilm formation of tissue using an epithelial tissue example. Microorganisms may invade through the epithelial barrier by either a paracellular or intracellular route (protected within endosomes). Biofilms may form on tissue surface or within the subepithelial tissues. Planktonic bacteria within the mucus blanket are readily eliminated in the normal flow of this material across the epithelial surface.

reaction that leads to further tissue destruction. Many bacterial pathogens may form tissue microabscesses and/or significant inflammation. Examples of such diseases are *Escherichia coli* prostatitis, *Yersinia enterocolitica* enteritis and *Staphylococcus aureus* mastitis.

HOST ELIMINATION OF BACTERIA

The ability of biofilms to persist in the presence of a normal host immune system has been the focus of study by a number of researchers. Chronic

Pseudomonas aeruginosa pneumonia associated with cystic fibrosis (CF), for example, has received much attention by researchers trying to understand how the host fails to eliminate the bacterial pathogen in spite of strong cellular and humoral immunity (Høiby *et al.* 1995). Antigenic proteins and peptides are released from both sessile and planktonic bacteria. However, we do not have sufficient understanding of the differences between the immunogenic antigens in planktonic and sessile bacteria. Planktonic organisms, however, often stimulate a more intense acute response than sessile bacteria. These antigens stimulate a humoral and cellular immune response and antibodies are produced that opsonize the invading pathogen. Planktonic bacteria are readily eliminated from the host, whereas bacteria existing within biofilms resist elimination despite being opsonized and the presence of activated phagocytes. This has been demonstrated by *in vitro* and *in vivo* models of *P. aeruginosa* lung infections and *E. coli* prostatitis (Høiby *et al.* 1995; Ceri *et al.* 1999c). Even immunization of laboratory animals with bacteria followed by experimental challenge induces a strong immune response that will readily eliminate planktonic bacteria but has no effect on sessile organisms (Ward *et al.* 1992; Høiby *et al.* 1995; Ceri *et al.* 1999c). The explanation for failure in the immune response has been attributed to the different composition of the biofilm. Biofilms consist of bacteria enmeshed in a glycocalyx containing LPS, exopolysaccharide (e.g. alginate) and a variety of secretory and degradation proteins. Within tissue the host also contributes a significant biomass to the biofilm (Buret *et al.* 1991). Phagocytes (polymorphonuclear neutrophils (PMNs) being the most predominant) are unable to phagocytize the biofilm effectively due to the size of the microcolony, the masking of binding sites by the exopolysaccharide and inhibition of activity to destroy phagocytized cells through an oxidative burst (Høiby *et al.* 1995; Costerton *et al.* 1999). Both planktonic and sessile bacteria are able to induce complement activation, but biofilm bacteria are able survive the lytic action of the complement cascade (Høiby *et al.* 1995). Bacterial biofilms are able to absorb complement fragments, thereby inhibiting the complement activation. Several strategies that biofilms use to resist elimination by the host have been identified. Certainly others will be identified as advances in genomics and proteomics continue to contribute to our understanding of pathogenesis.

Inflammation

The previous section described how microbial biofilms resist elimination by host defences. It is this failure in pathogen elimination and the continual attempt by the host to eliminate the biofilm bacteria that leads to much of the pathology associated with chronic bacterial infections. The pathogenesis of the inflammation associated with tissue biofilms is summarized in Figure 3.2.2. The bacteria and bacterial products (LPS, exopolysaccharide, proteases and toxins) recruit host phagocytes to the area. The bacterial toxins and proteases may also cause tissue necrosis that further attracts inflammatory cells (primarily PMNs) to the bacterial microcolonies. These neutrophils undergo necrosis and release their contents (proteolytic enzymes, such as elastase, acid hydrolases, lactoferrin and lysozyme), so inducing significant tissue damage and perpetuating the inflammatory reaction. The host main defence in blocking the sites of infection is to induce proliferation of extracellular matrix molecules, like collagen and fibrin. This, in itself, may also be harmful, as it further protects the microbial microcolonies from host or antibiotic elimination, and it can also induce adverse tissue structural alterations, such as vegetations on heart valves, pulmonary fibrosis and scarring of wounds. Chronic inflammatory processes can also lead to damage within other tissue sites, such as deposition of immune complexes, hypersensitivity and autoimmune disorders. Indeed, it may be the microbial biofilms that initiate the process but the host's own activated inflammatory process that causes most of the tissue damage.

Antimicrobial Resistance of Biofilm Bacteria

Antimicrobial agents have been highly effective in controlling and eliminating acute tissue infections involving microbes that have not had the opportunity to develop a protective biofilm. Many chronic tissue infections that fail to respond to chemotherapeutic medications are biofilm diseases. It is estimated that 65% of nosocomial infections are biofilm associated, costing greater than one billion dollars to the American health-care system annually (Potera 1999). It has been demonstrated in many studies that microbial biofilms are 10–1000 times more resistant to the effects of antimicrobial agents compared with their planktonic counterparts (Ceri *et al.* 1999b; Mah and O'Toole 2001). The mechanisms of resistance are still not understood, but it is multifactorial, as are the resistance strategies in planktonic organisms. It is believed that the exopolysaccharide matrix inhibits the penetration of the antimicrobial agent to the microorganism (Mah and O'Toole 2001). When biofilms form within tissues there is an added barrier of the host extracellular products (collagen, fibrin). Mathematical models suggest that this should not be a sufficient barrier, however, the bacterial and host products may bind the antibiotics, thus inhibiting their movement through the biofilm (Stewart 1996). This has been shown with such drugs as ciprofloxacin, piperacillin and tobramycin (Hoyle *et al.* 1992; Suci *et al.* 1994). Host and bacterial enzymes, such as β-lactamases, may reach high concentrations within the biofilm. The

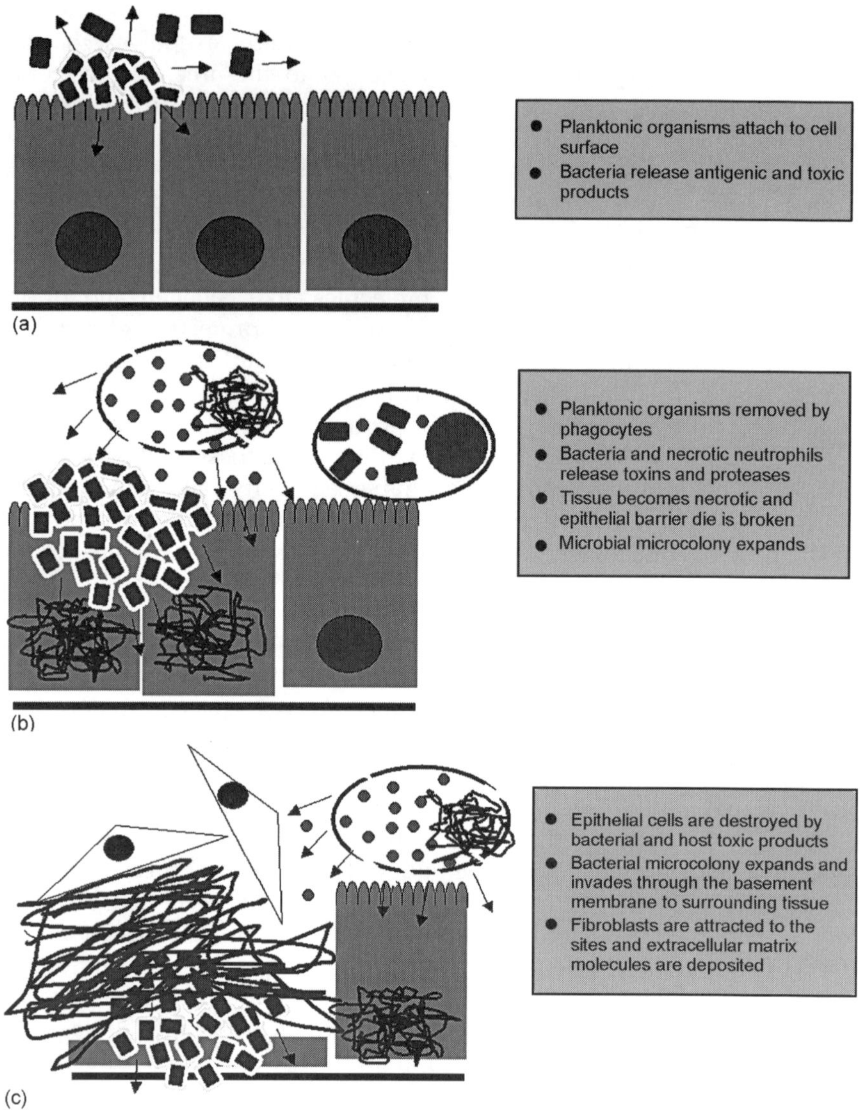

Figure 3.2.2. The pathogenesis of the inflammation associated with tissue biofilms. The bacteria and bacterial products (LPS, alginate, proteases, and toxins) recruit host phagocytes to the area. The bacterial toxins and proteases may also cause tissue necrosis, which attracts further inflammatory cells (primarily PMNs) to the bacterial microcolonies. These neutrophils undergo necrosis and release their contents (proteolytic enzymes, such as elastase, acid hydrolases, lactoferrin, and lysozyme), thus inducing significant tissue damage and perpetuating the inflammatory reaction. The host's main defence in walling off the sites of infections is to induce proliferation of extracellular matrix molecules like collagen and fibrin.

combination of inhibition of transport through the biofilm and a high concentration of degrading enzymes may effectively inactivate the antimicrobial agent. It has been suggested that bacteria within biofilms are slow growing, thereby making them resistant to uptake of the antimicrobial agent (Evans *et al.* 1991). In fact, microbial biofilms within tissue differ from environmental biofilms in that they are in a high-nutrient environment and are indeed rapidly multiplying. In some cases of biocide resistance in biofilms from industrial settings, where nutrient limitations exist, the decreased physiological activity of the cells may account for resistance, but this is unlikely for tissue biofilms. It has recently been suggested that slow growth of organisms within biofilms is induced as a stress response that is regulated by a σ factor, RpoS (Adams and McLean 1999; Cochran *et al.* 2000). This σ factor has been demonstrated in the sputum of CF patients with chronic *P. aeruginosa* infections, and RpoS-deficient *E. coli* mutants are unable to form biofilms. Quorum-sensing has recently been implicated in antimicrobial resistance within biofilms, as some authors have suggested that the *lasR–lasI* quorum-sensing systems might activate the σ factor, RpoS, to reduce the rate of growth within the biofilm. There is presently conflicting evidence on the role of quorum-sensing on antimicrobial resistance (Mah and O'Toole 2001).

Much of the recent data suggests that a resistant phenotype is induced within microbial biofilms (Pratt and Kolter 1999; Kuchma and O'Toole 2000). This phenotype is expressed when it attaches to a surface (foreign body or tissue) and a new set of genes are switched on and their corresponding products are formed. Researchers have yet to identify these gene products, but a σ factor, multi-drug efflux pump and porin pump may be involved. Pharmaceutical companies have developed drugs using the minimum inhibitory concentration (MIC) screening assay that targets microorganisms within a planktonic phenotype. Microorganisms that form biofilms have an entirely different phenotype with respect to physiology, therefore, it is logical that they will be more resistant than the planktonic phenotype. Certainly the development of chemotherapeutic agents directed against biofilm organisms must be a rational approach to new drug development (Ceri *et al.* 1999b). The minimum biofilm eradication concentration (MBEC) has been proposed as the standard for selection of chemotherapeutic drugs and biocides (Ceri *et al.* 1999b, 2000; Morck *et al.* 2000).

Presently, there is no clear explanation for antimicrobial resistance of biofilms within tissues. It is very clear that this resistance is very costly both in terms of the welfare of the patient and in terms of the economics. The pharmaceutical industry may need to re-examine their libraries of compounds and select compounds that are effective against the biofilm phenotype.

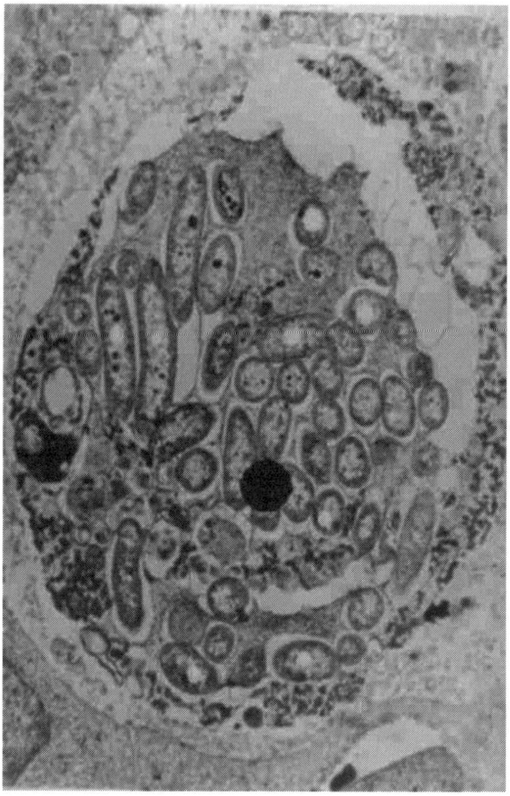

Figure 3.2.3. Transmission electron micrograph of a *P. aeruginosa* biofilm within the respiratory tissue of a CF rat animal model. In chronic pneumonia the bacteria exist within the tissue encapsulated in microcolonies that are protected from the host immune defences.

EXAMPLES OF BIOFILM TISSUE INFECTIONS

Biofilms in Lung Infections

Most bacterial lung infections cause an acute pneumonia (e.g. *Streptococcus* spp., *Haemophilus* spp.), however, there are many cases where microbial biofilms are involved in chronic pneumonias. In these cases, the host's immunity is compromised or there is chemical, viral or bacterial damage to the respiratory epithelium. Chronic lung infections involving bacterial biofilms have been demonstrated with a wide variety of bacteria (e.g. *S. aureus*, *Pseudomonas* spp., *Klebsiella* spp., *Pasteurella* spp. and *Actinomyces pyogenes*) in both humans and animals (Morck *et al.* 1989; Jones *et al.* 2000). The classic example of a biofilm lung infection has become the *P. aeruginosa* infections seen in the lungs of CF patients (Lam *et al.* 1980; Høiby *et al.* 1995; Kobayashi 1995). In CF, the host is compromised through a genetic defect involving the regulation of chloride ion transport. In those patients, the lungs are more susceptible to infection due an excessive build-up of thick dehydrated mucus in the airways. *P. aeruginosa*, a common environmental bacterium, is the most prevalent and severe chronic infection in CF patients. *P. aeruginosa* shows chemotaxis toward mucin-rich surfaces and readily adheres to injured respiratory epithelial cells. Once bacteria have colonized the respiratory epithelial surface as a biofilm (Figure 3.2.3) it produces toxins and virulence factors such as alginate, alkaline phosphatase, elastase, exotoxin A, exotoxin S, LPS and phospholipase. The presence of bacteria and bacterial toxins recruit host phagocytes (primarily PMNs) and immunoglobulins. These phagocytes are unable to eliminate the bacteria and their toxic products and lead to further tissue damage and immune complex disease. The existence of *P. aeruginosa* as a biofilm correlates with the chronic nature of the disease, the inability to clear the infection with antibiotics that display efficacy against the planktonic isolates, and the host's immune system failure to clear the infection.

The association of biofilms in non-CF bronchiectasis has not been established, however, similar changes are seen in the lungs in bronchiectasis (Ceri *et al.* 1991) that may also promote biofilm formation. Chronic biofilm infections of the lungs in humans and animals have similar pathogenesis but may vary depending on the virulence factors of the pathogen (or pathogens in mixed infections), the level of immunocompromisation, and the severity of inflammatory cell infiltration. It is evident to both human and veterinary clinicians that chronic pneumonia involving biofilm bacteria does not respond well to conventional antimicrobial therapy (Jones *et al.* 2000). Long-term administration of antibiotics has frequently been shown to lead to reinfection after the drug has been withdrawn, which indicates that only the planktonic forms have been eliminated. In addition, the emergence of antimicrobial-resistant forms have been observed following long-term therapy (Jones *et al.* 2000).

Wound Infections

A wound, which is defined as an interruption of tissue, can affect the skin, mucosa or organs. Wound repair is a complex and dynamic process that has three stages: inflammatory phase, proliferative phase and regenerative phase (Figure 3.2.4). The rate at which the wound passes through these phases influences the time it takes for a wound to heal. The inflammatory phase is mostly associated with bacterial colonization and lasts for 3 days or more (Davidson 1992; Eaglstein and Falanga 1997). During this phase the blood coagulation system is activated with the release of numerous

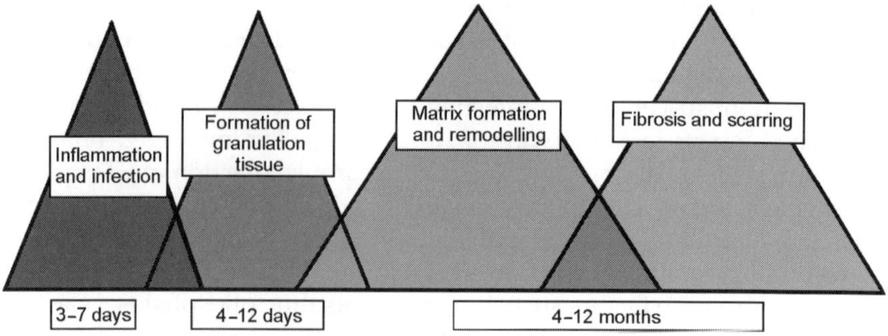

Figure 3.2.4. Pictorial description of the normal wound healing process. Bacterial biofilms impair wound healing by increasing the duration of the inflammatory phase.

activators (e.g. complement, platelet-activating factor). Inflammatory cells (primarily neutrophils) migrate to the wound site attracted by a number of chemotactic substances. They are involved in debridement and bacterial defence, as their granules contain a variety of proteolytic enzymes that are cytotoxic to bacteria, but they are also destructive to the host tissue. Neutophils and macrophages also ingest the bacteria and necrotic material at the wound site in order to establish the conditions for eventual wound closure. Macrophages also release factors that stimulate fibroblast proliferation, angiogenesis and migration and growth of keratinocytes. Once bacteria form biofilms by establishing protected microcolonies within the tissue, they are very difficult for the host to remove. If the phagocytes cannot remove the bacteria, then they will continue to be recruited to the wound site, causing damaging tissue destruction in an attempt to debride and disinfect the wound. Therefore, bacterial biofilms within wound tissue are responsible for extending the inflammatory phase and impairing wound healing. Bacteria themselves produce cytotoxic elements, such as coagulase, haemolysins, leucocidin, enterotoxins, catalase and a variety of proteases, that cause tissue destruction and impair the activity of inflammatory cells. This contributes to impairment in the wound healing process and maintains the wound in the inflammatory phase.

Bacteria exist in wounds as either a single or a mixed culture. *S. aureus* is the leading bacterial species in chronic wounds (63%), but other microorganisms (*E. coli, Bacteroides* spp., *Streptococcus faecalis, Staphylococcus epidermidis, Candida albicans* and *P. aeruginosa*) are frequently observed as pure or as a mixed cultures (Wright *et al.* 1998). In a pig model, we have demonstrated that, within 24 h of wounding/challenge, bacteria attach and form microcolonies within the wound bed (Figure 3.2.5). In the same model, we also demonstrated bacterial microcolonies deep within the wound bed

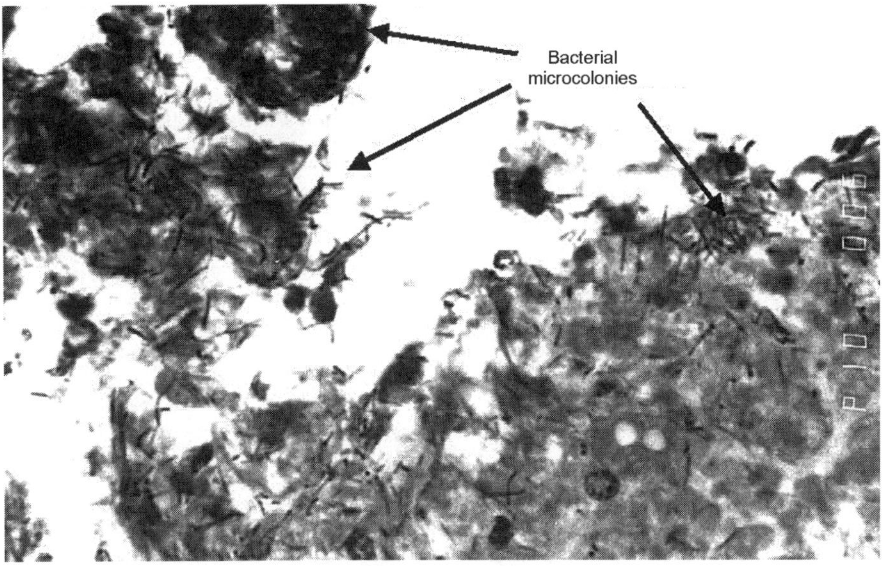

Figure 3.2.5. Histology of a 7-day-old infected wound. Bacterial biofilm is present on the surface and within a wound. Microcolonies within the tissue are surrounded by necrotic PMNs and fibroblasts.

after the tissue had re-epithelized. These localized deep tissue micro-abscesses may be responsible for recurrence of wounds, cellulitis and deep tissue infections. We have demonstrated that these cryptic bacteria are resistant to most systemic and topical antibiotics used in traditional wound therapies (Table 3.2.1). In fact, the chronic use of infective antimicrobial therapy may lead to highly resistant (planktonic and sessile) microorganisms, as is common in burn wards. Only recently have wound-care specialists recognized the importance of biofilm bacteria in the pathogenesis of delayed wound healing and, recently, novel strategies have been developed to control sessile organisms and microcolonies to enhance the rate of healing (Wright *et al.* 1998). For example a nanocrystalline silver band has been shown to eliminate a wide variety of wound bacteria rapidly, thereby improving the rate of healing and quality of the healed tissue (Tredget *et al.* 1998).

Biofilm in Urinary Tract Infections

Typically, when considering biofilms in the urinary tract, one invariably considers catheter-related infections. Without a doubt, one of the first recognized and most studied biofilm diseases has been catheter-associated

Table 3.2.1. MIC and MBEC (concentration required to kill bacteria growing as a biofilm) of antibiotics for burn wound isolates of *P. aeruginosa*. For methodology refer to Ceri *et al.* (1999b)

	P. aeruginosa 30299		*P. aeruginosa* ML56	
Antibiotic	MIC ($\mu g\,ml^{-1}$)	MBEC ($\mu g\,ml^{-1}$)	MIC ($\mu g\,ml^{-1}$)	MBEC ($\mu g\,ml^{-1}$)
Amikacin	256	512	128	>1024
Tobramycin	512	>1024	256	>1024
Imipenem	256	>1024	2	>1024
Piperacillin	>1024	>1024	>1024	>1024
Gentamicin	16	32	512	512
Aztreonam	128	>1024	8	>1024
Ciprofloxacin	32	512	2	16
Ceftazidine	512	>1024	2	512

infections (Nickel *et al.* 1994; Morris *et al.* 1999; Olson *et al.* 2000). It is clear, however, that biofilms are associated with urinary tract infections (UTIs) where indwelling devices are not the cause. Bacteria have been shown to form biofilms on bladder mucosal tissue (Reid *et al.* 1994 2000) and to associate in a lectin-dependent manner to sloughed urinary epithelial cells (Graham *et al.* 1992). Additionally, biofilm formation on the mucosal surface of the acini of prostate tissue has been clearly demonstrated in a rat model of bacterial prostatitis, both by direct staining and by immunofluorescence (Figure 3.2.6) (Ceri *et al.* 1999a). The lack of protection by specific antibodies to the initiation of bacterial prostatitis has also been shown (Ceri *et al.* 1999c). Recently, biofilm formation in biopsies from 3 of 12 patients presenting prostatitis has been visualized by electron microscopy (Arakawa *et al.* 1999). The formation of biofilms within the urinary tract is likely the best explanation for the recurrent and chronic infections that are the bane of urologists. One can imagine the recovery from the symptoms of cystitis following antibiotic treatment, which removes the planktonic bacteria, only to have the symptoms return as a result of regrowth of the planktonic population from a nidus of infection consisting of biofilm bacteria displaying a higher level of resistance to the antibiotic. The urinary tract is protected from pathogen colonization by the flushing action of sterile urine, the sloughing of the uroepithelial cells and a glycosaminoglycan layer (GAG). In the lower urinary tract, organisms like *Lactobacillus* spp. prevent pathogens from attaching and establishing a tissue infection. Any disturbance of the normal protective mechanisms of an organism, which can outcompete the normal microflora, can lead to a UTI. This disturbance can be in the form of a foreign body, such as a catheter, stent or inorganic deposit (struvite), which allows the bacteria or yeast to attach initially to the

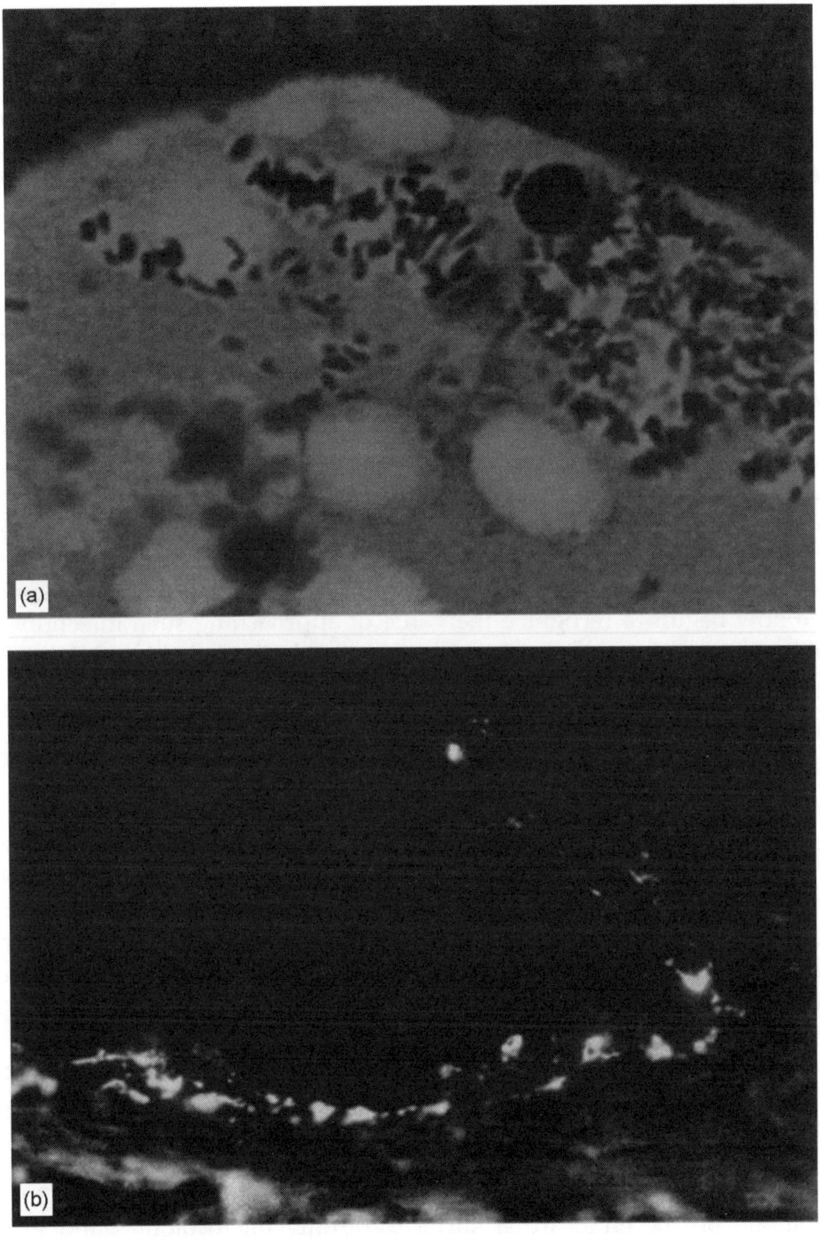

Figure 3.2.6. Biofilm formation on the mucosal surface of the acini of prostate tissue in a rat model of bacterial prostatitis by (a) direct silver staining and (b) immunofluorescence microscopy using fluorescent antibodies against *E. coli*.

inert material and then invade the tissue. Alteration in the host protection can also be due to a change in urine pH or trauma where the protective GAG layer is disturbed and the underlying tissue exposed. The prostate gland is susceptible to bacterial colonization, as it is immunocompromised through natural host secretions and a reduced flow. Once bacteria have attached to the uroepithelial surface they multiply and form biofilms on the tissue surface and are resistant to host and antibiotic elimination. We show that opsonized bacteria are not removed by the large numbers of phagocytes in the prostate acini (Ceri *et al.* 1999c). The recruitment of large numbers of PMNs to the sites of bacterial colonization causes further tissue damage and permits the site of infection to expand. In chronic bacterial prostatitis this can lead to fibrosis of the entire prostatic gland over years. Chronic biofilm infection with associated tissue destruction in the urethra, bladder, and ureter can result in fibrosis, strictures, stenosis and pyelonephritis. The colonization of the urinary tract with urease-producing bacteria, such as *Proteus mirabilis*, results in urinary calculi formation in the bladder and kidneys and in further significant health problems associated with infection and obstruction. Clearly, biofilm infections in urogenital tissue are associated with significant morbidity and mortality. Urologists have now begun to recognize the value of approaching such infections as biofilms in preventative and treatment protocols (Nickel *et al.* 1994).

Vegetative Endocarditis

Bacteria and yeast have been shown to colonize the inner lining of the heart (endocardium) with a predilection to the heart valves, leading to vegetative endocarditis. This is a life-threatening condition requiring long-term aggressive treatment with antibiotics. Vegetative endocarditis has been shown to be a classical example of a biofilm infection within tissue (Olson *et al.* 2000). Bacterial microcolonies are present within the proliferative fibrotic lesions on the damaged heart values (Figure 3.2.7). *S. aureus* and *Streptococcus* spp. are the most common pathogens associated with this disease, but a wide variety of microorganisms, either alone or as mixed infections, are associated with the vegetations (Brook 1999; Hoesley and Cobbs 1999). Bacteria are released into the circulation from an infection (GI, urinary tract), trauma (wound), colonized implant (vascular catheter) or medical and dental procedures (teeth cleaning). The circulating bacteria eventually attach to the endocardium or the heart valves (Chang 2000), proliferate on the endocardial surface, forming microcolonies and eventually invading and inducing localized inflammatory reaction with numerous neutrophils. The inflammatory response, which fails to clear the pathogen, leads to tissue damage and more inflammatory cell infiltration. The host responds to this inflammation with excessive fibrosis

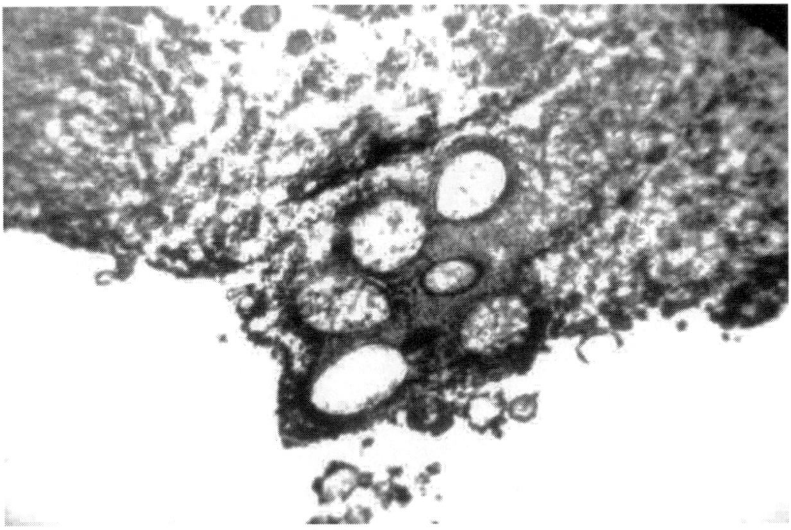

Figure 3.2.7. Transmission electron micrograph of *Streptococcus viridans* microcolonies within the proliferative fibrotic lesions on a heart valve. These microcolonies stimulate destructive changes to the heart valve by the host's own defence system.

(vegetations). These vegetations impair the normal function of the valves, leading to back flow of blood and eventually heart failure. The bacterial microcolonies are protected from host and antibiotic elimination by the glycocalyx and the excessive fibrosis. Heart valves can be colonized for weeks before the host becomes ill, and then it is even more difficult to treat the established infection (Brook 1999; Hoesley and Cobbs 1999; Chang 2000). Some bacterial microcolonies on heart valves can be released and then lodge in other organs, such as the lungs and the kidneys, leading to abscessation and infarction. Antimicrobial therapeutic agents have not been highly effective in the treatment of endocarditis (Chang 2000), but, by approaching this disease as a microbial biofilm infection, it may be possible to improve the therapeutic success using new drugs or combination therapy.

Osteomyelitis

Osteomyelitis is a chronic infection of the bone where bacteria have been shown to persist within glycocalyx-enclosed microcolonies (Power *et al.* 1990). A typical example of a *S. aureus* biofilm is represented in Figure 3.2.8. The exopolysaccharide is believed to provide protection against phagocytes

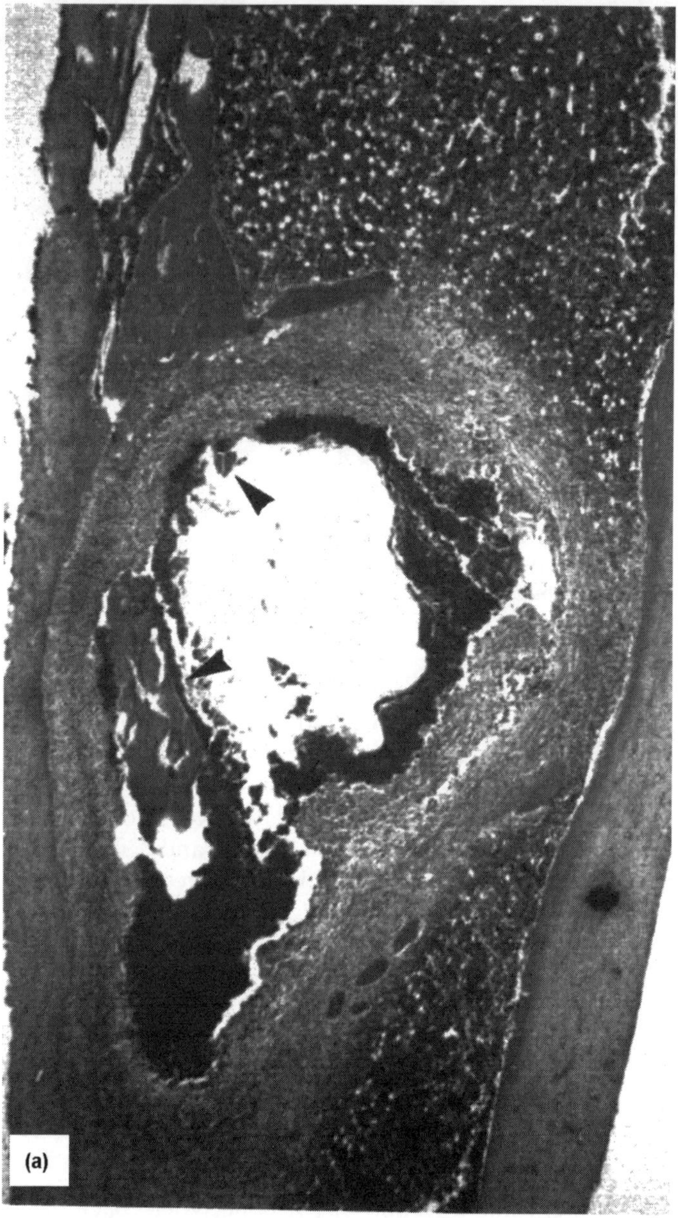
(a)

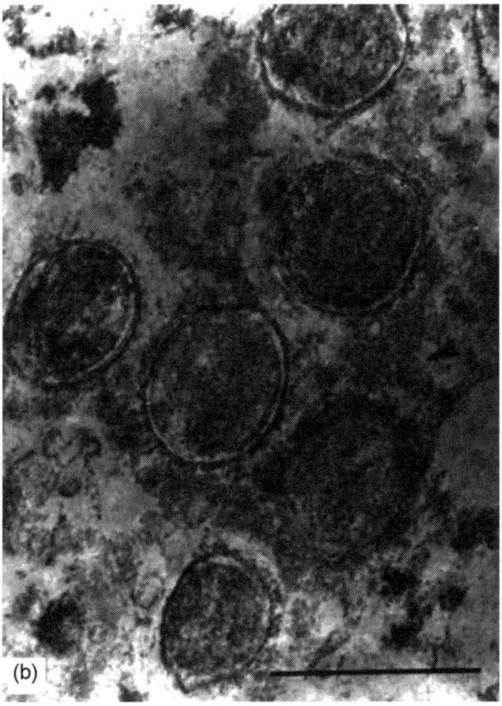

Figure 3.2.8. (a) Histological section through an infected bone of a rat model of osteomyelitis with extensive necrosis and remodelling of bone (arrows). (b) Transmission electron micrograph of *S. aureus* within glycocalyx-enclosed microcolonies inchronic osteomyelitis.

and chemotherapeutic agents. As in many chronic inflammatory infections, the host cellular response (mainly neutrophils) and bacterial toxins lead to both bone loss and soft tissue destruction (Power *et al.* 1990; Ciampoli and Harding 2000). There is also excessive fibrosis at the site of infection, leading to distortion of soft tissue. The infection often goes undiagnosed for weeks; by that time the biofilm is well organized and inflammation is severe, thereby making treatment difficult and often ineffective. Osteomyelitis is a serious and extremely difficult disease to treat. Frequently, the site of infection needs to be debrided and drained to remove microabscesses and permit direct treatment with antibiotics (Bamberger 2000; Ciampoli and Harding 2000). Successful treatment usually requires the use of high doses of one or more antibiotics over months, and surgical intervention is often necessary to remove the necrotic and contaminated bone tissue. Reappearance of the infection can also occur after apparently successful treatment, indicating that the bacterial biofilm was not entirely removed. Certainly, new antibiotics that are effective against the microcolonies within bone are needed (Bamberger 2000; Ciampoli and Harding 2000).

Gastrointestinal Biofilms

Microbial biofilms play an important role in the normal function of the gastrointestinal tract, and they can also be responsible for serious disease. In cattle, bacteria are essential for the digestion of plant material within the reticulo rumen (Cheng *et al.* 1995). Approximately 80% of the bacteria are associated as biofilms with the food particles and 1% with the epithelial surface. Within minutes of ingestion, amylolytic and cellulolytic bacteria attach to the food particles and establish a typical biofilm. The bacterial degradation products (butyric acid and propionic acid) and the bacteria themselves serve as the nutrient source for the animal. The mammalian gut is coated with a thick mucus blanket. Normally, the majority of the bacteria exist within the mucus blanket and fewer on the tissue surface itself (Figure 3.2.9). Disturbances of the mucus blanket can lead to the colonization of the gut surface with excessive numbers of autochthonous microflora, leading to bacterial overgrowth and intestinal disease. Indeed, this disturbance may be induced by simply eating a diet (i.e. containing plant lectins) that affects the mucus blanket (Banwell *et al.* 1988). Overgrowth of the gut surface with autochthonous bacteria may lead to impaired nutrient transport, gut inflammation and weight loss, as is seen in tropical sprue. Bacteria, such as *E. coli*, may attach to the intestinal surface, form microcolonies and release toxins, or they may translocate through the gut to form microabscesses within the lamina propria. Once the microorganisms have formed a protective biofilm, the host is then unable to remove the microbes through mucus flow or recruitment of inflammatory cells. The tissue damage is caused by the cytotoxic constituents of lysed neutrophils that have been recruited to the site of the microcolonies. Microbial biofilms that form on the gut surface and within gut tissue may be responsible for many of the chronic intestinal diseases that result in significant morbidity and mortality in animals and humans. These diseases are particularly serious in the young and elderly, who are unable to cope with the nutritional deficiencies and inflammation of chronic intestinal disease. Antibiotics are generally effective in treating acute intestinal diseases, but they are much less effective in eliminating tissue-associated bacteria once they have been established.

Mastitis

Microbial biofilm diseases are not restricted to humans. They play an important role in veterinary medicine. Numerous biofilm infections of

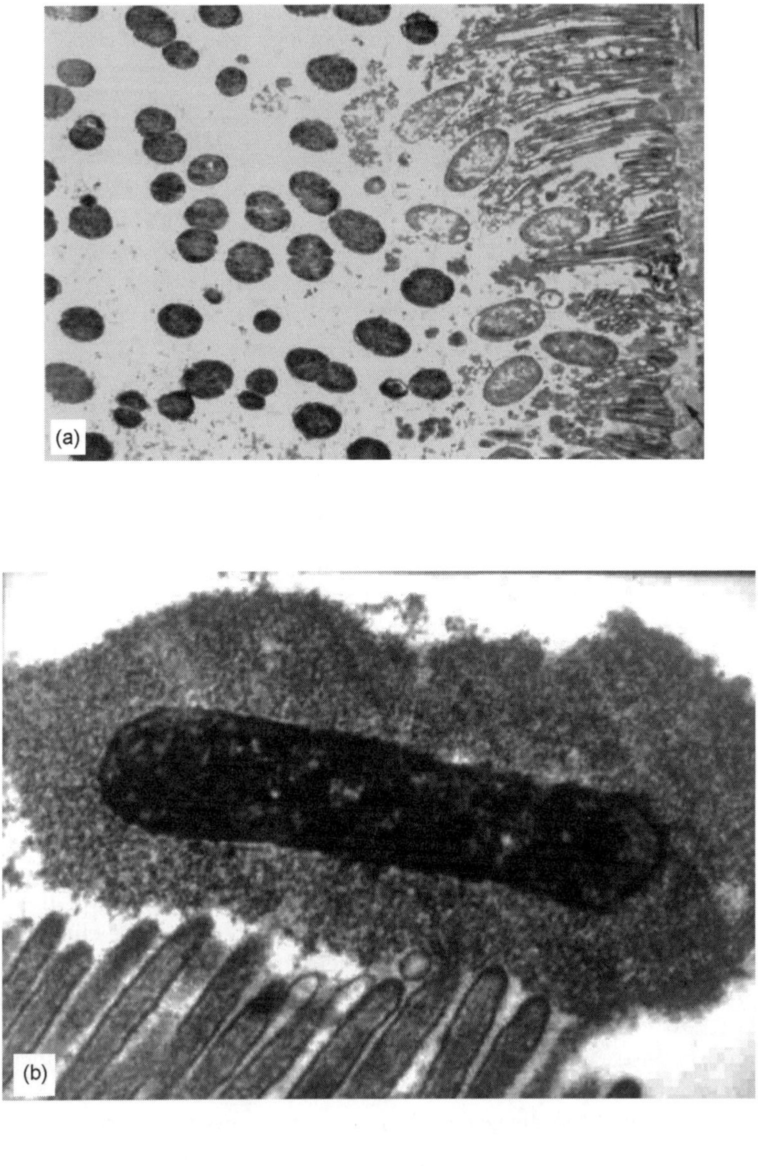

Figure 3.2.9. (a) Transmission electron micrograph of bacterial overgrowth and biofilm formation on the surface of the gut. This is a mixed population of microorganisms and the mucus blanket has been invaded, resulting in destruction of the microvilli. (b) Transmission electron micrograph of an *E. coli* attached to the enterocyte microvilli within the small intestine. Note the extensive glycocalyx surrounding the bacteria.

Table 3.2.2. MIC and MBEC of veterinary antibiotics for bovine mastitis isolates of *S. aureus* and *Streptococcus uberis*. For methodology refer to Ceri *et al.* (1999b)

	S. aureus		*S. uberus*	
Antibiotic	MIC (μg ml^{-1})	MBEC (μg ml^{-1})	MIC (μg ml^{-1})	MBEC (μg ml^{-1})
Amikacin	4	64	8	8
Gentamicin	<2	16	<2	<2
Tilmicosin	16	>1024	4	>1024
Pirlimycin	8	>1024	<2	64
Cephalothin	<2	>1024	<2	128
Erythromycin	128	>1024	<2	32
Penicillin G	<2	>1024	<2	<2
Novobiocin	8	16	<2	>1024
Tylocin	<2	1024	<2	512
Cloxacillin	<2	512	<2	512
Cephapirin	8	1024	<2	32
Oxytetracycline	<2	256	<2	128
Ceftiofur	4	1024	<2	128
Enrofloxacin	<2	256	<2	<2

tissues are responsible for animal disease; these have an economic impact on agriculture and affect the health and welfare of the animal. Bovine mastitis would be considered as one of the most significant biofilm infections of tissue in animals (Baselga *et al.* 1992, 1993). Mastitis is the inflammation of the mammary gland, which, in animals, is almost always due to the effects of infection by bacterial or mycotic pathogens. The most common bacterial pathogens are *S. aureus, Streptococcus* spp. and coliforms. The prevalence of bovine staphylococcal mastitis ranges from 7 to 40% of all dairy cattle and is the major reason for culling milking cows. It is also recognized that antibiotic therapy may temporarily eliminate clinical signs of mastitis, but the prognosis of a complete cure is poor (Sandholm *et al.* 1990; Bolourchi *et al.* 1995). Cows are usually treated at the drying-off period, when cows are no longer milking due to advanced pregnancy. Prolonged-release antibiotics (penicillin–streptomycin, cephalosporin, novobiocin and cloxacillin) are usually infused into the mammary canal. The pathogenesis of the disease is similar to other diseases described above. Bacteria usually enter the mammary gland through the teat from the environment. They adhere to the mammary epithelium, where they form microcolonies on the epithelial surface and invade the epithelial barrier to colonize the underlying tissue. PMNs are attracted to the site, where they induce a localized inflammatory reaction. The infection usually spreads throughout the mammary gland, leading to a chronic infection of the gland and a dramatic decrease in milk production. The massive infiltration of

PMNs into the gland is reflected by a large somatic cell count in the milk of cattle with chronic mastitis. Chronic Staphylococcal and Streptococcal mastitis in dairy cattle is considered an untreatable disease and has frustrated the dairy industry for decades. We compare the antibiotic susceptibility of planktonic and biofilm *Staphylococcus* and *Streptococcus* mastitis isolates in Table 3.2.2. Although planktonic organisms were highly susceptible to the antibiotics used for treatment of mastitis, the biofilm bacteria were highly resistant. Indeed, the dairy industry requires an effective treatment for chronic Staphylococcal and Streptococcal mastitis, to improve the quality and quantity of milk and to reduce the costs associated with culling infected animals. There is currently no product available.

CONCLUSIONS

Biofilm infections on tissues are as common and equally difficult to treat as device-related infections in medicine. However, this type of biofilm infection is not as well recognized, because of the slow evolution of understanding and knowledge accumulation regarding biofilms over the years (industrial biofilms, to medical device biofilms, to chronic tissue infections associated with biofilms). Biofilm infections involving tissues in the absence of foreign bodies have only recently been recognized as a significant cause of morbidity and mortality in human and veterinary medicine. The examples provided in this chapter only scratch the surface of the extensive nature of the problem. Methods for controlling and treating tissue biofilm infections must be developed in order for modern medicine to be successful in addressing one of the most problematic areas, i.e. chronic infectious diseases.

REFERENCES

Adams, JL and McLean, RJ (1999) Impact of RpoS deletion on *Escherichia coli* biofilms. *Applied and Environmental Microbiology* **65**:4285–4287.

Arakawa, S, Matsui, T, Gohji, K, Okada, H and Kamidono, S (1999) Prostatitis: the Japanese viewpoint. *International Journal of Antimicrobial Agents* **11**:201–203.

Bamberger, DM (2000) Diagnosis and treatment of osteomyelitis. *Comprehensive Therapy* **26**:89–95.

Banwell, J, Howard, R, Costerton, W, Ceri, H, Falk, P and Larson, G (1988) Chronic intestinal microbial infection in the rat induced by dietary intake of phytohemagglutin lectin. In: Inflammatory Bowel Disease: Current Status and Future Approach (Ed. MacDermott, RP), Elsevier, New York, pp. 609–614.

Baselga, R, Albizu, I, Aguilar, B, Iturralde, M and Amorena, B (1992) Hydrophobicity of ruminant mastitis *Staphylococcus aureus* in relation to bacterial aging and slime production. *Current Microbiology* **25**:173–179.

Baselga, R, Albizu, I, De la Cruz, M, Del Cacho, E, Barberan, M and Amorena, B (1993) Phase variation of slime production in *Staphylococcus aureus*: implications in colonization and virulence. *Infection and Immunity* **61**:4857–4862.

Bolourchi, M, Hovareshi, P and Tabatabyi, A (1995) Comparison of the effects of local and systemic dry cow therapy for staphyloccal mastitis control. *Preventative Veterinary Medicine* **25**:63–67.

Brook, MM (1999) Pediatric bacterial endocarditis, treatment and prophylaxis. *Pediatric Clinics of North America* **46**:275–287.

Buret, AG, Ward, KH, Olson, ME and Costerton, JW (1991) An *in vivo* model to study the pathobiology of infectious biofilms on bacterial surfaces. *Journal of Biomedical Materials Research* **25**:865–874.

Ceri, H, Hwang, HS and Rabin, H (1991) Structure, secretion, and bacterial specificity of an endogenous lectin from cystic fibrosis lung. *American Journal Respiratory Cellular and Molecular Biology* **5**:51–55.

Ceri, H, Olson, ME and Nickel, JC (1999a) Prostatitis: role of the animal model. In: Textbook of Prostatitis (Ed. Nickel, JC), Isis Medical Media, Oxford, pp. 109–114.

Ceri, H, Olson, ME, Stremick, C, Morck, DW, Read, RR and Buret, AG (1999b) The Calgary biofilm device: measurement of antimicrobial sensitivity of bacterial biofilms. *Journal of Clinical Microbiology* **37**:1771–1776.

Ceri, H, Schmidt, S, Olson, ME, Nickel, JC and Benediktsson, H (1999c) Specific mucosal immunity in the pathophysiology of bacterial prostatitis in a rat model. *Canadian Journal of Microbiology* **45**:849–855.

Ceri, H, Morck, DW and Ceri, H (2000) Biocide susceptibility of biofilms. In: Disinfection, Sterilization and Preservation (Ed. Block, SS), Lippincott Williams and Wilkins, Philadelphia, pp. 1329–1437.

Chang, FY (2000) *Staphylococcus aureus* bacteremia and endocarditis. *Journal of Microbiology, Immunology and Infection* **33**:63–68.

Cheng, K-J, McAllister, TA and Costerton, JW (1995) Biofilms of the ruminant digestive tract. In: Microbial Biofilms (Eds. Lappin-Scott, HM and Costerton, JW), Cambridge University Press, Cambridge, pp. 221–232.

Ciampoli, J and Harding, KG (2000) Pathophysiology of chronic bacterial osteomyelitis. Why do antibiotics fail so often? *Postgraduate Medical Journal* **76**:479–483.

Cochran, WL, Suh, SJ, McFeters, GA and Stewart, PS (2000) Role of RpoS and AlgT in *Pseudomonas aeruginosa* biofilm resistance to hydrogen peroxide and monochloramine. *Journal of Applied and Environmental Microbiology* **88**:546–553.

Costerton, JW, Lewandowski, Z, Caldwell, DE, Korber, DR and Lappin-Scott, HM (1995) Microbial biofilms. *Annual Reviews of Microbiology* **49**:711–745.

Costerton, JW, Stewart, PS and Greenberg, EP (1999) Bacterial biofilms: a common cause of persistent infections. *Science* **284**:1318–1322.

Davidson, JM (1992) Wound healing. In: Inflammation: Basic Principles and Clinical Correlation (Eds. Gallin, JI, Goldstein, IM and Snyderman, R), Raven Press Ltd, New York, pp. 809–919.

Eaglstein, WH and Falanga, V (1997) Chronic wounds. *Surgical Clinics of North America* **77**:575–586.

Evans, DJ, Allison, DG, Brown, MR and Gilbert, P (1991) Susceptability of *Pseudomonas aeruginosa* and *E. coli* biofilms towards ciprofloxacin: effect of specific growth rate. *Journal of Antimicrobial Chemotherapy* **30**:791–802.

Graham, LL, Ceri, H and Costerton, JW (1992) Lectin-like proteins from uroepithelial cells which inhibit *in vitro* adherence of three urethral bacterial isolates to uroepithelial cells. *Microbial Ecology in Health and Disease* **5**:77–86.

Habash, M and Reid, G (1999) Microbial biofilms: their development and significance for medical device-related infections. *Journal of Clinical Pharmacology* **39**:887–889.

Hoesley, CJ and Cobbs, CG (1999) Endocarditis at the millennium. *Journal of Infectious Diseases* **179**(Suppl. 2):S360–S365.

Høiby, N, Fomsgaard, A, Jensen, ET, Johansen, HK, Kromborg, G, Petersen, SS, Pressler, T and Kharazmi, A (1995) The immune response to bacterial biofilms. In: Microbial Biofilms (Eds. Lappin-Scott, HM and Costerton, JW), Cambridge University Press, Cambridge, pp. 233–250.

Hoyle, B, Alcantara, J and Costerton, JW (1992) *Pseudomonas aeruginosa* biofilm as a diffusion barrier to piperacillin. *Antimicrobial Agents and Chemotherapy* **36**:2054–2056.

Jones, RN, Croco, MA, Krugler, KC, Pfaller, MA and Beach, ML (2000) Respiratory tract pathogens isolated from patients hospitalised with suspected pneumonia: frequency of occurrence and antimicrobial susceptibility patterns from SENTRY Antimicrobial Surveilance Program (United States and Canada, 1997). *Diagnostic Microbiology and Infectious Disease* **37**:115–125.

Kobayashi, H (1995) Biofilm disease: its clinical manifestations and therapeutic possibilities of macrolides. *American Journal of Medicine* **99**:26s–30s.

Kuchma, SL and O'Toole, GA (2000) Surface-induced and biofilm-induced changes in gene expression. *Current Opinion in Biotechnology* **11**:429–433.

Lam, JR, Chan, J, Lam, K and Costerton, JW (1980) Production of mucoid microcolonies by *Pseudomonas aeruginosa* within infected lungs in cystic fibrosis. *Infection and Immunity* **28**:546–556.

Mah, T-FC and O'Toole, GA (2001) Mechanisms of biofilm resistance to antimicrobial agents. *Trends in Microbiology* **9**:34–39.

Morck, DW, Olson, ME, Acres, SD, Daoust, PY and Costerton, JW (1989) Presence of bacterial glycocalyx and fimbriae on *Pasteurella haemolyica* in feedlot cattle with pneumonic pasteurellosis. *Canadian Journal of Veterinary Research* **53**:167–171.

Morck, DW, Olson, ME and Ceri, H (2000) Biocide susceptibility of biofilms. In: Disinfection, Sterilization and Preservation (Ed. Block, SS), Lippincott Williams and Wilkins, Philadelphia, pp. 1329–1437.

Morris, NS, Stickler, DJ and McLean, RJ (1999) The development of bacterial biofilms on indwelling urethral catheters. *World Journal of Urology* **17**:345–350.

Nickel, JC, Costerton, JW, McLean, RJ and Olson, M (1994) Bacterial biofilms: influence on the pathogenesis, diagnosis and treatment of urinary tract infections. *Journal of Antimicrobial Chemotherapy* **33**:31–41.

Olson, ME, Morck, DW, Ceri, H, Read, RR and Buret, AG (2000) Animal models for the study of bacterial biofilms. In: Biofilms: Recent Advances in their Study and Control (Ed. Evans, L), Harwood Academic Publishers, Chur, Switzerland, pp. 133–147.

Potera, C (1999) Forging a link between biofilms and disease. *Science* **283**:1837–1838.

Power, ME, Olson, ME, Domingue, PAG and Costerton, JW (1990) A rat model of *Staphylococcus aureus* chronic osteomyelitis that proves a suitable system for studying the human infection. *Journal of Medical Microbiology* **33**:189–198.

Pratt, LA and Kolter, R (1999) Genetic analysis of biofilm formation. *Current Opinion in Microbiology* **2**:598–603.

Reid, G and Bruce, AW (1995) The role of the urogenital flora in probiotics. In: Microbial Biofilms (Eds. Lappin-Scott, HM and Costerton, JW), Cambridge University Press, Cambridge, pp. 274–281.

Reid, G, Dafoe, L, Delaney, G, Lacerte, M, Valvano, M and Hayes, KC (1994) Use of adhesion counts to help predict symptomatic infection and the ability of fluoroquinolones to penetrate bacterial biofilms on the bladder cells of spinal cord injured patients. *Paraplegia* **32**:468–472.

Reid, G, Potter, P, Delaney, G, Hsieh, J, Nicosia, S and Hayes, K (2000) Ofloxacin for the treatment of urinary tract infections and biofilms in spinal cord injury. *International Journal of Antimicrobial Agents* **13**:305–307.

Rosee, KR, Cooper, D, Lam, K and Costerton, JW (1982) Microbial flora of the mouse ileum mucous layer and epithelial surface. *Applied and Environmental Microbiology* **43**:1451–1463.

Sandholm, M, Kaartinen, L and Pyorala, S (1990) Bovine mastitis—why does antibiotic therapy not always work? An overview. *Journal of Veterinary Pharmacology and Therapy* **13**:248–260.

Stewart, PS (1996) Theoretical aspects of antibiotic diffusion into microbial biofilms. *Antimicrobial Agents and Chemotherapy* **40**:2517–2522.

Suci, PA, Mittelman, MW, Yu, FP and Geesey, GG (1994) Investigation of ciprofloxacin penetration into *Pseudomonas aeruginosa* biofilms. *Antimicrobial Agents and Chemotherapy* **38**:2125–2133.

Tredget, EE, Shankowsky, HA, Groeneveld, MN and Burell, R (1998) A matched-pair, randomized study evaluating the efficacy and safety of acticoat silver-coated dressing for the treatment of burn wounds. *Journal of Burn Care and Rehabilitation* **19**:531–537.

Ward, K, Olson, ME and Costerton, JW (1992) Mechanism of persistant infection associated with peritoneal implants. *Journal of Medical Microbiology* **36**:406–413.

Wright, JB, Hansen, CET and Burell, RE (1998) The comparative efficacy of two antimicrobial barrier dressings: *in vitro* examination of two controlled release silver dressings. *Wounds* **10**:179–188.

3.3 Control of Microbial Adhesion and Biofilm Formation on Tissue Surfaces

GREGOR REID, JAMES WATTERSON, PETER CADIEUX and JOHN DENSTEDT
Lawson Health Research Institute, and Departments of Surgery and Microbiology and Immunology, The University of Western Ontario, London, Ontario, Canada

INTRODUCTION

The focus of this chapter is to examine the current understanding of microbial interactions on tissues, including those that are found in healthy individuals and those associated with infections. Examples will be given for the gut, urogenital tract and wound sites, with some additional discussion of livestock. The section on wounds will be kept separate in order to retain the focus of the chapter. Control strategies will include antibiotics, prebiotics, probiotics and vaccines. In-depth discussion of the structure of biofilms is presented in Chapter 1 and will not be presented here.

GUT AND UROGENITAL TRACT

Life forms have existed on this planet for many years, in large part because of an ability to live alongside microorganisms. Such symbiotic or other associations are, for the most part, healthy, however, processes that take place within the microflora play a major role in the types of disease that disable or kill humans and animals. This is no more evident than with the biofilms found in the gut and urogenital tract, where as many as 400 and 50 species of microbes, respectively, have been recovered and identified. The production of carcinogens in the gut leading to cancer, the onset of some

Medical Biofilms. Edited by J. Jass, S. Surman, J. Walker

cases of heart disease, arthritis and autoimmune diseases can be linked to microorganisms. Given that cardiovascular disease, cancer and infectious diseases represent, by far, the major killers of humans, the burden placed on understanding microbial interactions is enormous. Furthermore, as approximately 65% of all infections are associated with microbial biofilms, it is incumbent upon scientists to encourage more research in these areas.

What Dictates Which Organisms Will Colonize Individuals?

Although there may be some transfer of flora from the mother to the foetus, it is upon vaginal birth that the process of microbial colonization begins. Unfortunately, premature babies, those born by caesarian section and undernourished babies are more susceptible to infection. Intestinal infections occur in around 16.5 million US children each year, and projections for developing nations in Africa and Asia bring this number to many hundreds of millions. Worldwide, over 2 million people die each year from diarrhoeal diseases. Even in developed countries, babies born under 1500 g are at high risk of a devastating disease called necrotizing enterocolitis, caused by a group of pathogens that are primary colonizers and which become enterotoxic to the gut. These pathogens include *Enterococcus faecalis, Escherichia coli, Staphylococcus epidermidis, Enterobacter cloacae, Klebsiella pneumoniae* and *Staphylococcus haemolyticus* (Gewolb *et al.* 1999). Signs and symptoms include abdominal distension, bilious emesis, bloody stools, lethargy, apnoea and bradycardia (Caplan and Jilling 2000). Mortality occurs in around 25–30% of cases, and in others a bowel resection is often needed, leading to major long-term health problems.

Much can be learned from the flora that colonizes babies. Clearly, most babies in developed countries survive and go on to lead a normal life. Apart from factors such as nutrition, birth size and weight, and genetic make-up, the primary colonization of the intestine by lactobacilli and bifidobacteria transferred from the breast-feeding mother appears to be critically important. In a review of the literature, Walker (2000) reported a correlation between normal gut flora at birth and protection against various infections. This apparent protective effect will be discussed later.

The microbial colonization of the gut at birth is really a model in which biofilms are formed, starting with the primary and secondary colonizers. The process likely involves adhesion to the intestinal epithelium and mucus (Reid *et al.* 1998; Tuomola *et al.* 1999). Adhesion by lactobacilli and bifidobacteria is not fully elucidated, but can involve electrostatic and hydrophobic interactions, as well as adhesins such as cell-wall teichoic acid and proteins (Chan *et al.* 1985; Reid *et al.* 1998). A collagen-binding protein, for example, has been isolated from strains of *Lactobacillus fermentum* and *Lactobacillus reuteri*, and is involved in adhesion to surfaces (Roos *et al.* 1996;

Turner *et al.* 1997). Other such proteins have been identified (Howard *et al.* 2000), and yet more have still to be found that facilitate establishment in the intestine.

In addition to bacterial adhesins, it is feasible that the host also selects which organisms it wants to colonize various sites. Studies on baby calves have shown that a small set of Gram-positive cocci (four to six species), which are facultative anaerobes and urea degraders, colonize within the first 4 days of birth and protect the rumen contents from the oxygen in the tissues and urea in the blood, converting the latter to ammonia, which then aids in further gut flora development (Cheng and Wallace 1979; Cheng and Costerton 1980; McCowan *et al.* 1980; Cheng *et al.* 1981). If this truly represents host selection, it might explain, in part, the finding of strains that are clearly persistent colonizers (McCartney *et al.* 1996; Tannock *et al.* 2000), and the uniqueness of each person's microflora. Some of this selection might be influenced by signals sent by the bacteria, such as lactobacilli, which turn on mucus expression in the gut and thus inhibit pathogen colonization (Mack *et al.* 1999). Studies that utilize molecular diagnostic techniques will certainly help advance our understanding of the selection process (Mackie *et al.* 1999). One outcome of the Human Genome Project could be to identify host receptivity to certain intestinal organisms. In preliminary findings, Tannock (unpublished data) has shown that animals susceptible to multiple sclerosis have a very stable, if somewhat unusual normal flora, perhaps indicating selectivity and an involvement of these organisms in the disease process.

In summary, our understanding of the flora that develops in each of us is scant, yet these organisms play a major role in health and disease. Studies on the basic microbial ecology of the gut are urgently needed if we are to develop new and effective interventions. Similar conclusions can be drawn from the urogenital tract and skin, where little is known about colonization and stability of microflora.

Diagnosis of Gut and Urogenital Tract Infections

The detection of infections in the gut is mostly diagnosed by the presence of diarrhoea, nausea and vomiting, and culture of stool specimens. In almost all cases of bacterial infection this reveals the offending pathogen, although use of strict anaerobic culture conditions may be needed. Wound and skin infections are diagnosed by culturing samples from the infected sites. For infections of the vagina and cervix, swabs are taken, but the resultant data produced by standard laboratories is often inadequate. If the purpose is to detect a sexually transmitted disease pathogen including *Trichomonas*, this is usually reliable. However, detection of yeast and bacterial vaginosis pathogens is too often missed. This may be due to the organisms' presence within select areas of the mucosa, the failure of their biofilms to be detached by the swab or to be broken down before plating, or their death upon exposure to air.

Various alternative detection methods have been explored experimentally, including the use of fluorescent antibodies to examine vaginal cells directly (Cook *et al.* 1989). Other possibilities exist, including the detection of specific bacterial by-products such as toxic-shock proteins or sialidases. In practical terms, such systems, and those emerging from proteomic and DNA/RNA research, are not yet available for general use. Therefore, the use of swabs and fluids such as urine will remain the standard methods. In the case of urine, the major problem is the failure of biofilm 'clumps' to be completely disrupted prior to plating, which produces counts much lower than are actually present. Also, many biofilms present on bladder cells are not detected using standard urine analysis, again giving false negative results (Reid *et al.* 1992). Unless laboratories return to microscopic analysis of urine sediment cells, something that is time consuming and not economically feasible for most centres, urine culture results must be treated with a degree of scepticism. This explains why many physicians do not culture urine, and it also explains why there has been a recommendation to use lower cut-off values (possibly even 100 cells ml^{-1} instead of 100 000 cells ml^{-1}) in the presence of symptoms and signs to diagnose urinary tract infections (UTIs) (Hooton 1999).

Another problem with existing detection systems is interpretation. For many years, the finding of *Staphylococcus* in urine was viewed as contamination. Now, some physicians believe it can reflect an infectious process, and they treat it accordingly. When a mixed culture is detected, it is difficult to determine which of the organisms is actually inducing the signs and symptoms, and thus the use of broad-spectrum drugs is recommended.

Management of Biofilms Within the Gut and Urogenital Tract

The ability of organisms in the gut and urogenital tract (as well as other sites) to form biofilms and survive, in some cases for the duration of the host's life (Tannock *et al.* 2000), creates an opportunity to use certain microorganisms to reduce the risk of disease. This intervention could occur at birth or at points thereafter, such as when antibiotic use destroys much of the normal flora and induce *Clostridium difficile* colitis. Intervention could be induced by prebiotics (nutrients, not metabolized by the host, that promote growth of normal flora) and probiotics (the use of living microbes administered to promote the health of the host).

In terms of the intestine, it is particularly difficult to manipulate microbial flora, which by adulthood contains 10^3–10^5 per gram of bacteria in the acidic stomach and first part of the small intestine, 10^8 per gram of bacteria in the upper bowel, and 10^{10}–10^{11} per gram of bacteria in the large intestine/colon, all contained within a biofilm structure. Studies on the penetration of biofilms by other bacteria have largely not been done. However, the urogenital tract provides an example of the process and outcome.

During the menstrual cycle the vaginal flora changes, perhaps on an hourly basis (Seddon *et al.* 1976), moving from a normal population dominated by lactobacilli to one that has more Gram-negative rods. Indeed, at any given time, in 50% of women, this flora may not be normal (Keane *et al.* 1997), with some women developing asymptomatic bacterial vaginosis. Currently, we have no insight into the biofilm dynamics that retain or lose a stable lactobacilli flora in the vagina. As urogenital infections (UTI, bacterial vaginosis (BV; an infection caused primarily by *Gardnerella* that causes fishy odour, some clear discharge, irritation and high pH in many patients) and yeast vaginitis) are so common (over 30 million cases in North America annually), and are invariably associated with reduced lactobacilli counts, it is imperative that studies be undertaken in this area. To date, the focal point of studies has been the investigation of virulence factors produced by pathogens, such as sialidases by *Gardnerella vaginalis*, toxins by *Staphylococcus aureus* and fimbriae and haemolysins by *E. coli* (Tierno and Hanna 1989; Cauci *et al.* 1998; Landraud *et al.* 2000). However, this only tells us how the host is affected, not how the pathogens interfere with, or resist, the action of the normal flora. Studies on vaginal lactobacilli and bifidobacteria have shown that they produce acids, bacteriocins, biosurfactants, and hydrogen peroxide that inhibit the binding and growth of pathogens (Reid *et al.* 1987; Eschenbach *et al.* 1989; Velraeds *et al.* 1998; Korshonov *et al.* 1999). Thus, there are clearly various microbial interactions, and perhaps even conflicts, taking place within the flora.

Antibiotics

Antimicrobial agents have been used as the primary intervention for both the treatment of and prophylaxis against UTIs. The general approach for UTIs in recent years has been to use broad-spectrum antibiotics because a physician and/or patient wants immediate treatment without culture, and because high levels of antibiotics, such as fluoroquinolones, can clear the bladder in 1 to 3 days. The problem is that antibiotics are too often given for 7 to 10 days and second- or third-line classes of drugs are prescribed as first line, all of which leads to rapid resistance and fewer third-line options when a patient is seriously ill. The emergence of, and increased concern over, multi-drug-resistant bacteria, including their link to livestock feed, makes alternative strategies to control biofilms ever more imperative (Gomez-Lus 1998; Cohen 2000; Garvin and Urban 2000; Hellinger 2000; van den Bogaard and Stobberingh 2000).

It was thought that the failure of antibiotics to eradicate biofilms was due to an inability to penetrate through them to the surface interface. However, studies with confocal microscopy have shown that water channels exist and that actual penetration of antibiotics to the surface occurs. Therefore, other factors around the biofilm, such as electrostatic repulsion, are involved in protecting the dense structure against complete killing. Reaching appropriate drug levels is not easy in the gut, and if achieved would destroy the normal flora and put the host at risk of *C. difficile* diarrhoea. Thus, a more strategic approach is needed to target pathogens only. In the urinary tract, drug levels can be achieved that are significantly higher than the minimum inhibitory concentration required to eradicate the pathogens; yet, in the presence of biofilms, complete cure is generally not possible (Reid *et al.* 2000).

A solution is not on the immediate horizon, but some cell signalling studies offer a possible new approach. Essentially, cell signals were found to be produced by bacteria within biofilms (Hellingwerf *et al.* 1998; Davies *et al.* 1998; Pesci *et al.* 1999; Slaughter 1999), some of which are sent out to detect the environmental conditions. Signals are then returned to the cells, for example 'telling' the bacteria that it is safe to multiply. One interventional concept is to 'fool' the bacteria by sending exogenously applied signals followed by a bactericidal agent. For this to work, the signals must be defined, produced in a safe and easily administered vehicle, and then tested extensively *in vivo*.

Vaccines

Vaccines are being developed against various intestinal and urogenital pathogens to reduce the risk of infection and morbidity. These approaches will likely find a place in disease prevention, but much remains to be determined. For example, would a vaccine against *Salmonella typhimurium* be effective if a large inoculum of the pathogen was ingested? Which class of antibody or other immune factor should be primed without damaging the host? Should vaccines be delivered on the cell wall of organisms such as lactobacilli? And if so, how is expression controlled and how can regulatory approval be obtained given the adverse public feelings towards genetically modified products?

The development of a FimH-adhesin-based vaccine against uropathogenic *E. coli* is quite advanced (Langermann *et al.* 1997) and clinical trials are now under way. The hope is that the vaccine will induce the host to generate antibodies that will block the pathogen's adhesion to bladder cells. In our view, this concept, while founded in excellent science, is narrow in its perspective. Firstly, *E. coli* has many tools with which to adhere to surfaces, including electrostatic and hydrophilic forces. Secondly, failure to mount an effective antibody response has not been associated with an increased incidence of UTIs. The vaccine may not have any effect on the vaginal colonization levels of *E. coli* and, therefore, not reduce the risk of infection. Indeed, it is hard to see how a vaccine could prevent *E. coli* from creating a vaginal biofilm that is dominated by potential pathogens, especially given the failure of host responses in general to prevent or eradicate biofilms. Lastly, many other uropathogens are problematic, particularly in the ageing population, and so the proposed vaccine will only benefit a portion of the patients at risk. Nevertheless, any scientifically based approach that provides an alternative to antibiotics is to be applauded.

Probiotics

The use of exogenous bacteria that are 'generally regarded as safe' (GRAS) by regulatory agencies, to reduce the risk of intestinal and urogenital infections, has been studied. A number of products, most of which are comprised of lactic acid bacteria, have been developed as probiotics designed to improve intestinal health. The problem is that few of the strains contained in the vast array of products have any scientific evidence in support of the claims being made by the distributors, and few of the products have reliable contents in terms of viable count and the presence of strains stated on the labels (Hamilton-Miller *et al.* 1999). Perhaps 7–15 strains exist with some degree of scientific and/or clinical evidence (Reid 1999; Sanders 1999). Of these, *Lactobacillus rhamnosus* GG (Valio, Finland), *Lactobacillus casei* Shirota (Yakult, Japan), *L. reuteri* MM53 (BioGaia, Sweden) and *Lactobacillus johnsonii* LJ1 (Nestlé, Switzerland) have arguably been studied the most.

There is good evidence that *L. rhamnosus* GG can colonize the gut for several days and hasten recovery from viral and bacterial diarrhoea (Isolauri *et al.* 1994; Guandalini *et al.* 2000), as well as prevent antibiotic-associated diarrhoea (Siitonen *et al.* 1990). Similar evidence is available for *L. reuteri* MM53 (Shornikova *et al.* 1997a,b). It is believed that the organisms adhere to the intestinal cells, multiply and produce substances (acid, bacteriocins, etc.) that outcompete the pathogens, but in truth none of this has been proven. Given the rapidity with which the organisms appear to be eradicated from the gut (based upon stool sampling, which will not detect organisms in low numbers within the gut), it is possible that the mechanisms of action involve other factors, such as adhesion to mucus (Tuomola *et al.* 1999) and stimulation of mucus production (Mack *et al.* 1999), which blocks pathogen access to the gut epithelia, thereby terminating signs and symptoms of infection.

The data concerning the potential probiotic mechanisms of *L. casei* Shirota and *L. johnsonii* LJ1 are concentrated on modulation of the immune system

(Donnet-Hughes *et al.* 1999; Matsuzaki and Chin 2000), although a recent study suggests that LJ1 also competes for carbohydrate binding sites in the gut (Neeser *et al.* 2000). However, the degree to which this occurs *in vivo* and leads to either prevention of infection or cessation of pathogenesis remains to be elucidated. Studies will likely have to be undertaken in animals, perhaps pigs, in order to comprehend the molecular and physiological events that take place at the bacterial-mucus–cell interface in the intestine. Biofilm models that combine detection of specific proteins, as well as peptide and DNA sequences, with sophisticated microscopy could rapidly move this field forward over the next 5 years.

The study of the vagina is much easier than that of the gut because the tract is more readily sampled. Although swabs and saline washes do not necessarily recover all adherent organisms, they have provided us with some insight into the vaginal flora. This flora clearly changes throughout the menstrual cycle, with much less stability during menses (Eschenbach *et al.* 2000). The dominant species in normal healthy women appear to be *Lactobacillus crispatus*, *Lactobacillus iners* and *Lactobacillus jensenii* (Reid *et al.* 1996; Antonio *et al.* 1999; Burton *et al.* 2003) followed by various others, including *L. casei*, *L. rhamnosus*, and *L. fermentum*. In order to provide a protective barrier population of lactobacilli in women prone to infection, or at times when the indigenous population is low, it might be tempting simply to install one of the dominant species. However, our viewpoint has been that any strain, regardless of the species, should have properties that improve its ability to antagonize urogenital pathogens. There is good evidence to show that hydrogen peroxide is one such characteristic (Gupta *et al.* 1998), along with an ability to adhere and inhibit growth and adhesion of pathogens (Reid *et al.* 1987). Furthermore, the ability to produce biosurfactants (Velraeds *et al.* 1998) and specific collagen-binding proteins (Howard *et al.* 2000) also appears important.

Two treatment approaches using probiotics have reached clinical testing stages. One is to use a single, hydrogen-peroxide-producing strain of *L. crispatus*, and the other uses a combination of strains that colonize the vagina, produce hydrogen peroxide, and biosurfactants, and resist the killing action of spermicide nonoxynol-9. Preliminary data on the former approach (Hillier, oral presentations) and extensive data on the latter are encouraging (Reid *et al.* 1995; 2000; Reid *et al.* unpublished data).

The concept of using prebiotics in the gut (Gibson and Roberfroid 1995) and vagina (Reid *et al.* 1998) to modify the flora in favour of one dominated by lactobacilli and bifidobacteria seems to be a potentially useful health maintenance tool. Oligosaccharides can be ingested safely, and appear to enhance selectively certain strains of commensals without necessarily increasing the total count of the gut flora. Application of skim milk to the vagina has also caused an increase in indigenous lactobacilli leading to reduced risk of UTI (Reid *et al.* 1995). A search for optimal prebiotics continues, and could lead to new approaches to the maintenance of a healthy flora (Gibson and Fuller 2000).

The lack of side effects with probiotic therapies in the gut and urogenital tract (Naidu *et al.* 1999; Reid *et al.* 2000), unlike antibiotic treatment, makes the use of probiotics greatly appealing to physicians and patients. As more confirmatory efficacy data are accumulated, there is hope that this approach to controlling disease and maintaining health can become widely available.

WOUNDS

A wound can be described as an 'injury or damage, usually restricted to those caused by physical means with disruption of the normal continuity of structures' (Dorland 2000) and can be classified as acute or chronic. Acute wounds can be either traumatic or iatrogenic, and generally involve healthy individuals possessing adequate self-healing capacity. Although medical intervention may be necessary to limit many of the infections associated with acute wounds, host defences are recognized to play an important role in eventual infection clearance. Chronic wounds, on the other hand, such as pressure, diabetic, vascular or arterial ulcers, have many possible causes and are generally potentiated by underlying disease. These wounds can take months to years to heal, since they are often incapable of completing all the events involved in the healing process. Acute and chronic wound-associated infection with any pathogenic organism (including opportunistic pathogens) poses a serious threat to the health and life of the patient if not promptly diagnosed and treated (Mertz and Ovington 1993).

In this setting, an infection can be defined as a series of events during which pathogens adhere to tissue (often damaged or necrotic), proliferate and invade viable tissue, and stimulate a host immune response. Infection rates in large surgical series are approximately 1.5 to 3.9% for clean wounds, 3.0 to 4.0% for clean-contaminated wounds, 8.5% for contaminated wounds, and 28 to 40% for dirty wounds (Howard 1994). Since the risk of developing any wound-associated infection is functionally dependent upon the initial inoculum of organisms at the wound site (Hong and Davis 1996), it is important both to limit the extent of contamination (as during surgical procedures) and to initiate cleaning and treatment strategies as soon as possible. This is especially important in the case of immunocompromised individuals, such as burn victims or those with underlying disease, as the inoculum required to cause infection is substantially lower. Additionally, strategies aimed at preventing and quickly dealing with these infections greatly reduce the likelihood of associated biofilm development.

Biofilms have been shown to be an integral part of many wound infections, and consideration of this phenomenon must be included in any search for preventive and therapeutic strategies (Costerton *et al.* 1999). Since biofilm-associated infections commonly occur on medical devices, they have been extensively studied on biomaterials. However, tissues taken from around devices and at the site of non-device-related infections also show the presence of biofilms, thus further establishing the importance of their study and clinical significance.

Among the vast list of wound infections, there exist common pathophysiological factors important in their development. Breakage of the skin or rupture of the gut represents portals of entry for microorganisms. The presence of inert prosthetic materials (such as central venous catheters, peritoneal dialysis catheters, or orthopaedic devices), devitalized tissue (burns, poor surgical technique, or bony sequestra), or foreign bodies (suture material, traumatic wounds) contribute to the development of biofilm-associated infections.

Mechanisms of Biofilm Formation in Wounds

Numerous species of both bacteria and fungi have been implicated as causal organisms in biofilm-associated wound infections, especially those involving prostheses and/or compromised patients (Table 3.3.1). However, the most commonly diagnosed are Gram-positive cocci, *Staphylococcus* and *Streptococcus*, and the Gram-negative bacillus *Pseudomonas*.

Of all the bacterial species isolated from wound infections, *S. aureus* is one of the most common and problematic (Villavicencio and Wall 1996). This pathogen is common to many chronic wound infections (i.e. diabetic ulcers) and is difficult to eradicate completely from patients due to multi-drug

Table 3.3.1. Common wound infections and associated pathogens

Infection	Pathogens
Musculoskeletal infections	Gram-positive cocci
Necrotizing fasciitis	Group A streptococci, polymicrobial
Osteomyelitis	Various bacterial and fungal species
Sutures	*S. epidermidis* and *S. aureus*
Exit sites	*S. epidermidis* and *S. aureus*
Arteriovenous shunts	*S. epidermidis* and *S. aureus*
Peritoneal dialysis (CAPD) peritonitis	A variety of bacteria and fungi
Central venous catheters	*S. epidermidis*
Orthopaedic devices	*S. aureus* and *S. epidermidis*
Burns	*Pseudomonas* species
Chronic ulcers (as in diabetes)	Polymicrobial, including anaerobic species

resistance, reinfection through contact with asymptomatic carriers, and multiple bacterial virulence factors. While coagulase production has been shown to be a major factor in its pathogenicity, additional cell-surface components and extracellular products have also been implicated as contributors to *S. aureus* infection. The possession of adhesins specific for a variety of host matrix proteins, such as fibronectin, fibrinogen and collagen, facilitate its adherence (Patti *et al.* 1994). Cell-wall peptidoglycan inhibits oedema production and the migration of leukocytes, delaying onset of the local host immune response, thus allowing additional time for bacterial proliferation and dissemination. Exopolysaccharide capsules function as ion-exchange resins controlling molecular uptake, inhibit opsonization and thus phagocytosis, and are utilized in organism adhesion, aggregation and interaction. Additionally, several strains of *S. aureus* have been shown to produce infection-associated biofilms, with increased virulence. Although fibrinogen is an important attachment protein for *S. aureus* during infection initiation, its conversion to fibrin is necessary for the formation of a biofilm (Akiyama *et al.* 1997).

The detection of *S. epidermidis* at an infection site was once viewed as a contaminant, as the organism is a common member of the commensal skin and mucous membrane flora. However, *S. epidermidis* is now regarded as the most common cause of nosocomial bacteraemia (especially in immunocompromised patients) and is the principal organism responsible for infections of implanted prosthetic medical devices (Rupp *et al.* 1999). The production of a biofilm mediated by polysaccharide intercellular adhesin/haemagglutinin is thought to be crucial in its pathogenesis of prosthetic-device infections. A number of studies have observed the elaboration of biofilms in clinically significant strains of *S. epidermidis* as opposed to contaminants or skin isolates (Christensen *et al.* 1983; Ishak *et al.* 1985).

Streptococcus pyogenes, Streptococcus pneumoniae and the viridans streptococcal group are important surgical infection pathogens. Group A streptococci have cell-surface components and extracellular products that inhibit host defences and/or promote spread of the bacterium (Howard 1994). The cell-surface M protein, along with the capsule, help streptococci resist phagocytosis. Hyaluronidase and streptokinase promote the spread of infection. Streptolysins O and S are cytotoxins that not only lyse erythrocytes but also leukocytes and other cell types, resulting in an impaired local immune response. Streptococci are responsible for many cellulitic wound infections and necrotizing soft tissue infections and abscesses.

Pseudomonas aeruginosa is an obligate aerobe and is another organism responsible for wound and surgical infections. It is a frequent cause of infection in immunologically compromised patients, especially if they have been hospitalized for any extended period of time. The virulence of *P. aeruginosa* is related to the production of extracellular polysaccharide and specialized pili involved in attachment to surfaces (Costerton *et al.* 1999). The importance of this pili-associated attachment was previously demonstrated using an experimental mouse model involving piliated and non-piliated mutants. It was observed that, in comparison with piliated *P. aeruginosa* mutants, a tenfold higher inoculum of non-piliated mutants was required to achieve a similar 50% infection rate (Sato and Okinaga 1987). *P. aeruginosa* utilizes Type IV pili in 'twitching', a form of surface-associated motility hypothesized as a potential requirement for the aggregation of cells into microcolonies (precursors to biofilms) (Davies and Geesey 1995). It has also been demonstrated that, following attachment to a solid substratum, *P. aeruginosa* genes involved in the synthesis of alginate, an extracellular polysaccharide involved in biofilm formation, are activated (Boyd and Chakrabarty 1995). Alginate serves to enhance the adhesion of the pathogen and protect it from potential detrimental environmental conditions.

Control of Wound-associated Biofilms

The development and progression of infections associated with biomaterials and wounds differ in several respects, including location, tissue involvement and immune system accessibility. However, since biomaterials and damaged tissue within wounds similarly provide sites for pathogen attachment, biofilm formation and ultimately infection development, many of the approaches discussed here will have potential application in the management of both wound and device-associated infections.

Biomaterial Modification

Bacterial colonization of implants and alloplastic devices is incurable by use of conventional systemic antimicrobial agents (Nickel *et al.* 1985; Radd and Bodey 1992). The limitations of this and other traditional methods used in the prevention and treatment of wound and biomaterial-related infections have forced researchers to seek out novel therapeutic strategies.

The modification of the physical–chemical properties of biomaterials is one approach to reduce nonspecific (electrostatic, hydrophobic) microbial attachment. This has led to conclusions that urinary catheters made from pure silicone could have lower rates of infection (Roberts *et al.* 1990). However, these conclusions, and certainly this one in particular, were based on *in vitro* studies of dubious rigour. The development of hydrogel coatings, i.e. hydrophilic polyurethane polymers that swell on contact with water, has been used to alter device surface properties, and some *in vitro* data is available on their ability to reduce platelet adhesion with a view to reducing thrombosis (Kulik and Ikada 1996). However, microorganisms are very diverse in nature and versatile in their ability to colonize biomaterials, and thus the likelihood of preventing biofilm growth in the long term by this approach alone, is limited.

Another strategy focuses on impregnating, coating, or otherwise incorporating leachable or non-leachable antimicrobial agents within or on the surface of the alloplast (Williams and Worley 2000). Antimicrobial substances such as heavy metals, silver oxide, and antibiotics (Sugarman 1980; Johnson *et al.* 1990; Reid *et al.* 1995c) or using silver-releasing polymers or antiseptics (Stickler *et al.* 1989; Gilchrist *et al.* 1991) could provide a degree of reduction in the incidence of infection, especially for patients requiring short-term insertion of a device. Central venous catheters impregnated with chlorhexidine and silver sulphadiazine have been demonstrated to prevent bacterial adherence and biofilm formation in swine (Greenfield *et al.* 1995). Use of such catheters reduces the risk of colonization by slime-producing *S. aureus* strains (Raad 1995). Given that the cost of treating an intravascular catheter infection is at least $6000 US (Garrison and Wilson 1994), prevention of infection has huge economic implications.

For permanent prosthetic devices, improvements in biomaterial design that allow a more rapid biointegration of host tissue into the alloplast could result in decreased device-related infection. Since biomaterials present novel, uninhabited surfaces that both microorganisms and host cells can potentially occupy, a primary research focus has been to alter surfaces in such a way that pathogen binding is halted while host cell integration is promoted. Although substantial progress has been made in this area, increased understanding of the attachment and adherence mechanisms of both microorganisms and host cells is imperative to achieving infection-free biomaterials.

Molecular and Physical Modalities

Jet lavage with saline, antibiotic and other additives such as Bacitracin, Neomycin, Polymyxin/Neomycin, and various detergents and surfactants have been used to remove biofilms of slime-producing *S. epidermidis* from stainless-steel screws used in orthopaedic devices (Moussa *et al.* 1996). Enzyme-based strategies targeting biofilm degradation and the enhancement of antibiotic activity are also being pursued. For example, the use of polysaccharide-hydrolysing enzymes in combination with bactericidal molecules has shown promise in removing biofilms of *S. aureus*, *S. epidermidis*, and *P. aeruginosa* on steel (Johansen *et al.* 1997), and ofloxacin activity against sessile cultures of numerous pathogens was greatly increased when used in tandem with a proteolytic enzyme serratiopeptidase (Selan *et al.* 1993). Fluoroquinolones, such as ofloxacin and ciprofloxacin, do appear capable of somewhat eradicating pathogenic biofilms *in vitro* and *in vivo* (Reid *et al.* 1994, 2000).

Physical modalities, such as ultrasound (Qian *et al.* 1996, 1997) and direct current (DC) fields (Costerton *et al.* 1993) have been proven to be useful adjuncts to antimicrobial therapy in reducing biofilm loads. The use of DC fields to generate this bioelectric effect, termed 'iontophoresis', has been shown to reduce biofilms, using both *in vitro* and animal models. The mode of action may involve alterations in ion exchange such that adherent organisms become susceptible to antibiotics or by increasing antimicrobial diffusion into the biofilm. The use of low-intensity ultrasound enhances the bactericidal activity of gentamycin against biofilms of *P. aeruginosa* (Qian *et al.* 1997) and *E. coli* (Rediske *et al.* 1999). Further development of these technologies could lead to improved treatment of clinical implant infections.

Traditional preventive strategies against wound infections include attention to surgical technique and tissue handling, adherence to proper sterile technique and draping of the surgical field, hand washing, gloves and other barriers, treatment of remote infections, limitation of traffic within the operating theatre, and control of environmental factors, such as air-handling systems. To diminish the patient's own skin flora, perioperative systemic antibiotics are administered prior to skin incision, if indicated. Hair removal and skin preparation, using a germicidal soap solution and antimicrobial solution, is performed in the operating theatre. Intraoperative irrigation with antibiotics has been used extensively in the orthopaedic literature to reduce the incidence of prosthetic-device-related infection (Dirschl and Wilson 1991). These irrigants are generally comprised of multiple broad-spectrum antibiotics, such as neomycin, polymyxin, and bacitracin, to target the majority of organisms most likely to cause infection.

The treatment of an established wound infection consists of systemic antimicrobial therapy, aggressive surgical debridement and wound care. Such treatment usually also requires the removal of any prosthetic device, because it contains the biofilms from which infections arise. The development of multi-drug-resistant organisms has forced researchers to investigate alternative therapies for the treatment of wound infections.

The creation of novel, antimicrobial wound dressings is being extensively investigated. Burn wound infection is a serious complication with a resultant high mortality. Thermal injury causes coagulation necrosis of the epidermis and varying levels of the dermis and subcutaneous tissue. Once necrosis occurs, the wound is essentially avascular, which prevents effective delivery of systemic antibiotics if infection occurs. Damage to the cutaneous barrier allows bacterial penetration into viable tissue and subsequent infection. Before the availability of penicillin, streptococci and staphylococci were the predominant organisms. By the late 1950s, Gram-negative bacteria, primarily *Pseudomonas* species, emerged as the dominant organisms causing fatal wound infections in burned patients (Goodwin *et al.* 1994). Topical antimicrobial chemotherapy, using agents such as silver nitrate, mafenide acetate, and silver sulphadiazine, is effective in controlling burn wound infection. However, problems with emerging microbial resistance, metabolic derangements, and superimposed fungal burn wound infection are limitations of such dressings. Advances in biomaterial development have expanded into the realm of wound dressings. Novel polymers are being investigated that have the capacity to release various substances, such as collagenases and growth factors, to improve wound healing.

Probiotics

The concept of using commensal microorganisms to prevent or treat wound infections might seem to have come from the Middle Ages. However, this was the basis for a series of studies undertaken in our laboratories. The hypothesis was that microorganisms, such as lactobacilli, quite effectively control the ability of pathogens to infect the intestine, and if these organisms or factors produced by them were antagonistic to pathogens, then they could be applied to wounds to prevent infection.

In the first series of experiments, several *Lactobacillus* strains were found to reduce significantly pathogen adhesion to biomaterials (Hawthorn and Reid 1990; Reid and Tieszer 1993, 1995; Reid *et al.* 1995c). Furthermore, biosurfactant substances were found to be produced by certain strains, in particular *L. fermentum* RC-14, and to inhibit pathogen binding significantly (Velraeds *et al.* 1996, 1998). Most recently, collagen-binding proteins have been identified within the biosurfactant mixture, and these too inhibit pathogen adhesion (Heinemann *et al.* 2000; Howard *et al.* 2000; Cadieux *et al.* 2003). Using an animal model for surgical implant infection with *S. aureus*, viable *L. fermentum* RC-14, its biosurfactant and a 29 kDa collagen-binding protein were found to prevent disease (Gan *et al.* 2003). This finding is potentially a new paradigm in disease management, where antibiotic killing of pathogens or immune stimulation are not principles upon which infections are prevented (Figure 3.3.1). Rather, simply by blocking the pathogens' spread, they prevent infection of the host. Whether this process involves cell-to-cell signalling remains to be determined. More studies are required, but if extracellular by-products of organisms like lactobacilli are proven to be effective in humans, the medical implications are enormous.

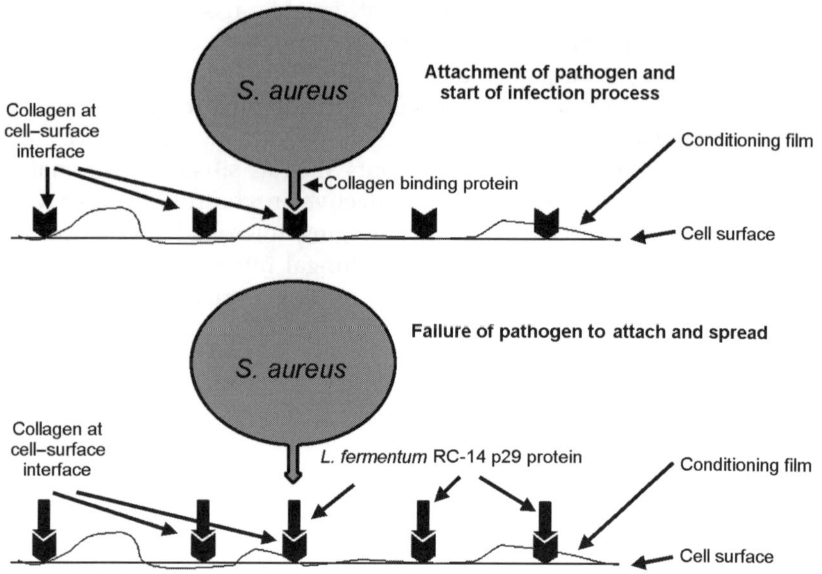

Figure 3.3.1. Effect of probiotic *Lactobacillus* biosurfactant protein on infection.

Quorum-sensing

Developments in molecular biology and refinements in microscopy techniques have resulted in an explosion of research into the biology of biofilms, focusing on their molecular and genetic bases of development. *P. aeruginosa* has been the most extensively studied biofilm-producing bacteria of clinical importance. This organism produces at least two extracellular signals involved in cell-to-cell communication and expresses many cell-density-dependent virulence factors that may be involved in the differentiation and maturation of its biofilms (Davies *et al.* 1998). This cell-to-cell signalling has been termed quorum-sensing and largely involves the regulation of genes responsible for growth, reproduction and pathogenicity (Davies and Geesey 1995). Targeting of specific gene products that are expressed when cells undergo changes from a planktonic to a sessile phenotype will most likely result in strategies to control bacterial spread.

Each step in the development of a biofilm, such as initial attachment and microcolony formation, maturation into a differentiated biofilm, and detachment of planktonic cells from biofilms, represents a potential target for therapies against biofilm infections. In time, it will be possible to mimic bacterial signalling and interfere with it in such a way that the infectious process will be prevented. Having stated that, we can equally expect that microbes will modify themselves and find new ways to overcome the host. Such is the continuing battle of humankind against microbes.

SUMMARY

In summary, microorganisms use a variety of mechanisms to adhere to tissue surfaces. Thereafter, in many cases, they form biofilms that can be infectious or represent a normal non-infectious flora. Attempts to manipulate biofilms using crude chemical (antimicrobial) interventions have succeeded in terms of saving lives, but largely failed in terms of eradicating biofilms and in creating a remedy that specifically attacks the offending organisms, while restoring the host's normal flora. Until the concerns over multi-drug-resistant bacteria, little attention was paid to the failures. Indeed, little attention has been paid to a better understanding of the microbial ecology of the intestine, skin and urogenital tract. Now, even pharmaceutical and biotech companies are paying attention to these issues, along with the Centers of Disease Control and other government public health bodies.

Remedies that control infectious biofilms and create non-infectious ones that promote well-being are on the horizon. These might include the use of adhesin blockers, such as the p29 collagen-binding protein described here, which interferes with *S. aureus* binding and sepsis. Other biogenesis approaches may emerge, but use of recombinant microbes and genetic manipulation of receptor sites on cells is a long way off, and raises a number of ethical considerations. In terms of probiotics, the current approach is relatively simplistic, in that a set composition of organisms is applied quite widely to large numbers of people. With the genome project complete, in due course, each of us will know our genetic make-up, making it possible to match specific probiotic strains with individuals. For example, certain lactobacilli might be transferred between families who have never experienced a urogenital infection; by propagating such strains, future generations might also benefit. The translation of such therapeutics into practical products will prove a major challenge logistically and economically. Nevertheless, although these concepts are futuristic, it is good to dream of what can be possible, then use scientific endeavour to make it happen!

ACKNOWLEDGEMENTS

Our research is supported by grants from the Natural Sciences and Engineering Research Council of Canada, the Kidney Foundation of Canada, and Urex Biotech Inc.

REFERENCES

Akiyama, H, Ueda, M, Kanzaki, H, Tada, J and Arata, J (1997) Biofilm formation of *Staphylococcus aureus* strains isolated from impetigo and furuncle: role of fibrinogen and fibrin. *Journal of Dermatological Science* **16**:2–10.

Antonio, MA, Hawes, SE and Hillier, SL (1999) The identification of vaginal *Lactobacillus* species and the demographic and microbiologic characteristics of women colonized by these species. *Journal of Infectious Diseases* **180**:1950–1956.

Boyd, A and Chakrabarty, AM (1995) *Pseudomonas aeruginosa* biofilms: role of the alginate exopolysaccharide. *Journal of Industrial Microbiology* **15**:162–168.

Burton, J, Cadieux, P and Reid, G (2003) Improved understanding of the bacterial vaginal microbiota of women before and after probiotic instillation. *Applied and Environmental Microbiology*. In Press.

Cadieux, P, Watterson, JD, Denstedt, J, Harbottle, RR, Puskas, J, Howard, J, Gan, BS and Reid, G (2003) Potential application of polyisobutylene polystyrene and a *Lactobacillus* protein to reduce the risk of device-associated urinary tract infections. *Colloids and surfaces B: Biointerfaces*. In Press.

Caplan, MS and Jilling, T (2000) Neonatal necrotizing enterocolitis: possible role of probiotic supplementation. *Journal of Pediatric Gastroenterology and Nutrition* **30**:S18–S22.

Cauci, S, Monte, R, Driussi, S, Lanzafame, P and Quadrifoglio, F (1998) Impairment of the mucosal immune system: IgA and IgM cleavage detected in vaginal washings of a subgroup of patients with bacterial vaginosis. *Journal of Infectious Diseases* **178**:1698–1706.

Chan, RCY, Reid, G, Irvin, RT, Bruce, AW and Costerton, JW (1985) Competitive exclusion of uropathogens from uroepithelial cells by *Lactobacillus* whole cells and cell wall fragments. *Infection and Immunity* **47**:84–89.

Cheng, KJ and Costerton, JW (1980) The formation of microcolonies by rumen bacteria. *Canadian Journal of Microbiology* 26:1104–1113.

Cheng, KJ and Wallace, RJ (1979) The mechanism of passage of endogenous urea through the rumen wall and the role of ureolytic epithelial bacteria in the urea flux. *British Journal of Nutrition* **42**: 553–557.

Cheng, KJ, Irvin, RT and Costerton, JW (1981) Autochthonous and pathogenic colonization of animal tissues by bacteria. *Canadian Journal of Microbiology* **27**:461–490.

Christensen, GD, Parisi, JT, Bisno, AL, Simpson, WA and Beachey, EH (1983) Characterization of clinically significant strains of coagulase-negative staphylococci. *Journal of Clinical Microbiology* **18**:258–269.

Cohen, ML (2000) Changing patterns of infectious disease. *Nature* **406**:762–767.

Cook, RL, Reid, G, Pond, DG, Schmitt, CA and Sobel, JD (1989) Clue cells and bacterial vaginosis: immunofluorescent identification of the adherent Gram negative bacteria as *Gardnerella vaginalis*. *Journal of Infectious Diseases* **160**:490–496.

Costerton, JW, Khoury, AE, Ward, KH and Anwar, H (1993) Practical measures to control device-related bacterial infections. *International Journal of Artificial Organs* **16**:765–770.

Costerton, JW, Stewart, PS and Greenberg, EP (1999) Bacterial biofilms: a common cause of persistent infections. *Science* **284**:1318–1322.

Davies, DG and Geesey, GG (1995) Regulation of the alginate biosynthesis gene *algC* in *Pseudomonas aeruginosa* during biofilm development in continuous culture. *Applied and Environmental Microbiology* **61**:860–867.

Davies, DG, Parsek, MR, Pearson, JP, Iglewski, BH, Costerton, JW and Greenberg, EP (1998) The involvement of cell-to-cell signals in the development of a bacterial biofilm. *Science* **280**:295–298.

Dirschl, DR and Wilson, FC (1991) Topical antibiotic irrigation in the prophylaxis of operative wound infections in orthopedic surgery. *Orthopedic Clinics of North America* **22**:419–426.

Donnet-Hughes, A, Rochat, F, Serrant, P, Aeschlimann, JM and Schiffrin, EJ (1999) Modulation of nonspecific mechanisms of defense by lactic acid bacteria: effective dose. *Journal of Dairy Science* **82**:863–869.

Dorland, WA (2000) Dorland's Illustrated Medical Dictionary, 29th edition, W.B. Saunders, Philadelphia, p. 1989.

Eschenbach, DA, Davick, PR, Williams, SJ, Klebanoff, SJ, Young-Smith, CM and Holmes, KK (1989) Prevalence of hydrogen peroxide-producing *Lactobacillus* species in normal women and women with bacterial vaginosis. *Journal of Clinical Microbiology* **27**:251–256.

Eschenbach, DA, Thwin, SS, Patton, DL, Hooton, TM, Stapleton, AE, Agnew, K, Winter, C, Meier, A and Stamm, WE (2000) Influence of the normal menstrual cycle on vaginal tissue, discharge and microflora. *Clinical Infectious Diseases* **30**:901–907.

Gan, BS, Kim, J, Reid, G, Cadieux, P and Howard, JC (2002) *Lactobacillus formentum* RC-14 inhibits *Staphylococcus aureus* infection of surgical implants in rats. *Journal of Infectious Diseases* **185**:1369–1372.

Garrison, RN and Wilson, MA (1994) Intravenous and central catheter infections. *Surgical Clinics of North America* **74**:557–570.

Garvin, KL and Urban, JA (2000) Emerging multiresistant strains: recommended precautions in the emergency room and surgical setting. *Instructional Course Lectures* **49**:605–614.

Gewolb, IH, Schwalbe, RS, Taciak, VL, Harrison, TS and Panigrahi, P (1999) Stool microflora in extremely low birthweight infants. *Archives of Disease Childhood. Fetal and Neonatal Edition* **80**:F167–F173.

Gibson, GR and Fuller, R (2000) Aspects of *in vitro* and *in vivo* research approaches directed toward identifying probiotics and prebiotics for human use. *Journal of Nutrition* **130**:391S–395S.

Gibson, GR and Roberfroid, MB (1995) Dietary modulation of the human colonic microbiota: introducing the concept of prebiotics. *Journal of Nutrition* **125**:1401–1412.

Gilchrist, T, Healy, DM and Drake, C (1991) Controlled silver-releasing polymers and their potential for urinary tract infection control. *Biomaterials* **12**:76–78.

Gomez-Lus, R (1998) Evolution of bacterial resistance to antibiotics during the last three decades. *International Microbiology* **1**:279–284.

Goodwin, CW, Finkelstein, JL and Madden, MR (1994) Burns. In: Principles of Surgery, 6th Edition (Eds. Schwartz, SI, Shires, GT and Spencer, FC), McGraw-Hill, New York, pp. 225–277.

Greenfield, JI, Sampath, L, Popilskis, SJ, Brunnert, SR, Stylianos, S and Modak, S (1995) Decreased bacterial adherence and biofilm formation on chlorhexidine and silver sulfadiazine-impregnated central venous catheters implanted in swine. *Critical Care Medicine* **23**:894–900.

Guandalini, S, Pensabene, L, Zikri, MA, Dias, JA, Casali, LG, Hoekstra, H, Kolacek, S, Massar, K, Micetic-Turk, D, Papadopoulou, A, de Soussa, JS, Szajewska, H and Weizman, Z (2000) *Lactobacillus* GG administered in oral rehydration solution to

children with acute diarrhea: a multicenter European trial. *Journal of Pediatric Gastroenterology and Nutrition* **30**:54–60.
Gupta, K, Stapleton, AE, Hooton, TM, Roberts, PL, Fennell, CL and Stamm, WE (1998) Inverse association of H_2O_2-producing lactobacilli and vaginal *Escherichia coli* colonization in women with recurrent urinary tract infections. *Journal of Infectious Diseases* **178**:446–450.
Hamilton-Miller, JM, Shah, S and Winkler, JT (1999) Public health issues arising from microbiological and labelling quality of foods and supplements containing probiotic microorganisms. *Public Health Nutrition* **2**:223–229.
Hawthorn, LA and Reid, G (1990) Exclusion of uropathogen adhesion to polymer surfaces by *Lactobacillus acidophilus*. *Journal of Biomedical Materials Research* **24**:39–46.
Heinemann, C, van Hylckama, Vlieg, JE, Janssen, DB, Busscher, HJ, van der Mei, HC and Reid, G (2000) Purification and characterization of a surface-binding protein from *Lactobacillus fermentum* RC-14 that inhibits adhesion of *Enterococcus faecalis* 1131. *FEMS Microbiology Letters* **190**:177–180.
Hellinger, WC (2000) Confronting the problem of increasing antibiotic resistance. *Southern Medical Journal* **93**:842–848.
Hellingwerf, KJ, Crielaard, WC, Joost, Teixeira de Mattos, M, Hoff, WD, Kort, R, Verhamme, DT and Avignone-Rossa, C (1998) Current topics in signal transduction in bacteria. *Antonie Van Leeuwenhoek* **74**:211–217.
Hong, J and Davis, JM (1996) Nosocomial infections and nosocomial pneumonia. *American Journal of Surgery* **172**(Suppl. 6A):33S–36S.
Hooton, TM (1999) Practice guidelines for urinary tract infection in the era of managed care. *International Journal of Antimicrobial Agents* **11**:241–245.
Howard, RJ (1994) Surgical infections. In: Principles of Surgery, 6th edition (Eds. Schwartz, SI, Shires, GT and Spencer, FC), McGraw-Hill, New York, pp. 145–173.
Howard, J, Heinemann, C, Thatcher, BJ, Martin, B, Gan, BS and Reid, G (2000) Identification of collagen-binding proteins in *Lactobacillus* spp. with surface-enhanced laser desorption/ionization-time of flight ProteinChip technology. *Applied and Environmental Microbiology* **66**:4396–4400.
Ishak, MA, Groschel, GL, Mandell, GL and Wenzel, RP (1985) Association of slime with pathogenicity of coagulase-negative staphylococci causing nosocomial septicemia. *Journal of Clinical Microbiology* **22**:1025–1029.
Isolauri, E, Kaila, M, Mykkanen, H, Ling, WH and Salminen, S (1994) Oral bacteriotherapy for viral gastroenteritis. *Digestive Disease Science* **39**: 2595–2600.
Johansen, C, Falholt, P and Gram, L (1997) Enzymatic removal and disinfection of bacterial biofilms. *Applied and Environmental Microbiology* **63**:3724–3728.
Johnson, JR, Roberts, PL, Olsen, RJ, Moyer, KA and Stamm, WE (1990) Prevention of catheter-associated urinary tract infection with a silver oxide-coated urinary catheter: clinical and microbiologic correlates. *Journal of Infectious Diseases* **162**:1145–1150.
Keane, FE, Ison, CA and Taylor-Robinson, D (1997) A longitudinal study of the vaginal flora over a menstrual cycle. *International Journal of STD and AIDS* **8**:489–494.
Korshonov, VM, Gudieva, ZA, Efimov, BA, Pikina, AP, Smeianov, VV, Reid, G, Korshonova, OV, Tyutyunnik, VL and Stepin, II (1999) Study of vaginal bifidoflora of women of reproductive age. *Zhurnal Microbiologii Epidemiologii i Immunobiologii* **4**:74–78.

Kulik, E and Ikada, Y (1996) *In vitro* platelet adhesion to nonionic and ionic hydrogels with different water contents. *Journal of Biomedical Materials Research* **30**:295–304.

Landraud, L, Gauthier, M, Fosse, T and Boquet, P (2000) Frequency of *Escherichia coli* strains producing the cytotoxic necrotizing factor (CNF1) in nosocomial urinary tract infections. *Letters in Applied Microbiology* **30**:213–216.

Langermann, S, Palaszynski, S, Barnhart, M, Auguste, G, Pinkner, JS, Burlein, J, Barren, P, Koenig, S, Leath, S, Jones, CH and Hultgren, SJ (1997) Prevention of mucosal *Escherichia coli* infection by FimH-adhesin-based systemic vaccination. *Science* **276**:607–611.

McCartney, AL, Wang, W and Tannock, GW (1996) Molecular analysis of the composition of the bifidobacterial and *Lactobacillus* microflora of humans. *Applied and Environmental Microbiology* **62**:4608–4613.

McCowan, RP, Cheng, KJ and Costerton, JW (1980) Adherent bacterial populations on the bovine rumen wall: distribution patterns of adherent bacteria. *Applied and Environmental Microbiology* **39**:233–241.

Mack, DR, Michail, S, Wei, S, McDougall, L and Hollingsworth, MA (1999) Probiotics inhibit enteropathogenic *E. coli* adherence *in vitro* by inducing intestinal mucin gene expression. *American Journal of Physiology* **276**:G941–G950.

Mackie, RI, Sghir, A and Gaskins, HR (1999) Developmental microbial ecology of the neonatal gastrointestinal tract. *American Journal of Clinical Nutrition* **69**(Suppl.):1035S–1045S.

Matsuzaki, T and Chin, J (2000) Modulating immune responses with probiotic bacteria. *Immunology and Cell Biology* **78**:67–73.

Mertz, PM and Ovington, LG (1993) Wound healing microbiology. *Dermatologic Clinics* **11**:739–747.

Moussa, FW, Gainor, BJ, Anglen, JO, Christensen, G and Simpson, WA (1996) Disinfecting agents for removing adherent bacteria from orthopaedic hardware. *Clinical Orthopaedics* **329**:255–262.

Naidu, AS, Biblack, WR and Clemens, RA (1999) Probiotic spectra of lactic acid bacteria (LAB). *Critical Reviews in Food Science and Nutrition* **39**:13–126.

Neeser, JR, Granato, D, Rouvet, M, Servin, A, Teneberg, S and Karlsson, KW (2000) *Lactobacillus johnsonii* La1 shares carbohydrate-binding specificities with several enteropathogenic bacteria. *Glycobiology* **10**:1193–1219.

Nickel, JC, Ruseska, I, Wright, JB and Costerton, JW (1985) Tobramycin resistance of *Pseudomonas aeruginosa* cells growing as a biofilm on urinary catheter material. *Antimicrobial Agents and Chemotherapy* **27**:619–624.

Patti, JM, Allen, BL, McGavin, MJ and Hook, M (1994) MSCRAMM-mediated adherence of microorganisms to host tissues. *Annual Reviews in Microbiology* **48**:585–617.

Pesci, EC, Millbank, JB, Pearson, JP, McKnight, S, Kende, AS, Greenberg, EP and Iglewski, BH (1999) Quinolone signaling in the cell-to-cell communication system of *Pseudomonas aeruginosa*. *Proceedings of the National Academy of Sciences* **96**:11 229–11 234.

Qian, Z, Stoodley, P and Pitt, WG (1996) Effect of low-intensity ultrasound upon biofilm structure from confocal laser microscopy observation. *Biomaterials* **17**:1975–1980.

Qian, Z, Sagers, RD and Pitt, WG (1997) The effect of ultrasonic frequency upon enhanced killing of *P. aeruginosa* biofilms. *Annals of Biomedical Engineering* **25**: 69–76.

Raad, I (1995) Antibiotics and prevention of microbial colonization of catheters. *Antimicrobial Agents and Chemotherapy* **39**:2397–2400.

Radd, II and Bodey, GP (1992) Infectious complications of indwelling vascular catheters. *Clinical Infectious Diseases* **15**:97–208.

Rediske, AM, Roeder, BL, Brown, MK, Nelson, JL, Robison, RL, Draper, DO, Schaalje, GB, Robison, RA and Pitt, WG (1999) Ultrasonic enhancement of antibiotic action on *Escherichia coli* biofilms: an *in vivo* model. *Antimicrobial Agents and Chemotherapy* **43**:1211–1214.

Reid, G (1999) The scientific basis for probiotic strains of *Lactobacillus*. *Applied and Environmental Microbiology* **65**:3763–3766.

Reid, G and Tieszer, C (1993) Preferential adhesion of bacteria from a mixed population to a urinary catheter. *Cells and Materials* **3**:171–176.

Reid, G and Tieszer, C (1995) Use of lactobacilli to reduce the adhesion of *Staphylococcus aureus* to catheters. *International Biodeterioration and Biodegradation* **34**:73–83.

Reid, G, Cook, RL and Bruce, AW (1987) Examination of strains of lactobacilli for properties which may influence bacterial interference in the urinary tract. *Journal of Urology* **138**:330–335.

Reid, G, Charbonneau-Smith, R, Lam, D, Lacerte, M, Kang, YS and Hayes, KC (1992) Bacterial biofilm formation in the urinary bladder of spinal cord injured patients. *Paraplegia* **30**:711–717.

Reid, G, Sharma, S, Advikolanu, K, Tieszer, C, Martin, RA and Bruce, AW (1994) Effect of ciprofloxacin, norfloxacin and ofloxacin *in vitro* on the adhesion and survival of *Pseudomonas aeruginosa* on urinary catheters. *Antimicrobial Agents and Chemotherapy* **38**:1490–1495.

Reid, G, Bruce, AW and Taylor, M (1995a) Instillation of *Lactobacillus* and stimulation of indigenous organisms to prevent recurrence of urinary tract infections. *Microecology and Therapy* **23**:32–45.

Reid, G, Busscher, HJ, Sharma, S, Mittelman, M and McIntyre, S (1995b) Surface properties of catheters, stents and bacteria associated with urinary tract infections. *Surface Science Reports* **21**:251–274.

Reid, G, Tieszer, C and Lam, D (1995c) Influence of lactobacilli on the adhesion of *Staphylococcus aureus* and *Candida albicans* to diapers. *Journal of Industrial Microbiology* **15**:248–253.

Reid, G, McGroarty, JA, Tomeczek, L and Bruce, AW (1996) Identification and plasmid profiles of *Lactobacillus* species from the vagina of 100 healthy women. *FEMS Immunology and Medical Microbiology* **15**:23–26.

Reid, G, Bruce, AW, Soboh, F and Mittelman, M (1998a) Effect of nutrient composition on the *in vitro* growth of urogenital *Lactobacillus* and uropathogens. *Canadian Journal of Microbiology* **44**:1–6.

Reid, G, van der Mei, HC and Busscher, HJ (1998b) Microbial biofilms and urinary tract infections. In: Urinary Tract Infections (Eds. Brumfitt, W, Hamilton-Miller, T and Bailey, RR), Chapman and Hall, London, pp. 111–118.

Reid, G, Potter, P, Delaney, G, Hseih, J, Nicosia, S and Hayes, KC (2000) Ofloxacin for the treatment of urinary infections and biofilms in spinal cord injury. *International Journal of Antimicrobial Agents* **4**:305–307.

Roberts, JA, Fussell, EN and Kaack, MB (1990) Bacterial adherence to urethral catheters. *Journal of Urology* **144**:264–269.

Roos, S, Aleljung, P, Robert, N, Lee, B, Wadstrom, T, Lindberg, M and Jonsson, H (1996) A collagen binding protein from *Lactobacillus reuteri* is part of an ABC transporter system? *FEMS Microbiology Letters* **144**:33–38.

Rupp, ME, Ulphani, JS, Fey, PD, Bartscht, K and Mack, D (1999) Characterization of the importance of polysaccharide intercellular adhesin/hemagglutinin of *Staphylococcus epidermidis* in the pathogenesis of biomaterial-based infection in a mouse foreign body infection model. *Infection and Immunity* **67**:2627–2632.

Sanders, ME (1999) Probiotics. *Food Technology* **53**:67–77.

Sato, H and Okinaga, K (1987) Role of pili in the adherence of *Pseudomonas aeruginosa* to mouse epidermal cells. *Infection and Immunity* **55**:1774–1778.

Selan, L, Berlutti, F, Passariello, C, Thaller, MC and Renzini, G (1993) Proteolytic enzymes: a new treatment strategy for prosthetic infections? *Antimicrobial Agents and Chemotherapy* **37**:2618–2621.

Seddon, JM, Bruce, AW, Chadwick, P and Carter, D (1976) Introital bacterial flora—effect of increased frequency of micturition. *British Journal of Urology* **48**:211–218.

Siitonen, S, Vapaatalo, H, Salminen, S, Gordin, A, Saxelin, M, Wikberg, R and Kirkkola, A-L (1990) Effect of *Lactobacillus* GG yoghurt in prevention of antibiotic associated diarrhoea. *Annals of Medicine* **22**:57–59.

Shornikova, A-V, Casas, IA, Isolauri, E, Mykkanen, H and Vesikari, T (1997a) *Lactobacillus reuteri* as a therapeutic agent in acute diarrhea in young children. *Journal of Pediatric Gastroenterology and Nutrition* **24**:399–404.

Shornikova, A-V, Casas, IA, Mykkanen, H, Salo, E and Vesikari, T (1997b) Bacteriotherapy with *Lactobacillus reuteri* in rotavirus gastroenteritis. *Pediatric Infectious Disease* **16**:1103–1107.

Slaughter, CJ (1999) The naturally occurring furanones: formation and function from pheromone to food. *Biological Reviews of the Cambridge Philosophical Society* **74**:259–276.

Stickler, D, Dolman, J, Rolfe, S and Chawla, J (1989) Activity of antiseptics against *Esherichia coli* growing as biofilms on silicone surfaces. *European Journal of Clinical Microbiology and Infectious Diseases* **8**:974–978.

Sugarman, B (1980) Effect of heavy metals on bacterial adherence. *Journal of Medical Microbiology* **13**:351–354.

Tannock, GW, Munro, K, Harmsen, HJM, Welling, GW, Smart, J and Gopal, PK (2000) Analysis of the fecal microflora of human subjects consuming a probiotic product containing *Lactobacillus rhamnosus* DR20. *Applied and Environmental Microbiology* **66**:2578–2588.

Tierno, PM and Hanna, BA (1989) Ecology of toxic shock syndrome: amplification of toxic shock syndrome toxin 1 by materials of medical interest. *Reviews of Infectious Diseases* **11**(suppl.):S182–S186.

Tuomola, EM, Ouwehand, AC and Salminen, SJ (1999) The effect of probiotic bacteria on the adhesion of pathogens to human intestinal mucus. *FEMS Immunology and Medical Microbiology* **26**:137–142.

Turner, MS, Timms, P, Hafner, LM and Giffard, PM (1997) Identification and characterization of a basic cell surface-located protein from *Lactobacillus fermentum* BR11. *Journal of Bacteriology* **179**:3310–3316.

Van den Bogaard, AW and Stobberingh, EE (2000) Epidemiology of resistance to antibiotics. Links between animals and humans. *International Journal of Antimicrobial Agents* **14**:327–335.

Velraeds, MC, van der Belt, B, van der Mei, HC, Reid, G and Busscher, HJ (1998) Interference in initial adhesion of uropathogenic bacteria and yeasts to silicone rubber by a *Lactobacillus acidophilus* biosurfactant. *Journal of Medical Microbiology* **49**:790–794.

Velraeds, MC, van der Mei, HC, Reid, G and Busscher, HJ (1996) Inhibition of initial adhesion of uropathogenic *Enterococcus faecalis* by biosurfactants from *Lactobacillus* isolates. *Applied and Environmental Microbiology* **62**:1958–1963.

Villavicencio, RT and Wall, MJ Jr (1996) The pathogenesis of *Staphylococcus aureus* in the trauma patient and potential future therapies. *American Journal of Surgery* **172**:291–296.

Walker, WA (2000) Role of nutrients and bacterial colonization in the development of intestinal host defense. *Journal of Pediatric Gastroenterology and Nutrition* **30**(suppl.):S2–S7.

Williams, JF and Worley, SD (2000) Infection-resistant nonleachable materials for urologic devices. *Journal of Endourology* **14**:395–400.

4 Dental Plaque and Bacterial Colonization of Dental Materials

4.1 Dental Plaque and Bacterial Colonization

DAVID SPRATT

Departments of Microbiology and Conservative Dentistry, Eastman Dental Institute for Oral Health Care Science, University College London, 256 Gray's Inn Road, London, UK

INTRODUCTION

Dental plaque is the term commonly used for the biofilm formed on teeth. It consists of a complex microbial community embedded in a matrix of polymers of bacterial and salivary origin. However, the oral cavity provides a numerous and varied range of both hard and soft tissue surfaces that act as substrata for biofilm development and so the term plaque has been extended to encompass biofilms on these surfaces. Most soft tissues in the oral cavity lack significant plaque accumulation, such as the lining and masticatory mucosa, due to the rapid rates of epithelial cell turnover and the desquamation of surface cells. The exception is the dorsum of the tongue, which is associated with a significant and characteristic microflora. In contrast, the hard non-shedding surfaces of the oral cavity (teeth) provide a far more stable substratum for the colonization of bacteria.

INITIAL COLONIZATION OF THE MOUTH

At birth, the oral cavity of the neonate is usually sterile, despite the encounters with the maternal resident flora of the uterus, cervix, vagina and perineum (Carlsson and Gothefors 1975). The subsequent acquisition of oral microflora is via passive transmission from a variety of sources, including food, milk, water and particularly saliva from the mother (Long and Swenson 1976; Li and Caufield 1995). The majority of these bacteria are transient, and only a limited number are actually acquired as oral flora.

Medical Biofilms. Edited by J. Jass, S. Surman, J. Walker

Table 4.1.1. Microflora associated with dentate infants

Facultative	Anaerobic
Streptococcus salivarius	*Actinobacillus* spp.
Streptococcus mitis Biovar 1 & 2	*Actinomyces* spp.
Streptococcus oralis	*Campylobacter* spp.
Streptococcus sanguis	*Fusobacterium* spp.
Streptococcus gordonii	*Lactobacillus* spp.
Streptococcus anginosus	*Leptotrichia* spp.
Capnocytophaga spp.	*Peptostreptococcus* spp.
Eikenella spp.	*Prevotella* spp.
	Selenomonas spp.
	Veillonella spp.

Adapted from Alaluusua and Asikainen (1988), Könönen *et al.* (1992, 1994, 1999), Pearce *et al.* (1995), Smith *et al.* (1993) and McCarthy *et al.* (1965).

Streptococci predominate the primary colonization, especially *Streptococcus salivarius* (McCarthy *et al.* 1965; Smith *et al.* 1993) and *Streptococcus mitis* biovar 1 (Pearce *et al.* 1995). The diversity of the flora increases rapidly with age, and by the time the infant becomes dentate, 6–18 months, numerous species are present (Table 4.1.1). The acquisition of flora is primarily due to the development of tissues with age, such as the eruption of teeth. For example, the acquisition of *Streptoccocus mutans* is during a discrete window of infectivity and starts at around 19 months and continues for a further 12 months (Caufield *et al.* 1993). Indeed, the probability of acquiring *S. mutans* increases with the increasing tooth surface area and the number of retentative sites (Catalanotto *et al.* 1975; Caufield *et al.* 1993).

The diversity of the oral microflora is significantly increased during puberty (12–16 years); for example, increases are seen in Gram-negative anaerobes (Wojicicki *et al.* 1986), e.g. *Prevotella intermedia* and *Prevotella nigrescens* (Nakagawa *et al.* 1994) and spirochaetes (Mombelli *et al.* 1995) among others. This is probably due to changes in the composition of gingival crevicular fluid (GCF; a tissue exudate that bathes the periodontal tissues) including increased levels of sex hormones brought about by puberty. Few changes are observed with further aging, except notable decreases in levels of *Actinobacillus actinomycetemcomitans* and corresponding increases in levels of *Porphyromonas gingivalis* (Rodenburg *et al.* 1990; Savitt and Kent 1991; Darby *et al.* 2000).

The normal flora associated with healthy mucosa consists predominantly of *S. mitis, S. sanguis, S. anginosus* group, *S. salivarius, Neisseria* spp. and *Haemophilus* spp. (Dahlén *et al.* 1982; Frandsen *et al.* 1991). In contrast, the dorsum of the tongue has been shown to harbour far more diverse microflora and is still dominated by viridans streptococci, particularly *S. mitis* (biovar 1) and *S. salivarius*, but with high numbers of *Neisseria* spp.,

Veillonella spp., *Actinomyces* spp., *Propionibacterium* spp., *Prevotella* spp., *Fusobacterium* spp. and *Haemophilus* spp. (Frandsen *et al.* 1991; Dahlén *et al.* 1992).

COLONIZATION OF TOOTH SURFACES

In a healthy mouth, the only tooth surface available for colonization is the enamel, a hard, highly calcified tissue. The enamel of teeth is covered with an acquired pellicle within seconds of cleaning and it is this surface that is colonized very rapidly by the bacteria present in saliva, which contains up to 10^8 CFU ml^{-1}. In fact, bacterial adherence to the tooth surface is detectable in minutes (Saxton 1973). A range of molecules present in saliva bind selectively to the tooth surface, e.g. proline-rich proteins, histatins and statherin (Schupbach *et al.* 2001), and these act to promote the adherence of some important oral bacteria (*Actinomyces naeslundii, S. mutans* and some black-pigmented anaerobes). Early work, carried out in the 1960s and supported by later studies, clearly indicates a progression of microorganisms dominated by streptococci followed by an increasing proportion of Actinomyces, culminating in a mature plaque biofilm containing a large proportion of Gram-negative anaerobes (Ritz 1967; Lai *et al.* 1975; Listgarten *et al.* 1975; Listgarten 1976).

The primary adhesion events can be broadly split into two processes involving separate mechanisms: (1) adsorption of cells to the pellicle, which requires specific adhesins to be present on the cell surface; and (2) the adherence of additional cells binding to cells already present. The *S. sanguis* group are good examples of primary colonizers, and adhere to the pellicle using two kinetically distinct steps. The first step involves a reversible interaction with the pellicle-coated enamel surface, mediated by electrostatic and hydrophobic forces. The second is a time-dependent shift to a higher affinity binding state; this involves multiple adhesins on the cell surface and is not hydrophobicity dependent (Cowan *et al.* 1986). The continued co-adhesion of bacteria over a period of time (weeks) eventually produces a climax community (mature plaque) and is termed succession. This community usually has a high species diversity: it has been estimated that the human oral cavity contains approximately 500 species (Paster *et al.* 2001) and contains numerous microenvironments with gradients of a range of nutrients, oxygen, Eh, and pH. For these reasons, the mature biofilm is an extremely complex and highly dynamic community. The variety of environments present in the dentate oral cavity is immense, and even biofilms associated with teeth are divided into numerous sub-categories depending on the location on the tooth surface. These include: supragingival plaque, that above the gingival margin, often giving rise to caries;

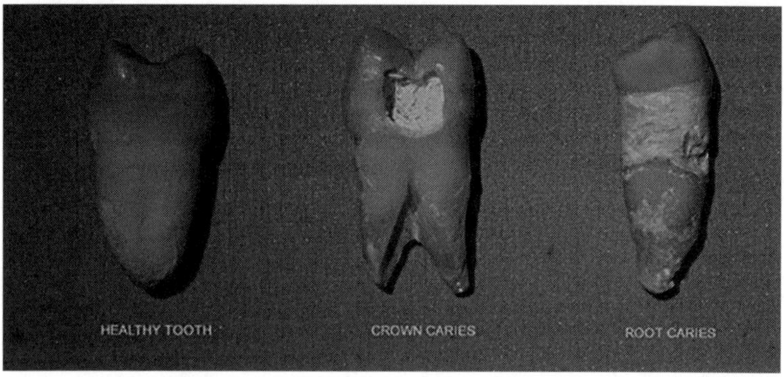

Figure 4.1.1. Carious lesions in teeth: (1) no caries; (2) crown caries, with large demineralized lesion in the enamel and dentine of the crown; and (3) root surface caries with demineralized cementum and enamel on root surface.

gingival margin plaque; subgingival plaque, that below the gingival margin and associated with periodontal disease; and aproximal plaque, that between teeth which is often very thick and gives rise to caries. The primary colonization of these surfaces is very similar, however, co-adhesion events and succession to a climax community are specific to the precise environmental conditions present during colonization. For a more comprehensive treatment of dental plaque formation, see the review by Rosan and Lamont (2000).

Oral Diseases Associated with Biofilms on Teeth

Caries

Dental caries can be defined as the localized demineralization of the tooth tissue by various acids produced by bacterial fermentation of dietary carbohydrates. Treatment of the disease involves removal of the damaged tooth tissue and its replacement with a restorative material. This disease is arguably the most common, chronic infectious disease in humans. Recently, it has been shown that 90% of all dentate adults in the UK have at least one restored tooth as a result of caries, with a mean frequency of seven per person (Pine *et al.* 2001). Caries can be simply and conveniently split into two categories: coronal (crown) caries and root surface caries. Coronal caries can occur on all surfaces of the crown where the supragingival plaque biofilm is allowed to develop and mature, however, it is most commonly associated with pits and fissures and aproximal sites (Figure 4.1.1).

A major question has been which bacteria, if any, are involved in the progression of disease. Over that last 30 years, a vast amount of research and debate has focused on this question (Loesche 1976; Theilade and Theilade 1976; Theilade 1986), with opinion polarizing into two camps: those supporting the 'specific plaque hypothesis' and those supporting the 'non-specific plaque hypothesis'. However, in the last 10 years, another concept that reconciles these two conflicting hypotheses has been suggested. The 'ecological plaque hypothesis' (Marsh 1991, 1994) suggests that small changes in the environment trigger shifts in the microbial community. In specific cases, this may predispose one to a more 'pathogenic' microbial community. In the case of coronal caries, a shift in the flora is brought about by increased amount and frequency of dietary fermentable carbohydrates. These substrates are fermented by the bacteria in the supragingival plaque biofilm, leading to production of acid end-products like lactic acid. This serves to lower the local pH and favour a shift in the microbial population to acid-tolerant bacteria such as mutans streptococci. Mutans streptococci is a collective term used to group the closely related species including *S. mutans*, *Streptococcus sobrinus*, *Streptococcus rattus*, *Streptococcus ferus* and *Streptococcus cricetus*. In addition to being aciduric, mutans streptococci are also highly acidogenic and can produce enough acid to lower the pH to levels that are inhibitory to many other bacteria within the biofilm (pH 4.5). The production of lactic acid is fundamental to the pathogenesis of mutans streptococci. It has been shown that mutant strains of *S. rattus* that lack lactate dehydrogenase activity fail to demineralize teeth in an animal model, despite colonization and plaque formation (Stashenko and Hillman 1989). Other bacteria associated with coronal caries include lactobacilli, non-mutans streptococci and *Actinomyces* spp. Lactobacilli are usually found in supragingival plaque biofilms in low numbers; conversely, however, they have been shown to be present in elevated proportions in established caries lesions and have, therefore, been associated more with progression of the lesion than with its initiation (Loesche 1986; Bowden 1991; van Houte 1994). Non-mutans streptococci are thought to be rarely involved with the disease progression of carious lesions, although they usually outnumber mutans streptococci and have been shown to produce acid in an acidic environment (Sansone *et al.* 1993).

Root surface caries, as the name implies, occurs on root cementum or dentine and is caused by a microbial biofilm. The disease is secondary to gingival recession, since, in a healthy mouth, cementum and dentine are not exposed to the microflora and, therefore, are unavailable for colonization. Gingival recession can be caused by a number of factors, including old age (the most common factor), mechanical injury (excessive tooth brushing) or periodontal treatment regimens. In industrialized countries, the proportion of the population over 65 years of age is increasing, additionally, the percentage of these remaining dentate is also increasing. A survey carried out by Steele *et al.* (1996) in 1991–92 showed, amongst other things, that in southern England 67% of patients over 60 years were dentate compared with an equivalent cohort in 1962, where only 15% remained dentate.

The microbiological nature of the associated plaque biofilm is different from that associated with crown caries, even though it is technically still a supragingival plaque. The lesion has been shown to have a definite progression, since changes in its clinical appearance are observed over time. Briefly, as the gingiva recedes new cementum/dentine is exposed and, in susceptible hosts, a lesion may start to form. The appearance of the lesion is described as 'soft' and consists of highly demineralized lesion replete with bacteria. Surrounding this, a further demineralized area is apparent and infiltrating bacteria can be observed. This lesion is actively carious. The progression of the lesion leads to a change in appearance and is categorized as 'leathery'. This is an intermediate stage and consists of a remineralized surface overlaying a heterogeneous mix of bacteria and demineralized and remineralized tissue. Further progression leads to a 'hard' lesion, which is fully remineralized and inactive with respect to caries. The microbiology of this biofilm has been the subject of numerous investigations over the years, however, only recently have the problems associated with sampling of the infected underlying dentine been identified and addressed (Beighton and Lynch 1995; Schupbach *et al.* 1996). Beighton and Lynch (1995) showed that the bacterial composition of the carious dentine biofilm associated with 'soft' lesions consists of significantly more lactobacilli and Gram-positive pleomorphic rods and, conversely, significantly fewer streptococci compared with the overlying plaque biofilm. Additionally, there is an increased number and/or proportion of *S. mutans* in 'soft' lesions compared with 'hard' lesions or sound surfaces. *Actinomyces* spp. have historically been associated with root surface caries, although the nomenclature of the species and genospecies in the literature confuses the matter greatly; see Johnson *et al.* (1990). However, in recent studies by Brailsford *et al.* (1998, 1999) *A. naeslundii* was shown not to be associated with active carious lesions and that *Actinomyces israelii* and *Actinomyces gerencseriae* predominated.

It is thought that the lesion occurs as a function of accumulation and subsequent stagnation of a plaque biofilm at the gingival margin. The nature of the flora is poorly understood, but there seems to be no single species responsible for disease initiation and progression. What may be important is the presence of particular strains of a defined but heterogeneous group of bacteria that are particularly suited to that environment. Indeed, Sansone *et al.* (1993) have shown that the acidogenic and aciduric flora associated with a carious lesion is far more diverse (approximately 25 taxa) than the corresponding acidogenic and aciduric flora associated with sound root surfaces (approximately eight taxa).

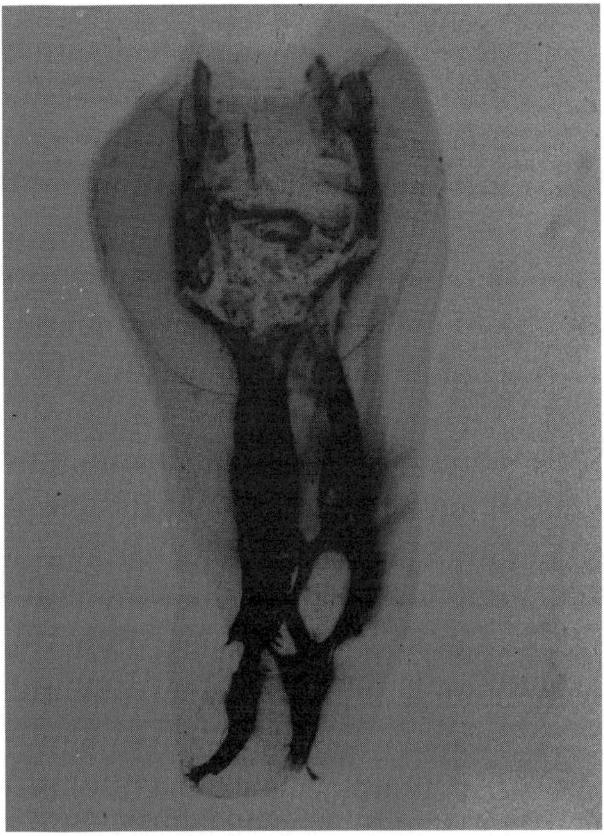

Figure 4.1.2. The complexity of the root canal system. Reprinted from Berkovitz *et al.* (1992) *Colour Atlas and Text of Endodontics*, published by Elsevier Health Sciences.

Endodontic Infections

All available surfaces in the oral cavity are colonized by different and diverse microbial biofilms. Structures present in the mouth but not exposed to the microflora are usually sterile, e.g. the endodontium, the pulp (neuro-vascular connective tissue that occupies the centre of the tooth in health and important for proprioception, nutrition and defence) and the root canal system within teeth. The root canals of teeth are complex systems of interconnecting channels (lateral and accessory) containing the blood vessels and nerve tissue leading from the tooth apex to the pulp chamber (Figure 4.1.2).

Endodontic infections are defined as infections of the pulp and periapical tissues. Miller (1894) suggested a bacterial cause for these diseases at the end of the 19th century when he demonstrated cocci, rods and spirochaetes in

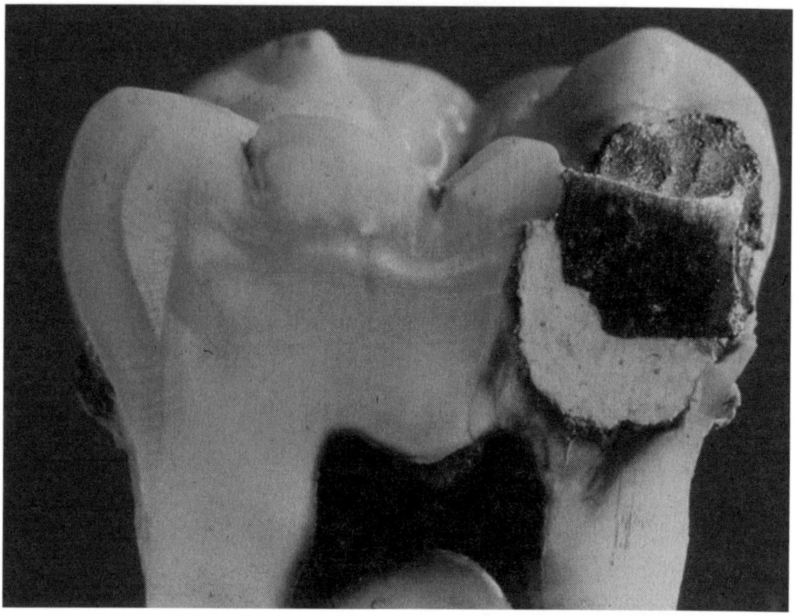

Figure 4.1.3. Leaking restoration, showing staining under the restoration caused by bacteria and their products, leading to pulp necrosis. Reprinted from Berkovitz *et al.* (1992) *A Colour Atlas and Textbook of Oral Anatomy*, published by Elsevier Health Sciences.

necrotic pulps. However, owing to other stronger arguments, namely the hollow tube theory (Rickert and Dixon 1931), a bacterial cause for these pulpal and periapical diseases has only been attributed since the mid-1960s, when pioneering work by Kakehashi *et al.* (1965) demonstrated the importance of bacteria as prerequisite to pulpal inflammation and subsequent necrosis. Bacteria and their products gain access to the pulp chamber in the majority of cases as a consequence of caries (Figure 4.1.3). Owing to significant demineralization of the enamel, cementum or dentine, the pulp can be directly exposed to insult by the biofilm associated with the lesion. Additionally, the pulp can be exposed by a number of other mechanisms, such as trauma, exposed dentinal tubules, congenital conditions, enamel lamellae and possibly anachoresis (Allard *et al.* 1979; Beynon 1982; Watts and Paterson 1990; Berkovitz *et al.* 1992). Prior to colonization, bacterial products, such as metabolic end-products and lipopolysaccharide, can elicit an inflammatory response from the pulp (Reeves and Stanely 1966) (Figure 4.1.4).

Once exposed, the pulp quickly becomes colonized; indeed, coronal pulp was shown to become colonized within 3 days of exposure to the oral cavity (Watts and Paterson 1990). The extent of the assault in terms of duration, number and specific virulence of the bacteria involved will determine the

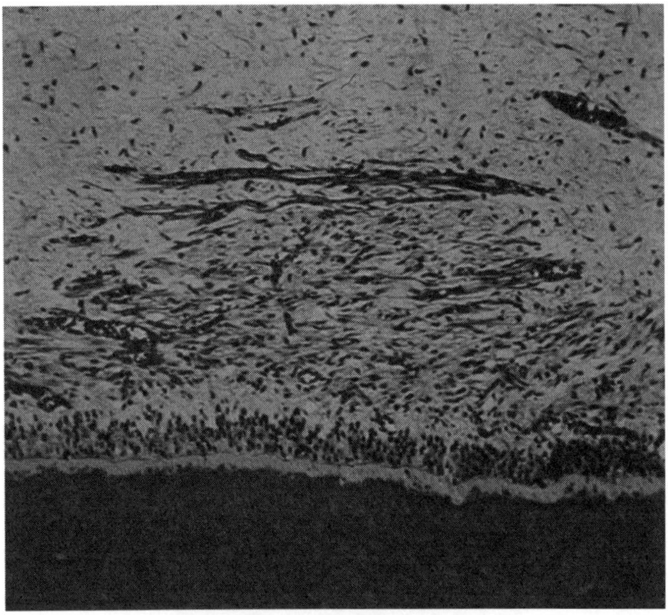

Figure 4.1.4. Localized inflammation of the pulp in response to bacterial assault on dentine. Reprinted from Berkovitz *et al.* (1992) *A Colour Atlas and Textbook of Oral Anatomy*, published by Elsevier Health Sciences.

outcome (Dahlén *et al.* 1982); persistent assault, and therefore inflammation, will lead to pulpal necrosis. Once the pulp is necrosed, an inflammatory response is elicited at the periapex and this leads to apical periodontitis. This can be visualized radiographically as a dark area at the root apex consistent with significant bone loss (Figure 4.1.5). Progression is often asymptomatic and only becomes apparent when it becomes acute, the pain being usually associated with the pressure exerted by pus in the supporting bone at the root apex. If untreated, the lesion will progress and further bone resorption will take place with concomitant tooth loss.

The bacteria associated with these lesions are surprisingly limited given the number of taxa potentially able to colonize the surface and the large number of taxa associated with periodontal lesions (Table 4.1.2). This reduced diversity implies special selective pressures operating within the root canal system. The bacterial composition of the biofilm within the root canal system is thought to change with time and also location within the system. A biofilm may line the interior surfaces of the tooth from the roof of the pulp chamber to beyond the root apex—a distance of up to approximately 15 mm. The difference in environmental conditions between the coronal and apical ends is thought to be significant. The coronal

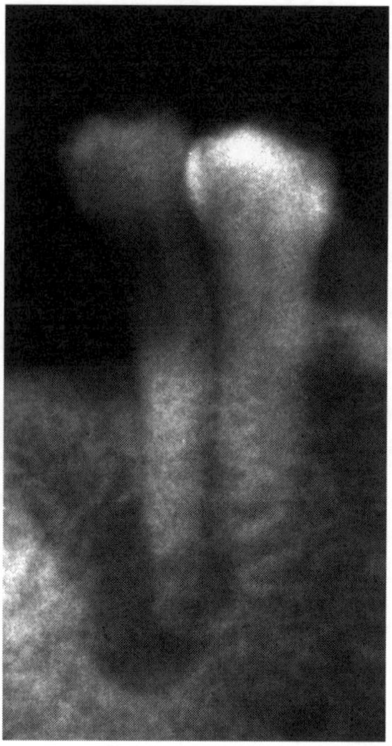

Figure 4.1.5. Radiograph showing a periapical lesion at the apex of the left-hand tooth. Visualized as a circular darkening in the alveolar bone.

environment is broadly more aerobic in contrast to the anaerobic apex, therefore, a gradient is formed between the two 'poles'. Similarly, nutrient gradients exist within the root canal system, however, the source of nutrients may be different in the coronal aspect than in the apex. Nutrition may be derived to some extent from the host diet in coronal areas via microleakage, while apical tissue fluid and breakdown products from the pulp are the major nutrient sources. Only recently have these differences been addressed with respect to the microbiology of the infection. Most studies use techniques based on sampling of the 'whole' root canal system, such as the paper point method, which aims to adsorb bacteria in the root canal system with small rolled cones of paper, often subsequent to limited filing. These techniques have shaped our understanding of the microbial nature of the biofilm associated with endodontic infection (Hirai *et al*. 1991; Sundqvist 1992; Brauner and Conrads 1995). A broad range of bacterial species are commonly isolated from endodontic infections, representing 20–30 genera, and this is dependent on the techniques used for isolation, identification and changes in nomenclature (Table 4.1.2). Of these, the most commonly occurring species are *F. nucleatum*, *Streptococcus* spp.,

Table 4.1.2. Diversity of species isolated for root canal infections

	Aerobic species	Facultative species	Anaerobic species
Gram-positive cocci		*Enterococcus faecalis* *Enterococcus faecium* *Staphylococcus warneri* *Staphylococcus lentus* *Streptococcus anginosus* *Streptococcus constellatus* *Streptococcus intermedius* *Streptococcus gordonii* *Streptococcus mitis* *Streptococcus mutans* *Streptococcus oralis* *Streptococcus salivarius* *Streptococcus sanguis*	*Peptostreptococcus micros* *Peptostreptococcus prevotii* *Peptostreptococcus magnus* *Peptostreptococcus asaccharolyticus*
Gram-positive rods		*Corynebacterium xerosis* *Lactobacillus acidophilus* *Lactobacillus catenaforme* *Lactobacillus fermentum* *Lactobacillus salivarius*	*Actinomyces naeslundii* *Actinomyces israelii* *Actinomyces meyeri* *Actinomyces odontolyticus* *Actinomyces viscosus* *Atopobium minutum* *Cryptobacterium curtum* *Eubacterium brachy* *Eubacterium lentum* *Eubacterium nodatum* *Mogibacterium timidum* *Propionibacterium acnes* *Propionibacterium granulosum* *Propionibacterium propionicus* *Pseudoramibacter alactolyticus* *Slakia exigua*
Gram-negative cocci		*Neisseria* spp.	*Veillonella parvula*
Gram-negative rods	*Pseudomonas aeruginosa*	*Campylobacter curvus* *Campylobacter rectus* *Campylobacter sputorum* *Capnocytophaga ochracea*	*Dialister pneumosinites* *Eikenella corrodens* *Fusobacterium nucleatum* *Fusobacterium necrophorum* *Porphyromonas gingivalis* *Porphyromonas endodontalis* *Prevotella oralis* *Prevotella oris* *Prevotella buccae* *P. intermedia* *Prevotella denticola* *Prevotella dentalis* *Prevotella melaninogenica* *Prevotella loescheii* *Selenomonas sputigena*

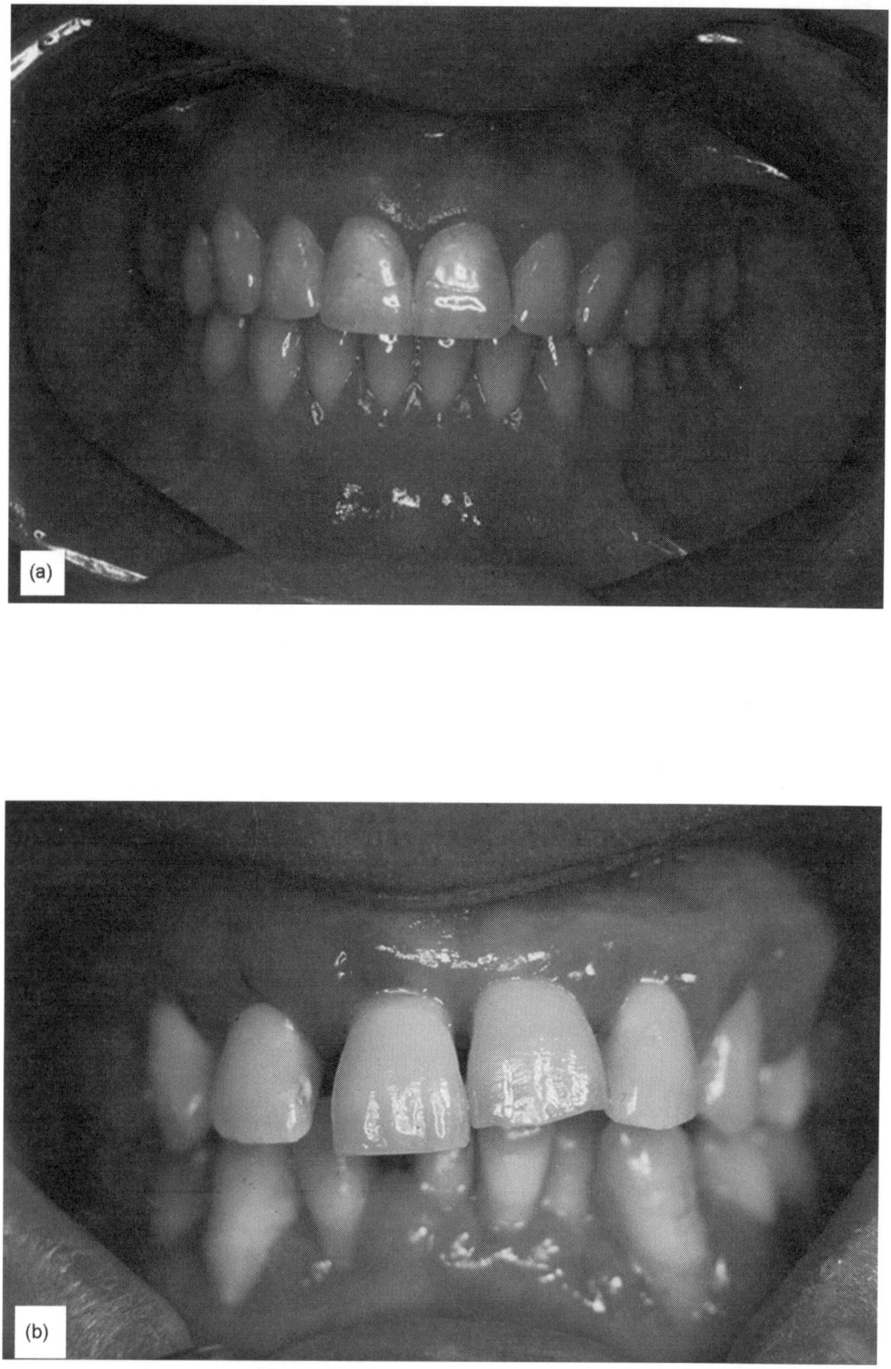
(a)
(b)

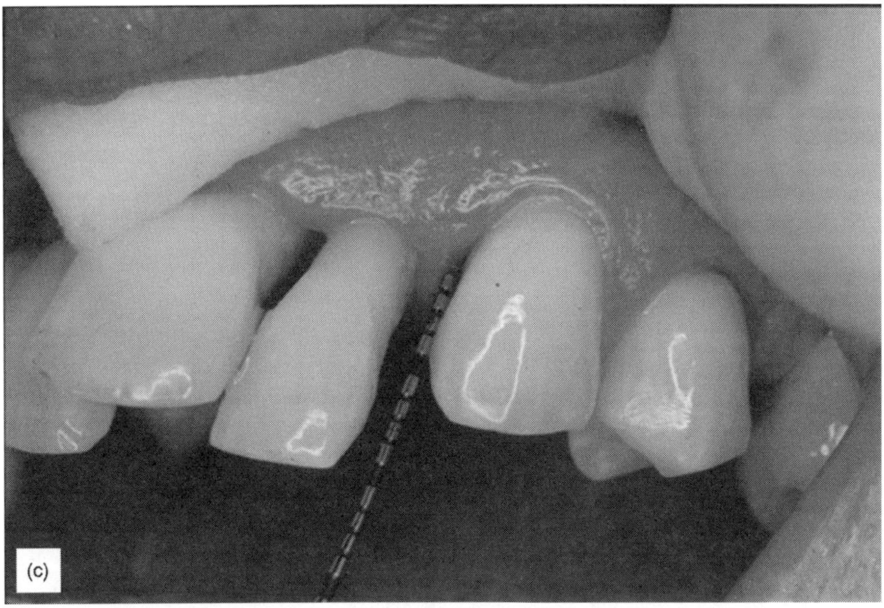

Figure 4.1.6. Clinical features of (a) a healthy periodontium and (b) chronic periodontitis. (c) Periodontal pocket depth measurement showing significant loss of tissue attachment to the teeth.

Porphyromonas spp., *P. intermedia*, *Peptostreptococcus* spp., *Actinomyces* spp. and *Eubacterium* spp. (the genus *Eubacterium* is very broad and is, at present, undergoing significant taxonomic revision).

Root canal infections are invariably polymicrobial in nature, however, mono-infections do occur (e.g. *Enterococcus* spp.). Typically, cultures include 4–12 bacterial isolates. This flora is often diverse with respect to growth atmosphere, nutritional needs and virulence determinants, and it may be regarded as an 'infection team'. For example, primary colonization and adherence to dentine is carried out by streptrococci, which utilize oxygen, thus making the environment more anaerobic for the colonization of less oxygen tolerant species. This colonization is analogous to tooth colonization, but, for some reason, only a limited number of bacterial species are involved in each case, suggesting a selection process is taking place (the nature of which is unknown). In addition, yeast species (e.g. *Candida albicans*) are present and, although reports differ, occur in about 10% of root canal infections (Egan *et al*. 2002). However, the relationship, both physically and nutritionally, between yeasts and bacteria in the root canal system biofilm is poorly understood.

The chronic asymptomatic periradicular lesion formed at the root apex (Figure 4.1.5) does not usually contain bacteria but is thought to be a host response to various bacterial products diffusing out of the root canal system. An acute exacerbation of a chronic lesion can occur; bacteria enter the periapical lesion and a pus-filled abscess forms. The size of the abscess and potential to spread is related to the diversity of the flora. A low diversity tends to lead to a small, non-spreading abscess, whereas a greater diversity leads to a larger, spreading and painful lesion.

Periodontal Diseases

This defines a broad group of diseases affecting the periodontal tissues, the most common are inflammatory processes of the gingiva and tissues attaching to the tooth (Figure 4.1.6). These diseases are usually associated with microbial infection due to accumulation of a plaque biofilm and calculus.

Gingivitis. The bacteria and their extracellular products present within the plaque biofilm on the surfaces of teeth at the gum margin can cause inflammation. This is termed gingivitis; it is the most common of the periodontal diseases, and is usually brought about by poor oral hygiene. It can be defined as 'a non-specific inflammatory process of the gingiva (gum) without destruction of the supporting tissues'. A complex range of gingival diseases are recognized and these have recently been reclassified into two main groups with 12 headings and numerous sub-headings (Table 4.1.3), however, only gingivitis associated with dental plaque will be discussed. This disease is usually reversible, and on removal of the biofilm (e.g. a return to good oral hygiene) the tissues revert to a healthy clinical state. It is likely that the entire population suffers to some extent from this disease. The microflora associated with dental plaque-induced gingivitis is different than that associated with health. In addition to an increase in biofilm mass, e.g. due to poor oral hygiene, the composition shifts from one dominated by streptococci (Slots 1977) to one where *Actinomyces* spp. dominate (Moore *et al*. 1987). Specifically, increased proportions of *A. naeslundii*, *E. corrodens*, *F. nucleatum* and *Capnocytophaga gingivalis* are detected (Savitt and Socransky 1984; Moore *et al*. 1987).

Periodontitis. This refers to a group of more advanced and related diseases within the broad heading of periodontal disease. It can be defined as 'an apical extension of gingival inflammation to involve the tissues supporting the tooth, including periodontal ligament and bone'. The destruction of the fibre attachment results in a periodontal pocket (Figures 4.1.6 and 4.1.7). This wide spectrum of diseases has recently been reclassified (Armitage 1999), and at least 48 specific periodontitis categories are now recognized (Table 4.1.3). By far the most common is chronic periodontitis (Table 4.1.3, section 2B) and this is the major cause of tooth loss in the adult population.

Table 4.1.3. Classification of periodontal diseases (adapted from Armitage 1999)

1. Gingival diseases
   - A. Dental-plaque-induced gingival disease
     - i Gingivitis associated with dental plaque
     - ii Gingival disease modified by systemic factors
     - iii Gingival diseases modified by medications
     - iv Gingival diseases modified by malnutrition
   - B. Non-plaque-induced gingival lesions
     - i Gingival disease of specific bacterial origin
     - ii Gingival diseases of viral origin
     - iii Gingival diseases of fungal origin
     - iv Gingival lesions of genetic origin
     - v Gingival manifestations of systemic conditions
     - vi Traumatic lesions, foreign body reactions
     - vii Other
2. Chronic periodontitis
   - A. Localized
   - B. Generalized
3. Aggressive periodontitis
   - A. Localized
   - B. Generalized
4. Periodontitis as a manifestation of systemic disease
   - A. Associated with haematological disorders
   - B. Associated with genetic disorders
   - C. Other
5. Necrotizing periodontal disease
   - A. Necrotizing ulcerative gingivitis
   - B. Necrotizing ulcerative periodontitis
6. Abscesses of the periodontium
   - A. Gingival abscess
   - B. Periodontal abscess
   - C. Pericoronal abscess
7. Periodontitis associated with endodontic lesions
8. Developmental or acquired deformities or conditions
   - A. Localized tooth-related factors that modify or predispose to plaque gingival disease/periodontitis
   - B. Mucogingival deformities and conditions around the teeth
   - C. Mucogingival deformities and conditions on edentulous ridges
   - D. Occlusal trauma

The disease is mediated by the microflora forming the plaque biofilm on the tooth surface (Figure 4.1.7). Additionally, as a consequence of the immune response elicited by the bacteria, further destruction may occur due to the host inflammatory response. A bacterial cause for these diseases has been shown with evidence arising from studies such as: longitudinal and cross-sectional studies (Löe *et al.* 1965); conventional and germ-free animal studies (Irving *et al.* 1978; Holt 1988); and antibiotic treatment studies (Garrett *et al.* 1999). The biofilm present in the gingival crevice, and later in

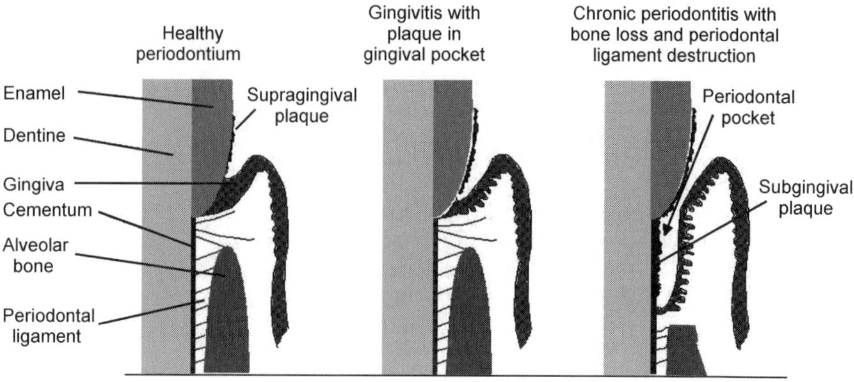

Figure 4.1.7. Stages in periodontal disease progression.

the periodontal pocket (Figure 4.1.7), is extremely diverse, with up to 100 culturable species from a single pocket (Haffajee and Socransky 1994). Since such a diverse flora is present, trying to identify the particular species responsible for disease initiation and progression is a very complex and difficult undertaking. This problem brings us back to the 'specific non-specific plaque hypothesis' debate (see Caries section) and, again, the most plausible explanation for the aetiology is supplied by the 'ecological plaque hypothesis'. Briefly, the primary inflammation events due to a large biofilm mass at the gingival margin increase the flow of GCF, thus changing the local environment and allowing proteolytic and anaerobic species to predominate. GCF is present in small quantities in healthy sites, but it is in much larger volumes in diseased sites and contains a wide range of complex molecules derived from a number of sources, including serum, connective tissue and epithelium (Lamster 1997). In addition, polymorphonuclear neutropathic granulocytes and monocytes are present in GCF in the pocket. Therefore, it is to be expected that a progressively more diverse and anaerobic flora will be isolated during disease progression. This premise is very well illustrated in Table 4.1.4, in which large numbers of anaerobes increase in their overall proportions during disease progression and, conversely, aerobes and facultative species decrease. The World Workshop on Clinical Periodontology (American Academy of Periodontology Consensus report 1996) has designated three species as aetiologic agents of periodontitis in susceptible hosts, namely *A. actinomycetemcomitans*, *P. gingivalis* and *Tannerella forsythensis* (formerly *Bacteroides forsythus*). The findings from the majority of the microbiology studies are based on data derived from the culturable flora. However, it has been estimated that only 50% of the oral flora is culturable (Socransky *et al.* 1963; Tanner *et al.* 1994). More recently, molecular techniques

Table 4.1.4. Changes in the microflora of the biofilm as a function of periodontitis severity (adapted from Moore and Moore (1994))

	Strict anaerobes	Aerobes/facultatives
Increase in proportions		
Gram-positive bacteria	*Atopobium rimae*	
	Eubacterium brachy	
	Eubacterium nodatum	
	Eubacterium saphenum	
	Mogibacterium timidum	
	Olsenella uli	
	P. anaerobius	
	P. alactolyticum	
Gram-negative bacteria	*Bacteroides gracilis*	*A. actinomycetemcomitans*
	C. rectus	
	C. curvus	
	Filfactor alocis	
	F. nucleatum	
	P. denticoa	
	P. intermedia	
	P. melaninogenica	
	P. oris	
	Prevotella tannerae	
	Prevotella veroralis	
	P. gingivalis	
	Selenomonas flueggei	
	Selenomonas inflex	
	Selenomonas noxia	
	Selenomonas sputigena	
Decrease in proportions		
Gram-positive bacteria	*Eubacterium saburreum*	*A. meryeri*
		A. naeslundii
		A. odontolyticus
		R. dentocariosa
		S. gordonii
		S. intermedius
		S. oralis
		S. salivarius
		S. sanguis
Gram-negative bacteria	*V. parvula*	*Capnocytophaga gingivalis*
		Haemophilus aphrophilus
		Haemophilus segnis
		Leptotrichia spp.
		Neisseria elongata
		Neisseria mucosa

have been used to detect and identify the unculturable portion of this highly diverse biofilm (Spratt *et al.* 1999; Paster *et al.* 2001).

A range of predisposing factors further complicates the aetiology of periodontal disease. The susceptibility of the host has been shown to be important. A specific genotype of the polymorphic interleukin-1 gene cluster (pro-inflammatory cytokine), a key regulator of the host responses to microbial infection and a major modulator of extracellular matrix catabolism and bone resorption, is associated with severity of periodontitis (Kornman *et al.* 1997). In addition, there is evidence that diabetes mellitus is a major risk factor for chronic periodontitis and the more severe and rapidly progressing forms (Cianciola *et al.* 1982; Shlossman *et al.* 1990; Oliver and Tervonen 1994). Smoking is also a significant risk factor (Holm 1994).

COLONIZATION OF EPITHELIAL SURFACES IN THE MOUTH

The microbial colonization of mucosal surfaces starts at birth. The normal flora associated with healthy mucosa consists of predominantly *S. mitis, S. sanguis, S. anginosus* group, *S. salivarius, Neisseria* spp., and *Haemophilus* spp. (Frandsen *et al.* 1991; Dahlén *et al.* 1992). In contrast, the dorsum of the tongue has been shown to harbour a far more diverse microflora still dominated by viridans streptococci, particularly *S. mitis* (biovar 1) and *S. salivarius*, but with high numbers of *Neisseria* spp., *Veillonella* spp., *Actinomyces* spp., *Propionibacterium* spp., *Prevotella* spp., *Fusobacterium* spp. and *Haemophilus* spp. (Frandsen *et al.* 1991; Dahlén *et al.* 1992). A large and complex microbial biofilm is associated with the oral mucosa, and under normal conditions this does not cause any disease. This is largely due to the balance of interactions between the microorganisms and the host defence system.

Diseases Associated with Epithelial Surfaces

A number of mucosal diseases have a microbial aetiology that can be associated with either the normal oral flora or microflora from an extra-oral environment. The majority of the oral mucosal infections are of a fungal origin, in particular *Candida* spp. These include thrush, angular cheilitis, denture stomatitis, *Candida* leucoplakia and median rhomboid glossitis. The majority of these diseases are rare in healthy adults and only affect the young, old and medically compromised. Fungal over-growth, especially by *Candida* spp., is the most common cause of these infections and is usually brought about by a lack of competitive microflora due to some sort of disruption (e.g. use of broad-spectrum antibiotics).

Acute Atrophic Candidosis (Thrush)

This condition is also a disease of the very young, old and medically compromised. It is characterized by white/yellow plaques, with associated acute inflammation, distributed over the mucosal surfaces of the oral cavity. The causative organisms are exclusively *Candida* spp., particularly *C. albicans*. The plaques consist of fungal hyphea, which are surrounding and invading into the mucosal cells, and the plaques can be wiped off to reveal a raw erythematous and often bleeding base.

Angular Cheilitis

Angular cheilitis is an infection at the corners of the lips and is characterized by a reddening of the tissue. It is most often associated with denture wearers with denture stomatitis, particularly those with badly fitting dentures that fail to support the face. The microorganisms associated with this disease are mainly *Candida* spp. and staphylococci, particularly *Staphylococcus aureus* (Warnakulasuriya *et al*. 1991; Dais and Samaranayake 1995).

Denture Stomatitis

This is characterized by mucosal inflammation and redness underneath a denture. It is caused by the microbial biofilm on the fitting surface of the denture rather than on the mucosal surface of, for example, the palate (Davenport 1970; Olsen 1974). Nearly 90% of cases are thought to be caused by yeast (Olsen 1974), typically *C. albicans,* although lesions have also been associated with extra-oral species (e.g. *S. aureus, Escherichia coli* and *Klebsiella* spp.). However, only a strong correlation has been shown for *C. albicans* and *S. aureus* (Palmqvist *et al*. 1984).

Candida Leucoplakia

This infection is presented in a similar fashion to acute atrophic candidosis, however, the lesions are usually more adherent to the mucosa. It is thought that *C. albicans* is the causative agent, since the grossly thickened hyperplastic epithelium is penetrated by fungal hyphea and systemic antifugal therapy often resolves the lesion. These lesions are clinically very important, since around 10% become cancerous.

Median Rhomboid Glossitis

Median rhomboid glossitis is characterized by a swelling of the tongue, which can be diamond shaped and red, nodular and depapillated or white.

It is mainly seen in adult males who smoke, and it is thought to be congenital. Infection by *Candida* spp. and/or bacteria is assumed, but this may be secondary to the lesion.

REFERENCES

Alaluusua, S and Asikainen, S (1989) Detection and distribution of *Actinobacillus actinomycetemcomitans* in the primary dentition. *Journal of Periodontology* **59**:504–507.

Allard, U, Nord, C-E, Sjöberg, L and Strömberg, T (1979) Experimental infections with *Staphylococcus aureus, Streptococcus sanguis, Pseudomonas aeruginosa* and *Bacteroides fragilis* in the jaws of dogs. *Oral Surgery, Oral Medicine and Oral Pathology* **48**:454–462.

Armitage, GC (1999) Development of a classification system for periodontal diseases and conditions. *Annals of Periodontology* **4**:1–6.

Beighton, D and Lynch, E (1995) Comparison of selected microflora of plaque and underlying carious dentine associated with primary root caries lesions. *Caries Research* **29**:154–158.

Berkovitz, BKB, Holland, GR and Moxham, BJ (1992) A Colour Atlas and Textbook of Oral Anatomy, 2nd Edition. Wolfe Medical Publishing, p. 122.

Beynon, AD (1982) Developing dens invaginatus (dens in dente). *British Dental Journal* **153**:255–260.

Bowden, GHW (1991) Which bacteria are cariogenic in humans? In: Risk Markers for Oral Diseases, Volume 1, Dental Caries (Ed. Johnson, NW), Cambridge University Press, Cambridge, pp. 266–286.

Brailsford, SR, Lynch, E and Beighton, D (1998) The isolation of *Actinomyces naeslundii* from sound root surfaces and root carious lesions. *Caries Research* **32**:100–106.

Brailsford, SR, Tregaskis, RB, Leftwich, HS and Beighton, D (1999) The predominant *Actinomyces* spp. isolated from infected dentin of active root caries lesions. *Journal of Dental Research* **78**:1525–1534.

Brauner, AW and Conrads, G (1995) Studies into the microbial spectrum of apical periodontitis. *International Endodontic Journal* **28**:244–248.

Carlsson, J and Gothefors, L (1975) Transmission of *Lactobacillus jensenii* and *Lactobacillus acidophilus* from mother to child at time of delivery. *Journal of Clinical Microbiology* **1**:124–128.

Catalanotto, FA, Shklair, IL and Keene, HJ (1975) Prevalence and localization of *Streptococcus mutans* in infants and children. *Journal of the American Dental Association* **91**:606–609.

Caufield, PW, Cutter, GR and Dasanayake, AP (1993) Initial acquisition of mutans streptococci by infants: evidence for a discrete window of infectivity. *Journal of Dental Research* **72**:37–45.

Cianciola, LJ, Park, BH and Bruck, E (1982) Prevalence of periodontal disease in insulin-dependent diabetes mellitus (juvenile diabetes). *Journal of the American Dental Association* **104**:653–660.

Cowan, MM, Taylor, KG and Doyle, RJ (1986) Kinetic analysis of *Streptococcus sanguis* adhesion to artificial pellicle. *Journal of Dental Research* **65**:1278–1283.

Dahlén, G, Fabricius, L, Heyden, G, Holm, SE and Moller, AJ (1982) Apical periodontitis induced by selected bacterial strains in root canals of immunized and nonimmunized monkeys. *Scandinavian Journal of Dental Research* **90**:207–216.

Dahlén, G, Jinsson, R, Öhman, S-C, Nielsen, R and Möller, AJR (1992) Infections of oral mucosa and submucosa. In: Contemporary Oral Microbiology and Immunology. Mosby Year Book (Eds. Slots, J and Taubman, MA), pp. 476–499.

Dais, AP and Samaranayake, LP (1995) Clinical, microbiological and ultrastructural features of angular chelitis lesions in southern Chinese. *Oral Disease* **1**:43–48.

Darby, IB, Hodge, PJ, Riggio, MP and Kinane, DF (2000) Microbial comparison of smoker and non-smoker adult and early-onset periodontitis patients by polymerase chain reaction. *Journal of Clinical Periodontology* **27**:417–424.

Davenport, JC (1970) The oral distribution of *Candida* in denture stomatitis. *British Dental Journal* **129**:151–156.

Egan, MW, Spratt, DA, Ng, Y-L, Lam, JA, Moles, DR and Gulabivala, K (2002) Prevalence of yeasts in saliva and root canals of teeth associated with apical periodontitis. *International Endodontic Journal* **35**:321–329.

Frandsen, EV, Pedrazzoli, V and Kilian, M (1991) Ecology of viridans streptococci in the oral cavity and pharynx. *Oral Microbiology and Immunology* **6**:129–133.

Garrett, S, Johnson, L, Drisko, CH, Adams, DF, Bandt, C, Beiswanger, B, Bogle, G, Donly, K, Hallmon, WW, Hancock, EB, Hanes, P, Hawley, CE, Kiger, R, Killoy, W, Mellonig, JT, Polson, A, Raab, FJ, Ryder, M, Stoller, NH, Wang, HL, Wolinsky, LE, Evans, GH, Harrold, CQ, Arnold, RM and Southard, GL (1999) Two multi-center studies evaluating locally delivered doxycycline hyclate, placebo control, oral hygiene, and scaling and root planing in the treatment of periodontitis. *Journal of Periodontology* **70**:490–503.

Haffajee, AD and Socransky, SS (1994) Microbial etiological agents of destructive periodontal diseases. *Periodontology 2000* **5**:78–111.

Hirai, K, Tagami, A and Okunda, K (1991) Isolation and classification of anaerobic bacteria from pulp cavities of nonvital teeth in man. *Bulletin of the Tokyo Dental College* **32**:95–98.

Holm, G (1994) Smoking as an additional risk for tooth loss. *Journal of Periodontology* **65**:996–1001.

Holt, SC (1988) Basic bacteriology. In: Contempary Oral Microbiology and Immunology. Mosby Year Book (Eds. Slots, J and Taubman, MA).

Irving, JT, Socransky, SS and Tanner AC (1978) Histological changes in experimental periodontal disease in rats mono infected with Gram-negative organisms. *Journal of Periodontal Research* **13**:326–332.

Johnson, JL, Moore, LV, Kaneko, B and Moore, WE (1990) *Actinomyces georgiae* sp. nov., *Actinomyces gerencseriae* sp. nov., designation of two genospecies of *Actinomyces naeslundii*, and inclusion of *A. naeslundii* serotypes II and III and *Actinomyces viscosus* serotype II in *A. naeslundii* genospecies 2. *International Journal of Systematic Bacteriology* **40**:273–286.

Kakehashi, S, Stanley, HR and Fitzgerald, RJ (1965) The effects of surgical exposures of dental pulps in germ-free and conventional laboratory rats. *Oral Surgery, Oral Medicine, Oral Pathology* **20**:340–349.

Könönen, E, Asikainen, S and Jousimies-Somer, H (1992) The early colonization of Gram-negative anaerobic bacteria in edentulous infants. *Oral Microbiology and Immunology* **7**:28–31.

Könönen, E, Asikainen, S, Saarela, M, Karjalainen, J and Jousimies-Somer, H (1994) The oral Gram-negative anaerobic microflora in young children: longitudinal

changes from edentulous to dentate mouth. *Oral Microbiology and Immunology* **9**:136–141.
Könönen, E, Kanervo, A, Takala, A, Asikainen, S and Jousimies-Somer, H (1999) Establishment of oral anaerobes during the first year of life. *Journal of Dental Research* **78**:1634–1639.
Kornman, KS, Crane, A, Wang, HY, di Giovine, FS, Newman, MG, Pirk, FW, Wilson, TG Jr, Higginbottom, FL and Duff, GW (1997) The interleukin-1 genotype as a severity factor in adult periodontal disease. *Journal of Clinical Periodontology* **24**:72–77.
Lai, CH, Listgarten, MA and Rosan, B (1975) Immunoelectron microscopic identification and localization of *Streptococcus sanguis* with peroxidase-labeled antibody: localization of *Streptococcus sanguis* in intact dental plaque. *Infection and Immunity* **11**:200–210.
Lamster, IB (1997) Evaluation of components of gingival crevicular fluid as diagnostic tests. *Annals of Periodontology* **2**:123–137.
Li, Y and Caufield, PW (1995) The fidelity of initial acquisition of mutans streptococci by infants from their mothers. *Journal of Dental Research* **74**:681–685.
Listgarten, MA (1976) Structure of the microbial flora associated with periodontal health and disease in man. A light and electron microscopic study. *Journal of Periodontology* **47**:1–18.
Listgarten, MA, Mayo, HE and Tremblay, R (1975) Development of dental plaque on epoxy resin crowns in man. A light and electron microscopic study. *Journal of Periodontology* **46**:10–26.
Löe, H, Theilade, E and Jensen, SB (1965) Experimental gingivitis in man. *Journal of Periodontology* **36**:177–187.
Loesche, WJ (1976) Chemotherapy of dental plaque infections. *Oral Science Review* **9**:65–107.
Loesche, WJ (1986) Role of *Streptococcus mutans* in human dental decay. *Microbiology Reviews* **50**:353–380.
Long, SS and Swenson, RM (1976) Determinants of the developing oral flora in normal newborns. *Applied and Environmental Microbiology* **32**:494–497.
Marsh, PD (1991) The significance of maintaining the stability of the natural microflora of the mouth. *British Dental Journal* **171**:174–177.
Marsh, PD (1994) Microbial ecology of dental plaque and its significance in health and disease. *Advances in Dental Research* **8**:263–271.
McCarthy, C, Snyder, MI and Parker, RB (1965) The indigenous flora of man I. The newborn to the 1-year-old infant. *Archives of Oral Biology* **10**:61–70.
Miller, WD (1894) An introduction to the study of bacterio-pathology of the dental pulp. *Dental Cosmos* **36**:505–528.
Mombelli, A, Rutar, A and Lang, NP (1995) Correlation of the periodontal status 6 years after puberty with clinical and microbiological conditions during puberty. *Journal of Clinical Periodontology* **22**:300–305.
Moore, LV, Moore, WE, Cato, EP, Smibert, RM, Burmeister, JA, Best, AM and Ranney, RR (1987) Bacteriology of human gingivitis. *Journal of Dental Research* **66**:989 995.
Moore, WEC and Moore, LVH (1994) The bacteria of periodontal disease. *Periodontology 2000* **5**:66–77.
Nakagawa, S, Fujii, H, Machida, Y and Okuda, K (1994) A longitudinal study from prepuberty to puberty of gingivitis. Correlation between the occurrence of *Prevotella intermedia* and sex hormones. *Journal of Clinical Periodontology* **21**:658–665.

Oliver, RC and Tervonen, T (1994) Diabetes—a risk factor for periodontitis in adults? *Journal of Periodontology* **65**(Suppl.):S530–S538.

Olsen, I (1974) Denture stomatitis. Occurrence and distribution of fungi. *Acta Odontologica Scandinavica* **32**:329–333.

Palmqvist, S, Unell, L and Lindquist, B (1984) Denture stomatitis in nursing home patients. *Swedish Dental Journal* **8**:73–80.

Paster, BJ, Boches, SK, Galvin, JL, Ericson, RE, Lau, CN, Levanos, VA, Sahasrabudhe, A and Dewhirst, FE (2001) Bacterial diversity in human subgingival plaque. *Journal of Bacteriology* **183**:3770–3783.

Pearce, C, Bowden, GH, Evans, M, Fitzsimmons, SP, Johnson, J, Sheridan, MJ, Wientzen, R and Cole, MF (1995) Identification of pioneer viridans streptococci in the oral cavity of human neonates. *Journal of Medical Microbiology* **42**:67–72.

Pine, CM, Pitts, NB, Steele, JG, Nunn, JN and Treasure, E (2001) Dental restorations in adults in the UK in 1998 and implications for the future. *British Dental Journal* **190**:48.

Reeves, R and Stanely, HR (1966) The relationship of bacterial penetration and pulpal pathos in carious teeth. *Oral Surgery* **22**:59–65.

Rickert, U and Dixon, CM (1931) The controlling of root surgery. In: Transactions of the eighth International Dental Congress, Section 111a. Paris, France. pp. 15–22.

Ritz, HL (1967) Microbial population shifts in developing human dental plaque. *Archives of Oral Biology* **12**:1561–1568.

Rodenburg, JP, van Winkelhoff, AJ, Winkel, EG, Goene, RJ, Abbas, F and de Graff, J (1990) Occurrence of *Bacteroides gingivalis*, *Bacteroides intermedius* and *Actinobacillus actinomycetemcomitans* in severe periodontitis in relation to age and treatment history. *Journal of Clinical Periodontology* **17**:392–399.

Rosan, B and Lamont, RJ (2000) Dental plaque formation. *Microbes and Infection* **2**:1599–1607.

Sansone, C, Van Houte, J, Joshipura, K, Kent, R and Margolis, HC (1993) The association of mutans streptococci and non-mutans streptococci capable of acidigenesis at a low pH with dental caries on enamel and root surfaces. *Journal of Dental Research* **72**:508–516.

Savitt, ED and Kent, RL (1991) Distribution of *Actinobacillus actinomycetemcomitans* and *Porphyromonas gingivalis* by subject age. *Journal of Periodontology* **62**:490–494.

Savitt, ED and Socransky, SS (1984) Distribution of certain subgingival microbial species in selected periodontal conditions. *Journal of Periodontal Research* **19**:111–123.

Saxton, CA (1973) Scanning electron microscope study of the formation of dental plaque. *Caries Research* **7**:102–119.

Schupbach, P, Osterwalder, V and Guggenheim, B (1996) Human root caries: microbiota of a limited number of root caries lesions. *Caries Research* **30**:52–64.

Schupbach, P, Oppenheim, FG, Lendenmann, U, Lamkin, MS, Yao, Y and Guggenheim, B (2001) Electron-microscopic demonstration of proline-rich proteins, statherin, and histatins in acquired enamel pellicles *in vitro*. *European Journal of Oral Science* **109**:60–68.

Shlossman, M, Knowler, W, Pettitt, DJ, and Genco, RJ (1990) Type-2 diabetes mellitus and periodontal disease. *Journal of the American Dental Association* **121**:535–536.

Slots, J (1977) Microflora in the healthy gingival sulcus in man. *Scandinavian Journal of Dental Research* **85**:247–254.

Smith, DJ, Anderson, JM, King, WF, van Houte, J and Taubman, MA (1993) Oral streptococcal colonization of infants. *Oral Microbiology and Immunology* **8**:1–4.

Socransky, SS, Gibbons, RJ, Dale, AC, Bortnik, L, Rosenthal, E and MacDonald, JB, (1963) The microbiota of the gingival crevice in man. 1 Total microscopic and viable counts and counts of specific organisms. *Archives of Oral Biology* **8**:275–280.

Spratt, DA, Weightman, AJ and Wade, WG (1999) *Eubacterium* species in periodontitis-identification of novel phylotypes representing uncultivated taxa. *Oral Microbiology and Immunology* **14**:56–59.

Stashenko, KP and Hillman, JD (1989) Microflora of plaque in rats following infection with and LDH deficient mutant of *Streptococcus rattus*. *Caries Research* **23**:375–377.

Steele, JG, Walls, AW, Ayatollahi, SM and Murray, JJ (1996) Major clinical findings from a dental survey of elderly people in three different English communities. *British Dental Journal* **180**:17–23.

Sundqvist, G (1992) Associations between microbial species in dental root canal infections. *Oral Microbiology and Immunology* **7**:257–262.

Tanner, A, Maiden, MF, Paster, BJ and Dewhirst, FE (1994) The impact of 16S ribosomal RNA-based phylogeny on the taxonomy of oral bacteria. *Periodontology 2000* **5**:26–51.

Theilade, E (1986) The non-specific theory in microbial etiology of inflammatory periodontal diseases. *Journal of Clinical Periodontology* **13**:905–911.

Theilade, E and Theilade, J (1976) Role of plaque in the etiology of periodontal disease and caries. *Oral Science Reviews* **9**:23–63.

van Houte, J (1994) Role of micro-organisms in caries etiology. *Journal of Dental Research* **73**:672–681.

Warnakulasuriya, KA, Samaranayake, LP and Peiris, JS (1991) Angular chelitis in a group of Sri Lankan adults: a clinical and microbiological study. *Oral Pathology and Medicine* **20**:172–175.

Watts, A and Paterson, C (1990) Detection of bacteria in histological sections of the dental pulp. *International Endodontic Journal* **23**:1–12.

Wojicicki, CJ, Harper, DS and Robinson, PJ (1986) Differences in periodontal disease associated microorganisms of gingival plaque in prepubertal, pubertal and postpubertal children. *Journal of Periodontology* **58**:219–223.

4.2 Detection of Microorganisms in Dental Plaque

DAVID DYMOCK
Department of Oral and Dental Science, Dental School, Lower Maudlin St., Bristol, UK

INTRODUCTION

This chapter summarizes the advances that have been made in understanding the diversity and complexity of microorganisms found in dental plaque. The emphasis is on determining which microorganisms are present as the association between various microorganisms becomes clearer with the development of new and existing technologies. Understanding the function of dental plaque in health and disease has occurred due to technological advances in a myriad of different fields from microscopy to molecular biology. However, in exploring the subject of 'detection', this chapter will focus on the techniques that have led to an understanding of microbial diversity and complexity with most potential impact for the clinician. Established culture techniques and their application, and more recent advances in molecular identification and detection of oral microorganisms, will be discussed. Intimate associations between bacteria will be considered briefly, although the reader is referred to a number of excellent reviews for further, more detailed, information on this subject matter (Whittaker *et al.* 1996; Kolenbrander 2000).

EARLY INDICATIONS OF BACTERIAL DIVERSITY IN DENTAL PLAQUE

Much of what is described in this chapter can be linked back through the centuries to the phenomenal observations of Antonie van Leeuwenhoek in 1683. The oral bacteria he observed using primitive microscopes illustrate a

Medical Biofilms. Edited by J. Jass, S. Surman, J. Walker

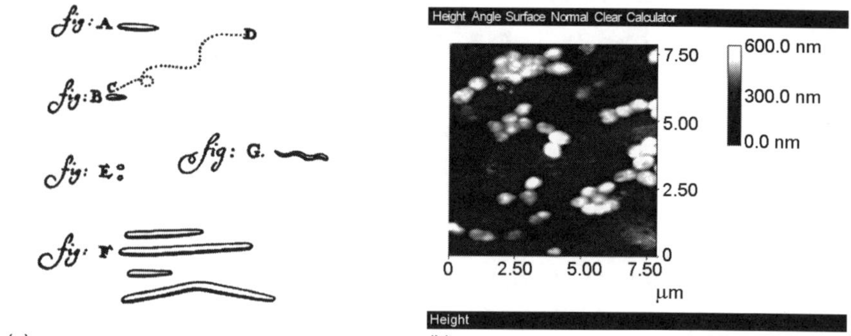

(a) (b)

Figure 4.2.1. (a) Van Leeuwenhoek's drawings of 'animalcules' he observed in dental plaque using a primitive microscope in 1683. (b) AFM image of *Streptococcus gordonii* cells adhering to saliva-conditioned human enamel (image kindly provided by ME Barbour and KD Jandt, Department of Oral and Dental Science, University of Bristol, UK).

fundamental concept of dental plaque, in that it contains a highly heterogeneous population of cell types. However, van Leeuwenhoek's observations (as shown in Figure 4.2.1) do not only suggest the diversity of organisms in plaque, they also suggest that he recognized the vital nature of these organisms by noticing that some organisms were motile and had numerous different properties. Indeed, van Leeuwenhoek recognized that these numerous microorganisms were indeed living organisms through the phraseology of his oft-quoted statement 'The number of these animalcules in the scurf of a man's teeth are so many that I believe they exceed the number of men in a kingdom'. Why should van Leeuwenhoek be interested in observations of dental plaque? Archaeological studies have indicated the prevalence of caries and periodontal disease in humans for thousands of years, and, moreover, there is evidence that mechanical methods were utilized by ancient cultures for cleaning of the tooth surface. This implies recognition of the link between dental plaque and oral disease for several millennia and explains why van Leeuwenhoek was stimulated to investigate plaque, particularly from older subjects with poor oral health. Since 1683 there have obviously been massive advances in microscopic observations of microflora in dental plaque. This chapter will not focus on visual observations of biofilms such as dental plaque, as excellent recent reviews on this subject are available (Surman *et al.* 1996; Beech *et al.* 2000). Possibly one of the most exciting innovations has been the use of the atomic force microscope (AFM) in examining interactions of individual cells with the pellicle-covered tooth surface (Figure 4.2.1). Like van Leewenhoek's observations of motility in oral microorganisms, in addition to the diversity of cell morphology the AFM allows greater understanding of the oral flora at the functional level rather than simply at the observational level.

MACROSCOPIC DETECTION OF DENTAL PLAQUE

All individuals grow dental plaque. However, there is massive variation in the rate of growth of plaque from one individual to another. This can be ascribed to a number of different variables, including constituents of saliva and differences in flow rate, diet, and microbial composition of plaque. Dyes such as erythrosin have been developed that enhance visualization of plaque on the tooth surface (Figure 4.2.2). Clinicians and hygienists use these dyes for a number of different purposes, for example to illustrate how to brush teeth correctly to a patient, or in trial studies to determine the effect of an oral hygiene product on dental plaque. Although macroscopic study of plaque has these particular uses, the 'detection' of dental plaque as described in this chapter has more bearing on ascertaining the microbiological content. Only by studying dental plaque at the level of understanding of which organisms are present, what properties they have and how they interact with one another and with the host is it possible to begin to understand how disease may develop in the oral cavity. A further complication to these studies is that dental plaque need not always lead to disease initiation and progression. Indeed, dental plaque plays a role in preventing colonization by exogenous, potentially dangerous, microorganisms. Thus, many individuals have perfect oral health with no indications of disease, but still grow dental plaque at varying rates from one individual to another. To understand why some individuals have disease, it has been important to isolate organisms from dental plaque and study their

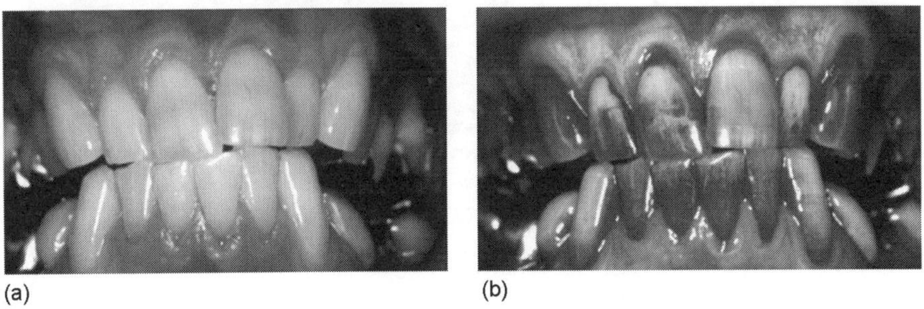

Figure 4.2.2. Dental plaque is not readily visible without the use of a disclosing dye. (a) A patient before disclosure and (b) the extent of plaque on tooth surfaces following a rinse with erythrosin disclosing solution.

properties under controlled conditions. Although van Leeuwenhoek could 'recognize' organisms by study of cellular morphology, this gave him no idea of the possible role of these organisms in disease. Indeed, recognition of this sort is limited, as there are only a finite number of cellular morphologies that can be accurately recognized. Clearly, as we now know, bacteria with vastly different properties can have similar microscopic morphologies. The next major breakthrough in the oral microbiology field was, therefore, the isolation and *in vitro* culture of oral microorganisms.

CULTURE OF ORAL MICROORGANISMS

Willoughby Dayton Miller (1890) described early microbial cultures from dental plaque. He describes a means to culture organisms from dental plaque obtained with a gelatine-based media and discovered that organisms could be subcultured to purity (Figure 4.2.3). A fascinating account of his understanding of the diverse flora in the human mouth is recommended to the interested reader (Miller, 1890). Based on the principles first proposed in the 19th century, commercial media are now used by most laboratories. Modern media are capable of growing a wide variety of organisms, but it is still estimated that approximately 50% of all oral flora may not yet have been cultured *in vitro* (Paster *et al.* 2001). This is

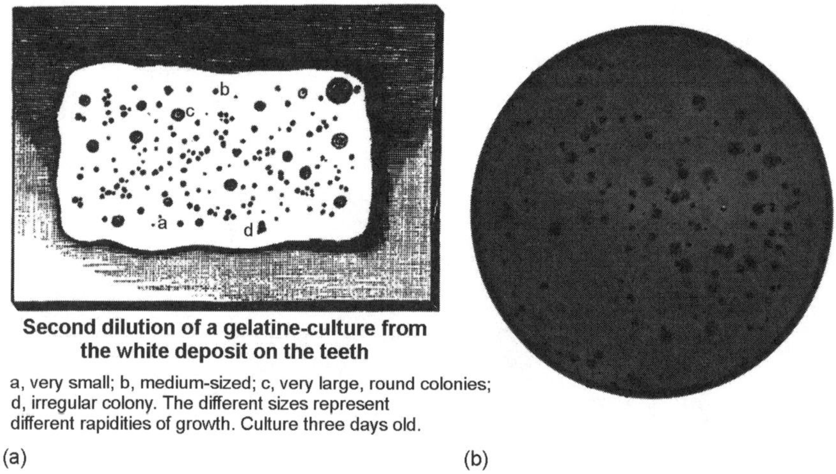

Figure 4.2.3. (a) Culture of oral microorganisms as illustrated by Miller (1890). (b) Modern primary culture of a subgingival plaque sample from a periodontal pocket. A diverse range of colony types is observed.

not to say that these organisms are unculturable, just that, as yet, the correct conditions have not been used for their culture. In developing culture media and conditions for growth of an organism the microbiologist is attempting to replicate the environmental conditions that are found at the site of growth of that particular organism. A variety of parameters can affect the ability of microorganisms to grow in the laboratory, such as nutritional dependency or atmospheric requirements. The replication of environmental conditions is difficult to produce because of the different physical localities of the mouth in which the bacteria proliferate, the interdependence of organisms upon one another in dental plaque, and the variations in individuals from whom samples are taken in saliva constituents and host defence components. Several culture media have been optimized for the isolation of groups of organisms, as in genus-specific media. A variety of these media are outlined in Table 4.2.1. These media have value in allowing relatively rapid isolation of selected groups of oral microorganisms and can be used to measure shifts in plaque microbial ecology. This is done by comparing numbers of organisms on selective media as a percentage of total, this latter figure being established from total numbers of microorganisms on a rich non-selective media such as Fastidious anaerobic agar or a Blood agar. This approach has been used in large-scale clinical trials where changes in the microbial ecology of dental plaque caused by oral hygiene products are monitored (Walker *et al.* 1994; Zambon *et al.* 1995). Advances in cultural techniques have played an important role in isolating bacteria for further study, for assessing shifts in bacterial populations in response to changing environmental conditions, and for the analysis of antibiotic resistance in oral bacteria. However, present-day culture does have many disadvantages. Although groups of bacteria may be isolated on different media, isolated bacteria often require further analysis for species-level identification, including biochemical and further phenotypic tests. Differentiation of strains within a species is even more problematical. Furthermore, bacteria that have been isolated and subcultured *in vitro* frequently change characteristics. This means that their potential for virulence may be miscalculated if conclusions are drawn from studies carried out in the laboratory. Comparisons of cultural versus other forms of microbial detection suggest that some strains of species considered to be relatively easy to culture could not, in fact, be grown in the laboratory.

MOLECULAR DETECTION AND ENUMERATION OF MICROORGANISMS

Despite the limitations of culture techniques, wide diverse ranges of oral bacteria have been isolated and are stored in culture collections around the

Table 4.2.1. Media for growth of oral bacteria

Target organism	Medium	Incubation conditions	Reference
Anaerobes	Fastidious anaerobe agar	Anaerobic, 37°C, 5–7 days	
Aerobes	Blood agar	Aerobic, 37°C, 2–3 days	
Actinobacillus actinomycetemcomitans	TSBV agar	10% CO_2, 37°C, 5–7 days	Slots *et al.* 1978
A. actinomycetemcomitans	Dentaid-1	10% CO_2, 37°C, 5–7 days	Alsina *et al.* 2001
Actinomyces	MMBA	Anaerobic, 37°C, 5–7 days	Lewis *et al.* 1995
Actinomyces	CFAT	Anaerobic, 37°C, 5–7 days	Zyther and Jordan 1982
Campylobacter rectus	*Wolinella* agar	Anaerobic, 37°C, 5–7 days	Hammond and Mallone 1988
Capnocytophaga	TPPB	Anaerobic, 37°C, 5–7 days	Mashimo *et al.* 1983
Eikenella corrodens	Clindamycin agar	Anaerobic, 37°C, 5–7 days	Walker *et al.* 1978
Enterics	McConkey's agar	Aerobic, 37°C, 2–3 days	MacFadden 1985a
Fusobacterium nucleatum	CVE agar	Anaerobic, 37°C, 5–7 days	Walker and Socransky 1979
Lactobacillus	Rogosa LS agar	Aerobic, 37°C, 5–7 days	Rogosa *et al.* 1951
Neisseria	*Neisseria* agar	Aerobic, 37°C, 5–7 days	Ritz 1967
Peptostreptococcus micros	PMM	Anaerobic, 37°C, 5–7 days	Turng *et al.* 1996
Streptococcus and *S. mutans*	Mitis–salivarius agar+tellurite	Anaerobic, 37°C, 5–7 days	Gold *et al.* 1973
Staphylococcus aureus	Mannitol salt agar	Aerobic, 37°C, 2–3 days	MacFadden 1985b
Veillonella	*Veillonella* agar	Anaerobic, 37°C, 5–7 days	Rogosa *et al.* 1958
Yeasts	Sabouraud dextrose agar	Aerobic, 37°C, 2–3 days	MacFadden 1985c

world. A novel approach to detecting these microorganisms was first developed by Socransky *et al.* (1994) at the Forsyth Dental Centre in Boston, USA. This technique, known as checkerboard hybridization, involves the extraction and labelling of total genomic DNA from a culturable microorganism for use as a probe in hybridization experiments with DNA extracted from plaque samples. By making probes for a number of oral microorganisms, and carefully optimizing the amount of each probe used in checkerboard experiments so that each probe gives an equal signal for the same number of bacteria, it is possible to obtain quantitative data for the presence of up to 40 oral species in 40 plaque samples. Thus, numerical data can be obtained for 1600 hybridizations from a single experimental membrane. Quantitation is obtained by measuring the strength of the signal obtained on the autoradiograph by computer-aided densitometry and comparing results with standards. The use of complex statistical analyses on the vast quantities of data generated by this technique allows confidence in the trends that have been observed. Checkerboard hybridization requires careful optimization, as the diverse range of bacteria found in the mouth has varying genome sizes and nucleotide content (GC ratios). A simple theoretical example of a checkerboard hybridization experiment is shown in Figure 4.2.4.

Papapanou *et al.* (1997b) assessed the relative merits of the checkerboard technique, as opposed to non-selective culture. They concluded that, for half of the ten species tested, the checkerboard technique was more

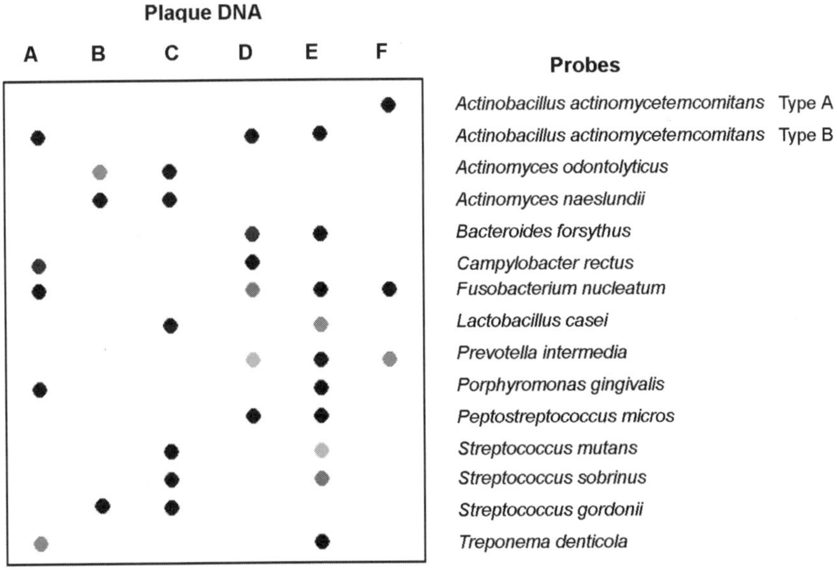

Figure 4.2.4. Graphical illustration of a checkerboard hybridization experiment. Six theoretical plaque samples are analysed in this experiment with extracted DNA fixed to the membrane in the vertical orientation. Labelled genomic DNA probes for 15 bacteria are applied to the membrane in the horizontal orientation. The strength of the signal is proportional to the number of cells of the microorganism tested in the plaque sample.

Table 4.2.2. Checkerboard hybridization in studies analysing the microbiota of periodontal disease. The size of each study is indicated to illustrate the vast amount of information that can be gained from this technique.

Study	Study size	Reference
Disease aetiology		
Subgingival microbiota in adult Chinese: prevalence and relation to periodontal disease progression	148 subjects, total of 1864 plaque samples, 18 bacterial taxa	Papapanou *et al.* 1997a
Microbiota of health, gingivitis and initial periodontitis	17 subjects (34 samples), 13 bacterial taxa, sampled once. Culture data also obtained	Tanner *et al.* 1998
Subgingival microbiota in healthy, well-maintained elder and periodontitis subjects	203 subjects, total of 5003 plaque samples, 40 bacterial taxa, sampled once	Haffajee *et al.* 1998
Microbial complexes in subgingival plaque	185 subjects, total of 13 261 plaque samples, 40 bacterial taxa	Socransky *et al.* 1998
Comparison of the microbiota of supra- and subgingival plaque in health and periodontitis	45 subjects, 2358 samples, 40 bacterial taxa	Ximenez-Fyvie *et al.* 2000a
Microbial composition of supra- and subgingival plaque in subjects with adult periodontitis	23 subjects, 1170 samples, 40 bacterial taxa	Ximenez-Fyvie *et al.* 2000b
Impact of surgical procedures		
Microbiota of successful osseointegrated dental implants	43 subjects, 220 samples, 23 bacterial taxa	Lee *et al.* 1999
The short-term effect of apically repositioned flap surgery on the composition of the subgingival microbiota	11 subjects, subgingival samples from each tooth, three visits, 29 subgingival taxa	Levy *et al.* 1999

Effect of antibiotic administration		
Systemic doxycycline administration in the treatment of periodontal infections (I). Effect on the subgingival microbiota	20 subjects, samples from up to 28 teeth at t_0 and 90 days, also samples from random two teeth at 3, 7 and 14 days, 40 bacterial taxa	Feres *et al.* 1999a
Systemic doxycycline administration in the treatment of periodontal infections (II). Effect on antibiotic resistance of subgingival species	20 subjects, t_0 saliva and six plaque samples, two more samples for three time points. Grow and enumerate bacteria on media containing doxycycline, remove from plates, checkerboard 40 bacterial taxa	Feres *et al.* 1999b
Change in subgingival microbial profiles in adult periodontitis subjects receiving either systemically-administered amoxicillin or metronidazole	17 subjects, samples from each tooth at baseline, 90, 180 and 360 days, and random posterior teeth at 3, 7 and 14 days post-treatment, 40 subgingival species	Feres *et al.* 2001
Effect of mechanical treatment regimes		
The effect of scaling and root planning on the clinical and microbiological parameters of periodontal diseases: 12-month results	32 subjects, 5 visits, 40 bacterial taxa	Cugini *et al.* 2000
The effect of repeated professional supragingival plaque removal on the composition of the supra- and subgingival microbiota	18 subjects, 1804 supragingival and 1804 subgingival plaque samples at four time points 3 months apart, 40 bacterial taxa	Ximenez-Fyvie *et al.* 2000c
Genetic factors		
Microbiological parameters associated with IL-1 gene polymorphisms in periodontitis patients	108 subjects, 2736 samples, 40 bacterial taxa	Socransky *et al.* 2000

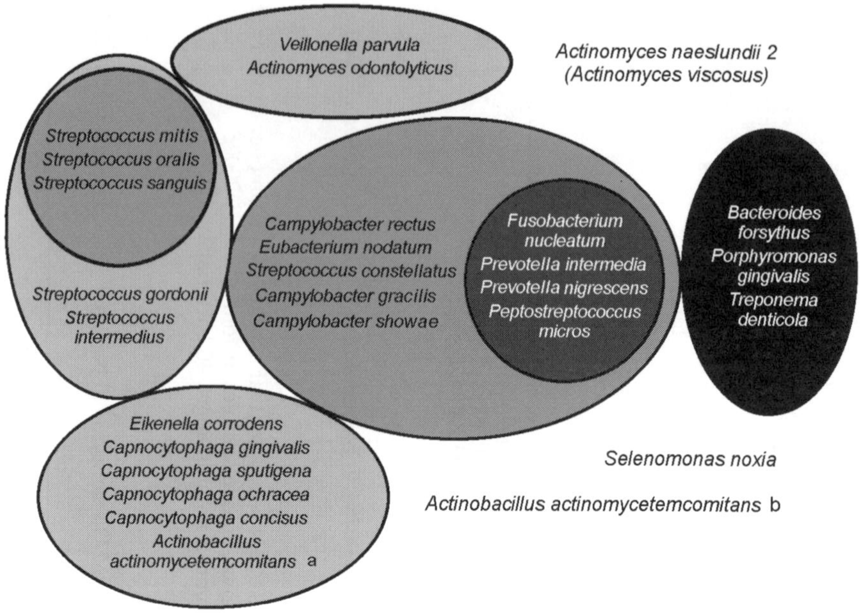

Figure 4.2.5. Groupings of bacteria associated with periodontal disease on the basis of checkerboard analysis. Adapted from Socransky *et al.* (1998). Published in the Journal of Clinical Periodontology.

sensitive, i.e. higher figures were obtained for the molecular study than those obtained from culture for those species. The technique has been optimized for use by several research groups in the world and has provided the opportunity to carry out microbiological analyses of high numbers of plaque samples, and it has proved particularly important in studying the particularly complex and diverse flora associated with periodontal disease. Table 4.2.2 lists a number of these studies subdivided into the areas of research to which the technique has been applied. Amongst the most important findings has been the complex association of different species of microorganisms in periodontal disease (Socransky *et al.* 1998). Five major complexes were found, and these are shown in Figure 4.2.5. The 'black' complex consisting of *Bacteroides forsythus*, *Porphyromonas gingivalis* and *Treponema denticola* appears to relate closely to clinical measures of periodontal disease, such as bleeding on probing and increasing pocket depth. Much discussion centres on whether these complexes result in a pathogenic flora that initiates the disease or whether they reflect a successful group of secondary invaders of the tissue. *P. gingivalis* is the best studied of these organisms. The role of the organism and its pathogenic potential are well established, and the interested reader should refer to a number of excellent recent reviews (Holt *et al.* 1999; Kadowaki *et al.* 2000; Lamont and Jenkinson 2000; Potempa *et al.* 2000). The genomes of both *P. gingivalis* and *T. denticola* have recently been sequenced, and this will, inevitably, lead to greater understanding of the roles of these organisms in disease. The second major group is the complex consisting of core microorganisms *Prevotella intermedia* and *Prevotella nigrescens*, *P. micros* and *F. nucleatum*. A number of other species may be associated with this complex, including *Campylobacter* spp. (Figure 4.2.5). The role of these organisms in disease initiation and/or progression is much less defined than that of the 'black' complex (Figure 4.2.5). The significance of other complexes is also not yet understood.

CHECKERBOARD ANALYSES OF PERIODONTAL TREATMENT REGIMES

Interestingly, commonly used mechanical treatment regimes for periodontal disease appeared to have most effect on the 'black' complex organisms (Figure 4.2.5). For example, repeated professional plaque removal resulted in a return to plaque composition normally associated with periodontal health when 40 bacterial taxa were monitored in 18 adult subjects (Ximenez-Fyvie *et al.* 2000c; Table 4.2.2) over a period of 1 year, and 57 subjects for 6 months with further data for 32 subjects for up to 1 year (Cugini *et al.* 2000). Clinical parameters, such as pocket depth and attachment level, also improved during these studies, coincident with the recurrent removal of subgingival plaque and reduction in pre-eminence of 'black' complex organisms. Checkerboard hybridization has also allowed studies of effects of antibiotic treatment regimes in periodontal therapies. Surprisingly, systemic administration of doxycycline appeared not to reduce proportions of *B. forsythus*, *P. gingivalis* or *T. denticola* in subgingival plaque (Feres *et al.* 1999a). However, in a further study (Feres *et al.* 2001) with systemically administered amoxicillin or metronidazole, 'black' complex organisms reduced in number and proportion in subgingival plaque samples even 1 year after treatment. *Campylobacter* species, *Eubacterium nodatum*, *F. nucleatum* subspecies, *Fusobacterium periodonticum* and *P. nigrescens* all reduced through the 14 day period of antibiotic treatment but gradually increased through the following year. These studies are of enormous size in terms of the number of samples analysed and the variety of bacterial taxa enumerated using the checkerboard hybridization technique.

There is little doubt that advances in the use of techniques such as checkerboard hybridization have made tremendous leaps in our understanding of microbial ecology of dental plaque, in particular subgingival plaque from periodontal pockets. The technique can also be applied to provide information on the effectiveness of antibiotics or other treatment regimes in eradicating the pathogens. However, the checkerboard hybridization is not without its drawbacks. Firstly, it is important that one has access to genomic DNA from a desired organism in order to make a labelled probe. Many oral microorganisms have not yet been cultured, and obtaining genomic DNA is difficult without a pure isolate in the laboratory. Secondly, the signal obtained from the experiment is likely to come from hybridization of the probe to DNA extracted from both dead and living cells. This could potentially provide problems in assessing data on the effectiveness of antibiotic treatment regimes, particularly when samples are obtained during administration of the antibiotics. Thirdly, this is a technique that requires a high degree of laboratory expertise, and experiments require several days before results are known. It is unlikely, therefore, to be of value in the clinical setting, where the practitioner may require knowledge of the microbial content of the plaque from a patient before deciding on the best treatment plan. The clinician would preferably require this information at the chair-side. The best prospect for identification and detection of specific microorganisms in dental plaque is likely to come through advances in polymerase chain reaction (PCR) methodology.

PCR AND UNDERSTANDING OF PLAQUE ECOLOGY

The PCR involves the specific amplification of DNA dependent on the sequence of the oligonucleotide primers in the reaction mix. Following heat denaturation of DNA strands (e.g. bacterial genomic DNA from a plaque sample), lowered temperature of the reaction allows primer annealing to complementary sequences on template strands. The temperature is raised to allow the catalytic action of a temperature-stable DNA polymerase, which results in the synthesis of a new strand of DNA. Each cycle of denaturation, primer annealing and DNA synthesis results in a doubling of the number of newly synthesized molecules of DNA, and this amplification will, after sufficient cycles, result in a DNA product that can be detected, usually by agarose gel electrophoresis. Environmental microbiologists carried out the first investigations into microbial ecology using PCR (Weller *et al.* 1991). The strategy taken was to target a molecule that is present in all living cells and to compare the sequences of DNA of these molecules with one another to achieve a phylogeny for unknown or unculturable organisms. By far the most widely used molecule for these studies is the 16S rRNA molecule that is absolutely required for protein synthesis in all living organisms (the larger 18S rRNA is the equivalent molecule in the Eukarya). The 16S rRNA molecule must have a specified conformation in order to be biologically active. Most mutations in the sequence will result in a catastrophic alteration in conformation and result in a non-viable mutant. Thus, some regions of the molecule are absolutely conserved. These regions have been particularly useful for the design of oligonucleotide primers from all microorganisms, and for aligning sequences. However, other regions of the molecule are less functionally active and there is, therefore, opportunity for mutation. In comparing 16S rRNA molecules from different microorganisms it is clear that they share regions of identity where function is preserved, but these regions are interspersed by nucleotide sequences that differ between organisms. In fact, these regions reflect evolutionary significance, since closely related organisms have 16S rRNA molecules that are similar, whereas distantly related bacteria have 16S rRNA molecules that, although identical in the functional conserved regions, are very different in other regions and, therefore, are quite different overall. These variable regions, reflecting evolution, can therefore be used to design much more specific oligonucleotide primers that will only amplify from a chosen species or even strain within a species (Stackebrandt *et al.* 1992).

When analysing a diverse and complex microbial population, such as found in dental plaque, the high level of specificity that can be obtained through oligonucleotide primer design for PCR can be a distinct advantage. The specificity of the primers does, in fact, allow for the proverbial needle in the haystack to be found. Thus, by using oligonucleotide primers designed from variable regions for known sequences of oral bacteria, it is possible to detect their presence in plaque. The sensitivity is such that as few as 100 cells in a sample can be detected, the technique is very quick and can be even more specific than species level. The first major study using PCR oligonucleotide primers directed to 16S rRNA molecules for detection of selected bacteria in plaque was done by Ashimoto *et al.* (1996). In that study, primers were designed for eight suspected periodontal pathogens. Detection was found to be more sensitive than culture for most of the species tested. A number of different primer sets directed to 16S rRNA of other oral bacteria have now been designed and validated (Table 4.2.3). Not all PCR detection of oral bacteria is directed to 16S rRNA. Some detection systems are targeted to virulence factors of important pathogens such as *A. actinomycetemcomitans* and *P. gingivalis* (Table 4.2.3). Commercial laboratories now provide services for clinicians using the PCR technique to detect a panel of selected microorganisms. Results are generally qualitative, rather than quantitative, but more sophisticated approaches are improving bacterial enumeration in plaque samples using PCR. The day when clinicians are able to analyse subgingival plaque samples for microbial content in PCR-based assays in computerized equipment is not too far away. Thus, in the future the clinician can expect to be able to choose a

Table 4.2.3. PCR primers used for detection of oral bacteria in clinical samples

Species	Target	Forward primer sequence (5′–3′)
A. actinomycetemcomitans	16S rRNA	AAACCCATCTCTGAGTTCTTCTTC
A. actinomycetemcomitans	leucotoxin	CAGCAAGCTGCACAGTTTGCAAA
A. actinomycetemcomitans	16S rRNA	ATTGGGGTTTAGCCCTGGTG
B. forsythus	16S rRNA	GCGTATGTAACCTGCCCGCA
B. forsythus	16S rRNA	TACAGGGGAATAAAATGA
C. rectus	16S rRNA	TTTCGGAGCGTAAAACTCCTTTTC
E. corrodens	16S rRNA	CTAATACCGCATACGTCCTAAG
P. micros	16S rRNA	TCGAACGTGATTTTTGTGGA
P. gingivalis	16S rRNA	AGGCAGCTTGCCATACTGCG
P. gingivalis	fimbriae	ATAATGGAGAACAGCAGGAA
P. gingivalis	16S rRNA	TGTAGATGACTGATGGTGAAAACC
P. intermedia	16S rRNA	TTTGTTGGGGAGTAAAGCGGG
P. intermedia	16S rRNA	CAAAGATTCATCGGTGGA
P. nigrescens	16S rRNA	ATGAAACAAAGGTTTTCCGGTAAG
P. nigrescens	16S rRNA	CAAAGGTTTTCCGGTAAG
Streptococcus salivarius	Dextranase	AACGTTGACCTTACGCTAGC
T. denticola	16S rRNA	TAATACCGAATGTGCTCATTTACAT
T. denticola	*tdpA*	AAGGCGGTAGAGCCGCTCA

treatment regime based on the microbial population found within a plaque sample, and to assess the effect of the treatment using clinic-based equipment.

PCR detection of microorganisms is just a single aspect of the vast amount of sequence information available for 16S rRNA molecules. Tens of thousands of these sequences can be accessed in computer databases, and unknown sequences may be rapidly compared with known, aligned sequences to obtain an 'identity' for the host cell from which the rRNA gene sequence was derived. 'Universal' oligonucleotide primers designed from conserved regions of 16S rRNA molecules have been used to amplify the gene from all bacteria in a sample. Amplified genes are separated in a cloning step, and individual cloned rRNA genes may be sequenced very rapidly. Comparisons with databases allow the diversity of bacteria within a sample to be assessed. One of the main advantages of this approach has been to overcome the difficulties of culturing only approximately 50% of the bacteria in dental plaque. Providing that the 16S rRNA gene of an unculturable microorganism can be amplified by PCR using 'universal' primers, then it is possible to establish where in the phylogenetic tree of life it should fit. This does not, however, provide information on potential virulence factors or possible antibiotic resistances of this organism. A recent comprehensive study using PCR and sequence analysis 16S rRNA sequences from bacteria in subgingival plaque suggests that approximately

Table 4.2.3. Continued

Reverse primer sequence (5′–3′)	Size	Reference
ATGCCAACTTGACGTTAAAT	557	Ashimoto *et al.* 1996
CATTAGTTAATGCCGGGCCGTCT	238	Watanabe and Frommel 1996
ACGTCATCCCCACCTTCCTC	360	Tran and Rudney 1999
TGCTTCAGTGTCAGTTATACCT	641	Ashimoto *et al.* 1996
ACGTCATCCCCACCTTCCTC	745	Tran and Rudney 1999
TTTCTGCAAGCAGACACTTTT	598	Ashimoto *et al.* 1996
CTACTAAGCAATCAAGTTGCCC	688	Ashimoto *et al.* 1996
TCCAGAGTTCCCACCTCT	1074	Riggio *et al.* 2001
ACTGTTAGCAACTACCGATGT	404	Ashimoto *et al.* 1996
TCTTGCCAACCAGTTCCATTGC	131	Watanabe and Frommel 1996
ACGTCATCCCCACCTTCCTC	197	Tran and Rudney 1999
TCAACATCTCTGTATCCTGCGT	575	Ashimoto *et al.* 1996
GCCGGTCCTTATTCGAAG	296	Stubbs *et al.* 1999
CCCACGTCTCTGTGGGCTGCGA	804	Ashimoto *et al.* 1996
GCCGGTCCTTATTCATGA	291	Stubbs *et al.* 1999
GATTCTGTCAAAGAAGCCAC-3′	2271	Igarashi *et al.* 2001
TCAAAGAAGCATTCCCTCTTCTTCTTA	316	Ashimoto *et al.* 1996
AGCCGCTGTCGAAAAGCCCA	311	Watanabe and Frommel 1996

415 species are likely to be present (Paster *et al.* 2001). That study compared subgingival plaque samples from healthy subjects with those from patients with periodontal diseases such as refractory periodontitis, adult periodontitis, human immunodeficiency periodontitis and acute necrotizing ulcerative gingivitis. Some species or phylotypes were identified from only patients with disease, and, conversely, some were found only in healthy subjects and never associated with disease. The former group clearly requires further attention as potential pathogens. In total, nine bacterial phyla were identified, including six phyla for which oral species have been cultured: the *Spirochaetes, Fusobacteria, Actinobacteria, Firmicutes, Proteobacteria* and *Bacteroidetes*. The other three phyla include the *Deferribacteres*, for which uncultured clones from oral samples had already been characterized, and phyla known as Obsidian pool OP11 and TM7. Neither of these latter two have, as yet, any cultured representatives. The Paster *et al.* (2001) study was the first occasion on which oral samples had yielded 16S rRNA genes that derived from these latter two phyla, suggesting that even larger studies in the future are likely to extend the diversity of classified oral microorganisms.

The study described above takes a shotgun approach to sequencing as many different 16S rRNA genes as can be amplified from selected plaque samples. Inevitably, a number of 'not-yet-cultured' species or phylotypes were identified. Earlier studies targeted these uncultured organisms in a

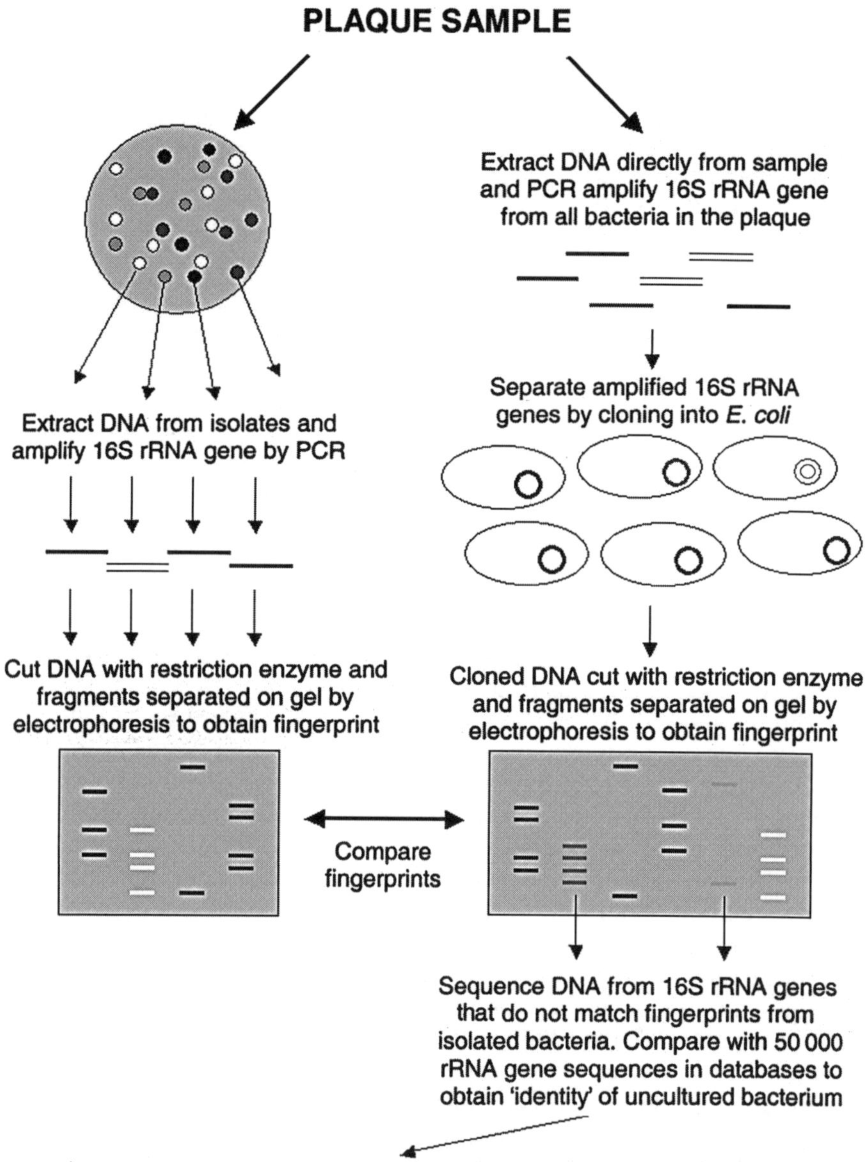


```
Species 1 ATTCAGCGCTGGGTGCTCCTGAAAGGCAGTTGAGTCTCGGTGGTTTATAGCCGCG
Species 2 ATTCAGCGCTGGGTGCTCCTGAAAGTCTGCTGAGTCTTCGTGGTTTATAGCCGCG
Species 3 ATTCAGCGCTGGGTGCTCCTGAAAGTCAGTTCAGTCTACGTGGCATATAGCCGCG
Unknown X ATTCAGCGCTGGGTGCTCCTGAAAGTCAGTTGAGTCTACGTGGTTTATAGCCGCG
                                    ::::::::::::::::::::::
Specific oligonucleotide for X      GTCAGTTGAGTCTACGTGGTTT
```

more direct way. Choi *et al.* (1994) used universal primers to amplify bacterial genes in subgingival plaque samples from a patient with severe periodontitis. A treponema-specific probe was used to identify clones derived from these notoriously culture-resistant organisms. One of the surprising conclusions of their study was that 23 different species or phyla of oral treponemes could be identified from a single subgingival plaque sample, suggesting a massive diversity of spirochaetes in this patient. Dymock *et al.* (1996) reported direct comparisons of cultured bacteria with not-yet-cultured bacteria by investigating the less-diverse flora associated with dentoalveolar abscesses. The approach taken is outlined in Figure 4.2.6 and could be applied to any sample that was to be investigated. In the Dymock *et al.* (1996) study, several clones were sequenced from bacteria that were not cultured, including some that were from the *Bacteroides* phylum. On the basis that some species in this phylum were difficult to culture, PCR primers specific for *Prevotella,* and for both *Prevotella* and *Bacteroides* genera, were designed. A simple validation experiment for specificity is shown in Figure 4.2.7 indicating the power of the approach. Unculturable phyla from the Dymock *et al.* (1996) study were later detected by PCR directed to 16S rRNA genes in subgingival plaque from patients with periodontal disease (Harper-Owen *et al.* 1999). Recently, these approaches have been applied to the detection of *T. denticola* in atherosclerotic lesions (Okuda *et al.* 2001). Thus, difficulties in culture may be overcome using a directed PCR approach based on 16S rRNA sequences for rapid detection of organisms from the oral cavity in other disease conditions. Advances in these studies will continue to improve our understanding of the role of oral microbes in human disease.

Figure 4.2.6. Identification of unculturable bacteria from a dental plaque sample. DNA is extracted from cultured organisms and from the original plaque sample and all samples are subject to PCR reactions using 'universal' primers directed to 16S rRNA genes. PCR products from cultured organisms are digested with a restriction enzyme to obtain a characteristic 'fingerprint' as indicated by the pattern of fragments obtained on an agarose gel. The PCR product obtained from the plaque sample is cloned into a vector and individual plasmid clones are 'fingerprinted' with the same restriction enzyme used for cultured microorganisms. Fingerprints that do not match those of cultured organisms are likely to be derived from 16S rRNA genes of bacteria that could not be cultured. In practice, most research laboratories would obtain a fingerprint from several restriction enzyme digests. Increasingly, as sequencing technologies improve and costs fall, the fingerprinting step is not carried out, but clones are subject to a single sequencing reaction that is then compared with 16S rRNA sequences deposited in databases. Oligonucleotide primers can then be designed that are specific for unculturable microorganisms, and these can then be detected by PCR in dental plaque samples.

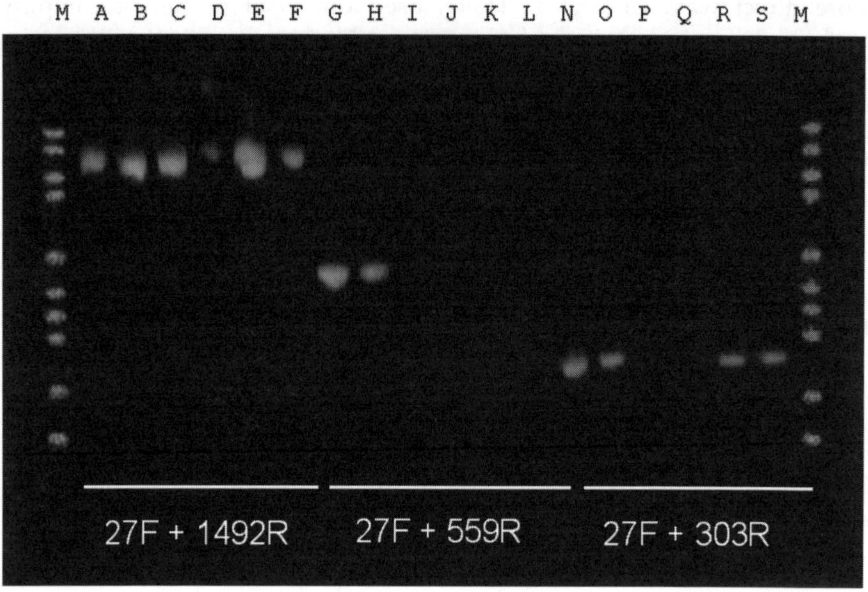

Figure 4.2.7. Validation of genus-specific PCR primers. M: molecular weight markers. The universal forward oligonucleotide primer, 27F, is used in all reactions. A control set of reactions is carried out with the universal reverse primer, 1492R (lanes A–F). A *Prevotella*-specific forward primer, 559R (lanes G–L) and *Prevotella*- and *Bacteroides*-specific forward primer 303R (lanes N–S) have been validated in PCR reactions with selected type strains. These are *Prevotella bivia* (lanes A, G, N), *Prevotella corporis* (lanes B, H, O), *Porphyromonas asaccharolytica* (lanes C, I, P), *Porphyromonas endodontalis* (lanes D, J, Q), *B. forsythus* (lanes E, K, R) and *Bacteroides fragilis* (lanes F, L, S).

CONCLUSIONS

Culturing organisms remains an important tool for the detection of bacteria from dental plaque. Cultured microorganisms are required for antibiotic resistance data and for elucidation of virulence mechanisms. However, cultured bacteria may rapidly alter their phenotypic characteristics *in vitro*, and 50% of oral microorganisms have not yet been cultured. Genus-specific media may be used to monitor shifts in microbial populations in dental plaque. The checkerboard hybridization technique is a useful tool to monitor populations in dental plaque, and has been important in understanding complexes of microorganisms associated with periodontal disease. Analyses of PCR-amplified 16S rRNA genes have been important for understanding microbial diversity, including microorganisms that are difficult to culture. It is estimated that approximately 400–500 species of bacteria are found in the oral cavity, the majority being in dental plaque. Rapid detection of selected microbes in dental plaque can be accomplished by PCR when the 16S rRNA gene sequence (or another specific target) is known. Developments in PCR technologies will lead to rapid, clinic-based, detection of oral pathogens in the near future, thus aiding the practitioner in the choice of treatment. Culture and molecular approaches should be considered as complementary.

REFERENCES

Alsina, M, Olle, E and Frias, J (2001) Improved, low-cost selective culture medium for *Actinobacillus actinomycetemcomitans*. *Journal of Clinical Microbiology* **39**:509–513.

Ashimoto, A, Chen, C, Bakker, I and Slots, J (1996) Polymerase chain reaction detection of 8 putative periodontal pathogens in subgingival plaque of gingivitis and advanced periodontitis lesions. *Oral Microbiology and Immunology* **11**:266–273.

Beech, IB, Tapper, RC and Gubner, RJ (2000) Microscopy methods for studying biofilms. In: Biofilms: Recent Advances in Their Study and Control (Ed. Evans, LV), Harwood Academic Publishers, Chur, Switzerland, pp. 51–70.

Choi, BK, Paster, BJ, Dewhirst, FE and Gobel, UB (1994) Diversity of cultivable and uncultivable oral spirochetes from a patient with severe destructive periodontitis. *Infection and Immunity* **62**:1889–1895.

Cugini, MA, Haffajee, AD, Smith, C, Kent, RL Jr and Socransky, SS (2000) The effect of scaling and root planning on the clinical and microbiological parameters of periodontal diseases: 12-month results. *Journal of Clinical Periodontology* **27**:30–36.

Dymock, D, Weightman, AJ, Scully, C and Wade, WG (1996) Molecular analysis of microflora associated with dentoalveolar abscesses. *Journal of Clinical Microbiology* **34**:537–542.

Feres, M, Haffajee, AD, Goncalves, C, Allard, KA, Som, S, Smith, C, Goodson, JM and Socransky, SS (1999a) Systemic doxycycline administration in the treatment of periodontal infections (I). Effect on the subgingival microbiota. *Journal of Clinical Periodontology* **26**:775–783.

Feres, M, Haffajee, AD, Goncalves, C, Allard, KA, Som, S, Smith, C, Goodson, JM and Socransky, SS (1999b) Systemic doxycycline administration in the treatment of periodontal infections (II). Effect on antibiotic resistance of subgingival species. *Journal of Clinical Periodontology* **2**:784–792.

Feres, M, Haffajee, AD, Allard, K, Som, S and Socransky, SS (2001) Change in subgingival microbial profiles in adult periodontitis subjects receiving either systemically-administered amoxicillin or metronidazole. *Journal of Clinical Periodontology* **28**:597–609.

Gold, OC, Jordon, HV and Van Houte, J (1973) A selective medium for *Streptococcus mutans*. *Archives of Oral Biology* **18**:1357–1361.

Haffajee, AD, Cugini, MA, Tanner, A, Pollack, RP, Smith, C, Kent, RL Jr and Socransky, SS (1998) Subgingival microbiota in healthy, well-maintained elder and periodontitis subjects. *Journal of Clinical Periodontology* **25**:346–353.

Hammond, BF and Mallone, D (1988) A selective/differential medium for *Wolinella recta*. *Journal of Dental Research* **67**:1712.

Harper-Owen, R, Dymock, D, Booth, V, Weightman, AJ and Wade, WG (1999) Detection of unculturable bacteria in periodontal health and disease by PCR. *Journal of Clinical Microbiology* **37**:1469–1473.

Holt, SC, Kesavalu, L, Walker, S and Genco, CA (1999) Virulence factors of *Porphyromonas gingivalis*. *Periodontology 2000* **20**:168–238.

Igarashi, T, Yano, Y, Yamamoto, A, Sasa, R and Goto, N (2001) Identification of *Streptococcus salivarius* by PCR and DNA probe. *Letters in Applied Microbiology* **32**:394–397.

Kadowaki, T, Nakayama, K, Okamoto, K, Abe, N, Baba, A, Shi, Y, Ratnayake, DB and Yamamoto, K (2000) *Porphyromonas gingivalis* proteinases as virulence determinants in progression of periodontal diseases. *Journal of Biochemistry (Tokyo)* **128**:153–159.

Kolenbrander, PE (2000) Oral microbial communities: biofilms, interactions, and genetic systems. *Annual Review of Microbiology* **54**:413–437.

Lamont, RJ and Jenkinson, HF (2000) Subgingival colonization by *Porphyromonas gingivalis*. *Oral Microbiology and Immunology* **15**:341–349.

Lee, KH, Maiden, MF, Tanner, AC and Weber, HP (1999) Microbiota of successful osseointegrated dental implants. *Journal of Periodontology* **70**:131–138.

Levy, RM, Giannobile, WV, Feres, M, Haffajee, AD, Smith, C and Socransky, SS (1999) The short-term effect of apically repositioned flap surgery on the composition of the subgingival microbiota. *International Journal of Periodontics and Restorative Dentistry* **19**:555–567.

Lewis, R, Mackenzie, D, Bagg, J and Dickie, A (1995) Experience with a novel selective medium for isolation of *Actinomyces* species from medical and dental specimens. *Journal of Clinical Microbiology* **33**:1613–1616.

MacFadden, JF (1985a) MacConkey (MC, MAC) agar. In: Media for Isolation–Cultivation–Identification–Maintenance of Medical Bacteria. Williams and Wilkins, Baltimore, MD, pp. 471–478.

MacFadden, JF (1985b) Mannitol salt agar (MSA). In: Media for Isolation–Cultivation–Identification–Maintenance of Medical Bacteria. Williams and Wilkins, Baltimore, MD, pp. 483–487.

MacFadden, JF (1985c) Sabouraud culture media. In: Media for Isolation–Cultivation–Identification–Maintenance of Medical Bacteria. Williams and Wilkins, Baltimore, MD, pp. 687–691.

Mashimo, PA, Yamamoto, Y, Nakamura, M and Slots, J (1983) A selective medium for the recovery of *Capnocytophaga* species from the oral cavity. *Journal of Clinical Microbiology* **17**:187–191.

Miller, WD (1890) The Micro-organisms of the Human Mouth. Graphische Anstalt Schuler AG, Biel, Switzerland.

Okuda, K, Ishihara, K, Nakagawa, T, Hirayama, A, Inayama, Y and Okuda, K (2001) Detection of *Treponema denticola* in atherosclerotic lesions. *Journal of Clinical Microbiology* **39**:1114–1117.

Papapanou, PN, Baelum, V, Luan, WM, Madianos, PN, Chen, X, Fejerskov, O and Dahlen, G (1997a) Subgingival microbiota in adult Chinese: prevalence and relation to periodontal disease progression. *Journal of Periodontology* **68**:651–666.

Papapanou, PN, Madianos, PN, Dahlen, G and Sandros, J (1997b) "Checkerboard" versus culture: a comparison between two methods for identification of subgingival microbiota. *European Journal of Oral Science* **105**:389–396.

Paster, BJ, Boches, SK, Galvin, JL, Ericson, RE, Lau, CN, Levanos, VA, Sahasrabudhe, A and Dewhirst, FE (2001) Bacterial diversity in human subgingival plaque. *Journal of Bacteriology* **183**:3770–3783.

Potempa, J, Banbula, A and Travis, J (2000) Role of bacterial proteinases in matrix destruction and modulation of host responses. *Periodontology 2000* **24**:153–192.

Riggio, MP, Lennon, A and Smith, A (2001) Detection of *Peptostreptococcus micros* DNA in clinical samples by PCR. *Journal of Medical Microbiology* **50**:249–254.

Ritz, HL (1967) Microbial population shifts in developing human plaque. *Archives of Oral Biology* **12**:1561–1568.

Rogosa, M, Mitchell, JA and Wiseman, RF (1951) A selective medium for the isolation and enumeration of oral and fecal lactobacilli. *Journal of Bacteriology* **62**:132–133.

Rogosa, M, Fitzgerald, RJ, MacKintosh, ME and Beaman, AJ (1958) Improved medium for selective isolation of *Veillonella*. *Journal of Bacteriology* **76**:455–456.

Slots, J, Moenbo, D, Langebaek, J and Frandsen, A (1978) Microbiota of gingivitis in man. *Scandinavian Journal of Dental Research* **86**:174–181.

Socransky, SS, Smith, C, Martin, L, Paster, BJ, Dewhirst, FE and Levin, AE (1994) "Checkerboard" DNA–DNA hybridization. *Biotechniques* **17**:788–792.

Socransky, SS, Haffajee, AD, Cugini, MA, Smith, C and Kent, RL Jr (1998) Microbial complexes in subgingival plaque. *Journal of Clinical Periodontology* **25**:134–144.

Socransky, SS, Haffajee, AD, Smith, C and Duff, GW (2000) Microbiological parameters associated with IL-1 gene polymorphisms in periodontitis patients. *Journal of Clinical Periodontology* **27**:810–818.

Stackebrandt, E, Liesack, W and Witt, D (1992). Ribosomal-RNA and rDNA sequence analyses. *Gene* **115**:255–260.

Stubbs, S, Park, SF, Bishop, PA and Lewis, MA (1999) Direct detection of *Prevotella intermedia* and *P. nigrescens* in suppurative oral infection by amplification of 16S rRNA gene. *Journal of Medical Microbiology* **48**:1017–1022.

Surman, SB, Walker, JT, Goddard, DT, Morton, LHG, Keevil, CW, Weaver, W, Skinner, A and Kurtz, J (1996) Comparison of microscope techniques for the examination of biofilms. *Journal of Microbiological Methods* **25**:57–70.

Tanner, A, Maiden, MF, Macuch, PJ, Murray, LL and Kent, RL Jr (1998) Microbiota of health, gingivitis, and initial periodontitis. *Journal of Clinical Periodontology* **25**:85–98.

Tran, SD and Rudney, JD (1999) Improved multiplex PCR using conserved and species-specific 16S rRNA gene primers for simultaneous detection of *Actinobacillus actinomycetemcomitans, Bacteroides forsythus*, and *Porphyromonas gingivalis*. *Journal of Clinical Microbiology* **37**:3504–3508.

Turng, BF, Guthmiller, JM, Minah, GE and Falkler, WA Jr (1996) Development and evaluation of a selective and differential medium for the primary isolation of *Peptostreptococcus micros*. *Oral Microbiology and Immunology* **11**:356–361.

Walker, C, Borden, LC, Zambon, JJ, Bonta, CY, DeVizio, W and Volpe, AR (1994) The effects of a 0.3% triclosan-containing dentifrice on the microbial composition of supragingival plaque. *Journal of Clinical Periodontology* **21**:334–341.

Walker, CB and Socransky, SS (1979) A selective medium for *Fusobacterium nucleatum* from human periodontal pockets. *Journal of Clinical Microbiology* **10**:844–849.

Walker, CB, Tanner, ACR, Smith, C and Socransky, SS (1978) Selective medium for *Eikenella corrodens*. *Journal of Dental Research* **57**:961.

Watanabe, K and Frommel, TO (1996) *Porphyromonas gingivalis, Actinobacillus actinomycetemcomitans* and *Treponema denticola* detection in oral plaque samples using the polymerase chain reaction. *Journal of Clinical Periodontology* **23**:212–219.

Weller, R, Weller, JW and Ward, DM (1991) 16S ribosomal-RNA sequences of uncultivated hot-spring cyanobacterial mat inhabitants retrieved as randomly primed cDNA. *Applied and Environmental Microbiology* **57**:1146–1151.

Whittaker, CJ, Klier, CM and Kolenbrander, PE (1996) Mechanisms of adhesion by oral bacteria. *Annual Review of Microbiology* **50**:513–552.

Ximenez-Fyvie, LA, Haffajee, AD and Socransky, SS (2000a) Comparison of the microbiota of supra- and subgingival plaque in health and periodontitis. *Journal of Clinical Periodontology* **27**:648–657.

Ximenez-Fyvie, LA, Haffajee, AD and Socransky, SS (2000b) Microbial composition of supra- and subgingival plaque in subjects with adult periodontitis. *Journal of Clinical Periodontology* **27**:722–732.

Ximenez-Fyvie, LA, Haffajee, AD, Som, S, Thompson, M, Torresyap, G and Socransky, SS (2000c) The effect of repeated professional supragingival plaque removal on the composition of the supra- and subgingival microbiota. *Journal of Clinical Periodontology* **27**:637–647.

Zambon, JJ, Reynolds, HS, Dunford, RG, DeVizio, W, Volpe, AR, Berta, R, Tempro, JP and Bonta, Y (1995) Microbial alterations in supragingival dental plaque in response to a triclosan-containing dentifrice. *Oral Microbiology and Immunology* **10**:247–255.

Zyther, L and Jordan, H (1982) Development of a partially selective medium for *Actinomyces viscosus* and *A. naeslundii*. *Journal of Clinical Microbiology* **15**:253–259.

4.3 Control of Dental Plaque

RACHEL SAMMONS
University of Birmingham, School of Dentistry, St Chad's Queensway, Birmingham, UK

WHY SHOULD WE CONTROL ORAL BIOFILMS?

There is experimental evidence of the benefit of controlling supragingival plaque in order to prevent the onset of gingivitis and the risk and progression of periodontal disease. Studies in humans have shown that, if plaque is allowed to accumulate without toothbrushing, signs of gingival inflammation and lesions develop within about 4 days (Egelberg 1964). If toothbrushing is resumed, then gingivitis quickly subsides (Löe *et al.* 1970), whereas if it is untreated, then gingivitis can proceed to periodontitis (Lindhe *et al.* 1975). In susceptible adults, loss of attachment of the periodontal ligament (a measure of periodontitis) can be halted almost completely by self-performed plaque control combined with professional cleaning to remove calculus three to six times a year (Axelsson *et al.* 1991).

The loss of teeth through periodontal disease is extremely distressing in itself, but until recently it was not thought to be a serious health risk. However, there is increasing evidence that periodontal disease and poor oral hygiene may be associated with a number of important systemic diseases, including atherosclerosis and coronary heart disease (Beck *et al.* 1996; de Stefano *et al.* 1993; Meyer and FivesTaylor 1998; Seymour and Steele 1998; Offenbacher *et al.* 1999). The reasons for this are not yet entirely clear, but transient bacteraemia may be at least a contributing factor. Patients with more plaque and gingivitis have more frequent and severe bacteraemias following dental manipulations than those with better oral hygiene (Beck *et al.* 1996). Therefore, the control of dental plaque, in order to avoid the onset of gingivitis and its progression to periodontal disease and to decrease the risk of severe bacteraemia, may be important for the maintenance of our general health.

Medical Biofilms. Edited by J. Jass, S. Surman, J. Walker

Studies by Löe *et al.* (1970) and Syed and Loesche (1978) have indicated that there is a threshold level of bacteria that is compatible with gingival health, and when this threshold is exceeded by two to three orders of magnitude then gingival inflammation is initiated (Gaffar *et al.* 1997). The aim of dental plaque control therefore, is to bring the numbers of bacteria down to a level sufficient to prevent the onset or progression of disease whilst, if possible, avoiding changes in the oral microbial ecology so that homeostasis is maintained.

POTENTIAL ROUTES TO THE CONTROL OF ORAL BIOFILMS

There are a number of potential routes by which oral biofilms can be controlled (modified from Cummins (1991)):

1. By reducing or preventing colonization of a surface.
2. By inhibiting the growth of microorganisms.
3. By disrupting or preventing the build up of the extracellular plaque matrix.
4. By modifying plaque biochemistry, in order to prevent formation of cytotoxic or cariogenic products.
5. By modifying plaque ecology to a less pathogenic flora.

The efficiency of a plaque control method or agent is measured by its effect on plaque, gingivitis, periodontitis or calculus formation. There are several standard methods for determining these effects. These include measurement of the thickness or extent of plaque or calculus, indications of inflammatory changes in the gingiva, such as bleeding from the sulcus or gingiva on probing, and loss of periodontal tissues, as indicated, for example, by pocket probing depth. The results are expressed as an index, plaque index, gingival index, bleeding index, etc. For a comprehensive description of how the different measurements are obtained, the reader is referred to Wilkins (1999a), Lang (1998) or Papapanou and Lindhe (1998).

MECHANICAL CONTROL OF SUPRAGINGIVAL PLAQUE

Mechanical removal of supragingival plaque, by brushing with a toothbrush, reduces the total number of colonizing bacteria on the tooth surface and disrupts the process of biofilm formation. It prevents or hinders the accumulation and maturation of plaque, and thus the establishment of anaerobic conditions, which may favour the growth of pathogenic organisms. Removal of supragingival plaque also prevents the inflammatory changes that would otherwise take place in the gingiva and favour the establishment of anaerobic species and the build up of a largely anaerobic bacterial population in the subgingival region (Listgarten 1999).

The Evolution of Toothbrushes

Oral hygiene products have a long history, going back over 6000 years (reviewed by Fischman (1997)). The first record of a mechanical aid for cleaning the teeth is probably in Chinese literature of about 1600 BC. This was a 'chewing stick', which was probably similar to the 'miswak' or 'siwak' (Figure 4.3.1), which is still recommended for cleaning teeth before prayers by some Moslems today in preference to a conventional toothbrush. The modern-style toothbrush evolved from this.

The first toothbrushes were probably developed by the Chinese in the late 15th century and were made of hog bristles set in ox bone. The modern-design toothbrush is believed to have been invented by a London tradesman, William Addis (1734–1808). The earliest mass-produced brushes were made of horsehair and wild boar bristles, which were imported to England from China until World War II prevented this. Fortunately, by then, nylon had been invented, and this was used for the manufacture of both bristles and toothbrush handles. Nylon is still universally used for bristles, whereas the handles of modern toothbrushes are made of thermoplastic materials such as cellulose acetate, styrene acrylonitrile or cellulose proprionate. Nickel silver (an alloy of copper and nickel) is used to anchor the bundles of bristles in the toothbrush head (Glass and Lare 1986).

Toothbrush Design and Methods of Brushing

There are many designs of the manual toothbrush on the market, and new ones are constantly being developed. Because of individual differences, an ideal brush does not exist (Iacono *et al.* 1998). According to the Consensus of the Proceedings of the European Workshop on Mechanical Plaque Control (Jepsen 1998), an ideal tooth brush should have the following attributes:

- Handle size appropriate to user age and dexterity.
- Head size appropriate to the size of the patient's mouth.
- Use of end-rounded nylon or polyester filaments not larger than 0.22 mm in diameter.
- Use of soft-bristle configurations, as defined by the acceptable international industry standards (ISO).
- Bristle patterns that enhance plaque removal in the interdental spaces and along the gum line.

Figure 4.3.1. Miswaks. The miswak is made from the fibrous root or twig of a bitter-tasting tree, especially the peelo (*Salvadora persica* or 'toothbrush tree'), olive or walnut. It is used without toothpaste and is said to '*Strengthen gums, effectively remove plaque and yellowness of teeth, prevent tooth decay, clean and brighten teeth and remove bad odour, improve the sense of taste, ease headache and tension, sharpen the memory and intelligence, assist the digestion, improve the eyesight, the health and the lustre of the face of a continual user*' (originally from Kitabut Tahaarah and Sunnats, quoted on commercial packaging). Photograph by R. Sammons. Lifesize.

Most people are not taught to brush their teeth correctly, and few do it effectively. A study by de la Rosa *et al.* (1979) on teenage boys brushing daily for 28 days showed that up to 60% of plaque was left on the teeth after brushing (Jepsen 1998). Several standard methods of brushing exist, and these are classified according to the type of motion that the brush performs, such as rolling, vibratory, circular, vertical or horizontal. These are all illustrated in Wilkins (1999b). The Bass technique (also called sulcular cleansing), in which the bristles are directed at the gum line (into the gingival sulcus) at an angle of 45° to the tooth, is widely accepted as the most effective method of removing plaque adjacent to and immediately beneath the gingival margin. Time and frequency of toothbrushing are also key factors in determining the efficiency of supragingival plaque removal (Cummins 1997). Most people brush for less than 1 min (Hawkins *et al.* 1986), although they may think that they do so for longer (Cancro and Fischman 1995). Recognizing that it is difficult to change people's behaviour, Oral B developed a 'CrossAction' toothbrush, which has tufts of bristles arranged at an angle of 16° to the brushhead in a criss-cross design, to 'maximize plaque removal regardless of how the user brushes' (Beals *et al.* 2000). Clearly, efficiency of brushing is also dependent upon patient skill and dexterity. For some patients, particularly the elderly, and patients with cerebral palsy, toothbrush handles may be individually adapted for an easier grip. Use of an electric toothbrush by the patient or a care worker may also be beneficial.

Interdental Cleaning Aids

The filaments of the toothbrush usually penetrate to some extent between the teeth, but even with good brushing technique it is impossible to clean the interdental areas thoroughly. Tools to aid interdental cleaning include woodsticks, small brushes and dental floss. Of these, dental floss is the most widely used aid and has the advantage that it can slide under the gingiva to remove subgingival plaque; it is also highly effective for cleaning crowns and bridges. However, when used incorrectly it can cause trauma to the gingival margin and it requires considerable manual dexterity to manipulate. Some types of floss incorporate a section of sponge and are helpful for cleaning irregularly shaped concavities between the teeth. These can also be used to deliver antiseptics, such as chlorhexidine, and fluoride into the interdental area. A small, round 'bottle-brush' called an interdental brush can be useful for cleaning under bridges and between implants. Single-tufted brushes may also be useful for some patients, particularly for reaching regions at the back of the mouth (Galgut 1991).

Electric Toothbrushes

Awareness of the benefits of good oral hygiene is also a very important factor in motivation for dental hygiene. The use of electric toothbrushes may improve plaque control for persons with poor motivation or dexterity. For example, they have been shown to improve the control of plaque in periodontal patients with poor compliance to oral hygiene (Hellstadious *et al.* 1993). Electric toothbrushes function by movement of the brushes or tufts of bristles in an oscillating lateral, rotatory or pulsating, in and out, motion. These motions may be combined in a single brush, with the rotatory design varying the angle of rotation. The frequency of oscillation varies from 40 Hz for battery-powered brushes to 250 Hz for magnetostrictive ones (Walmsley 1997). Several designs have interchangeable heads of different sizes and some have timers to aid thorough cleaning. The cleaning action of an electro-mechanical toothbrush is the result of mechanical forces (scrubbing action), as in manual brushing. However, it is at least theoretically much more efficient, as the cleaning action is produced by rapid oscillation of the bristles at up to 7000 times per minute (Cobb 1999). For example, in the Braun Oral-B Plak Control toothbrush, tufts of bristles oscillate at over 2800 rotations per minute with a circular angle of 70° (Iacono *et al.* 1998).

In a sonic and ultrasonic toothbrush the bristles oscillate even faster, producing up to 500 brush strokes per second. In addition, ultrasonic toothbrushes have a transducer embedded in the head, which creates low-intensity ultrasonic waves intended to break up and dislodge plaque. In both sonic and ultrasonic brushes, acoustic microstreaming forces produced by alternating pressure fields, created by the movement of the bristles and/or the sound waves, cause the rapid displacement of water around the vibrating bristles. These are accompanied by hydrodynamic shear stresses that are designed to be strong enough to dislodge plaque but not to disrupt biological cells and tissues (Walmsley 1997; Cobb 1999). Sonic and ultrasonic vibrations are intended to remove plaque beyond the bristle tips, in the interdental regions, and this does occur *in vitro*. For example, in one study, decreased bacterial viability and adhesion on a titanium surface was demonstrated at a distance of several millimetres beyond the end of the bristles (McInnes *et al.* 1992). Theoretically, bacteria in plaque may be damaged by shear stresses, possibly aided by abrasive particles in toothpaste, resulting in loss of fimbrae, which thus inhibits adhesion. The alternating pressures may also force oxygen into subgingival regions, thus changing the environment to more aerobic conditions (Cobb 1999). All these factors would seem to suggest that electric toothbrushes, in particular sonic or ultrasonic toothbrushes, should result in a marked decrease in plaque and a reduction in gingival and bleeding indices, compared with manual brushes. However, there are conflicting reports of their efficiency for all tooth surfaces (see reviews by Walmsley (1997) and Cobb (1999)). They are more effective than manual brushes in supervised trials and a small rotating brush is more effective than one with side-to-side motion (Walmsley 1997). A sonic brush was shown to be more effective in reducing the plaque index, whereas a manual one was more effective at reducing the bleeding and gingival indices (Johnson and McInnes 1994; Iacono *et al.* 1998). Cobb (1999) has suggested several reasons why sonic and ultrasonic electric toothbrushes do not perform clinically as well as might be expected: contact of the bristles with the tooth surface and soft tissue will dampen the acoustic transmission, and the volume of liquid in which the brush performs is very small. In the gingival sulcus the volume of crevicular fluid is only about 0.5 μl, and this may be insufficient for transmission of the long wavelength and amplitude of the sound waves.

Toothbrushes and Infection

Most dental practitioners recommend that patients change their toothbrushes every 3 months (Abraham *et al.* 1990), although there is no evidence to suggest that new toothbrushes remove plaque more efficiently than used ones (Daly *et al.* 1996). There is some concern about the potential of toothbrushes to harbour microorganisms and thus to be a potential source of infection. Svanberg (1978) first suggested that toothpaste and toothbrushes could become contaminated with *Streptococcus mutans*. Glass and Lare (1986) observed that patients with oral inflammatory disease tended to respond better to therapy if their toothbrushes were replaced regularly every 2 weeks. These authors and others have cultured organisms from used toothbrushes and detected a great variety of both oral and intestinal bacterial species, including *Staphylococcus, Enterobacter, Klebsiella, Clostridium, Serratia, Propionibacterium, Streptococcus* and *Candida*. *Herpes simplex* virus has also been shown to remain viable on a dried toothbrush for at least 48 h (Glass and Jensen 1988). Single brushes can harbour up to 10^6–10^8 organisms (Kozai *et al.* 1989; Verran and Leahy-Gilmartin 1996). Figure 4.3.2 shows a biofilm around bristles of a toothbrush that had been in use for about 5 months. Minor dental procedures, including toothbrushing, frequently cause bacteraemia (Beck *et al.* 1996; Roberts *et al.* 2000). In the case of oral disease and for patients at risk from bacteraemia, it would seem sensible to change toothbrushes frequently and to disinfect them with a proprietary toothbrush disinfectant after each use.

Care of Dentures

Dentures should be regularly disinfected to avoid build up of plaque and calculus. *Candida* (a frequent cause of denture stomatitis) can penetrate the

30kV X50
500µm 530000
(a)
30kV X5,000
5µm 530000
(b)

depths of the pores in the methylmethacrylate portion of a denture surface (Verran and Maryan 1997), and it is sometimes necessary to remake this part of the denture to eliminate persistent or recurrent symptoms of the stomatitis. A denture can be cleaned by brushing using a denture brush to remove plaque and disinfection by soaking the denture for 10–15 min in a mixture of 5% sodium hypochlorite and household detergent, or in a proprietary cleaner, in which hydrogen peroxide is the active ingredient. Exposure to microwave energy is also an effective disinfectant (Baysan *et al.* 1998). Dilute acids, such as 3% or 5% hydrochloric acid or acetic acid (white vinegar), may be used to remove calculus, although corrosion of metal parts of the denture may be a disadvantage (Wilkins 1999c).

Mechanical Control of Plaque

It is necessary to remove subgingival plaque and calculus in order to prevent the onset and progression of periodontal disease. This is done by scaling and root planning and is normally carried out by the dental surgeon. Surgical removal of the gingiva to expose the root surface may be required in cases of severe periodontal disease. Scaling is the mechanical removal of plaque, calculus and stains from the tooth crown or root surfaces in which there is no deliberate attempt to remove root substance. Scaling is commonly performed after routine dental surgery to remove small amounts of supragingival and subgingival calculus that normally accumulate even in healthy persons. Root planning is the process by which residual calculus and portions of cementum are removed from the roots to yield a smooth, hard and clean surface (Pattison and Pattison 1996). Root planning is performed in periodontal disease to remove diseased tissue, which may harbour microorganisms and be a reservoir for reinfection. In periodontal disease therapy, the task has to be performed periodically because pathogens such as *Actinobacillus actinomycetemcomitans* and *Porphyromonas gingivalis* may evade eviction by invasion of the dentine tubules or cementum of diseased teeth (Adriaens *et al.* 1988; Giuliana *et al.* 1997; Cobb 1999).

Traditional hand instruments used for scaling and root planing include small spoon-shaped blades with two curved cutting edges (curettes), pointed blades of triangular cross-section ('sickles') and a tool with one 45° angular cutting edge (hoe) (Rylander and Lindhe 1998). Scaling and root planing by manual instrumentation is effective but it is time consuming and sometimes difficult, particularly in root furcation areas. The task has been

---

Figure 4.3.2. (a) Scanning electron micrograph of a section of the head of a toothbrush, showing a biofilm around a tuft of bristles as they enter the head. (b) A magnified image of the biofilm in the area indicated by the arrow. The biofilm is porous and appears to consist predominantly of cocci. Micrographs by R. Sammons.

made much simpler and quicker since the introduction of piezoelectric, sonic and ultrasonic scalers. These all consist of a vibrating rod activated by compressed air that oscillates in an elliptical pattern or horizontal movement. The frequency of oscillation is in the range 25–30 kHz in the case of sonic and ultrasonic scalers to 40–42 kHz for piezoelectric scalers. These frequencies are sufficient potentially to cause acoustic cavitations and microstreaming with bactericidal effects. However, there is no evidence that cavitation actually occurs in the subgingival pocket, and experimental studies by O'Leary *et al.* (1997) suggested that, if a reduction in microorganisms does occur, it may be due to the associated rise in temperature (Cobb 1999). The tips of ultrasonic scalers can be modified according to function. For example, a flow-through design allows delivery of liquids containing antimicrobial agents, such as chlorhexidine (Cobb 1999), although such irrigation procedures during scaling and root planing have not been shown to produce significant clinical benefits (reviewed by Hastings-Drisko 1999). Efficiency of removal of subgingival plaque depends on the depth of the pocket, root anatomy, tooth position and ease of access. Neither manual nor sonic ultrasonic instrumentation alone, or in combination, will remove all plaque. However, it may only be necessary to remove sufficient plaque and calculus to achieve a shift in the composition of the microbial flora from predominantly Gram-negative organisms to predominantly Gram-positive, associated with gingival health (Listgarten 1976; Haffajee *et al.* 1998; Cobb 1999).

Lasers

The use of lasers in subgingival periodontal therapy is controversial, and Cobb (1999) states that they have no advantages over manual instrumentation and frequently damage the root surface. Low-power lasers, however, may have application in the control of subgingival plaque and pathogens (see below).

CHEMICAL METHODS OF PLAQUE CONTROL

There are a number of situations when it is impossible to achieve adequate plaque control by manual methods alone. There are a number of chemical agents that can be used, either alone or as an adjunct to mechanical methods. The chemical control of plaque has been the subject of several excellent reviews (Cummins 1991, 1997; Addy and Moran 1997; Gaffar *et al.* 1997; Jackson 1997; Jones 1997; Addy 1998; Iacono *et al.* 1998; Eley 1999; Hastings-Drisko 1999). The most commonly used substances are described here, and the reader is referred to the references listed for a more comprehensive account.

Agents that Inhibit the Growth of Microorganisms: Antiplaque Agents

By definition (according to The European Federation of Periodontology, second workshop, 1997), an antiplaque agent is one that produces a prolonged and profound reduction in plaque sufficient to prevent the development of gingivitis (Eley 1999). Antiplaque agents should have the following merits (Van der Ouderaa 1991; Gaffar *et al.* 1997; Iacono *et al.* 1998):

- non-toxic
- non-allergenic
- non-irritating
- safe
- effective at causing clinically significant reductions in plaque and gingivitis
- substantive
- have selective and specific effects on pathogenic microorganisms
- pleasant tasting
- economical
- easy to use
- compatible with toothpaste ingredients
- no disturbance of oral ecology.

Substantivity (the ability to be retained in the mouth for long enough to be effective) is a critical requirement of all antimicrobial agents used in the oral cavity. Not all antiplaque agents that are used are sufficiently substantive on their own but are used in conjunction with carriers or delivery systems that improve retention.

Vehicles for Antiplaque Agents

Vehicles for antiplaque agents include mouthrinses, toothpastes, gels, gums, varnishes and irrigants. Common ingredients of toothpaste and their functions are listed in Table 4.3.1. A gel is a thickened version of a mouthwash, often consisting of a water or water–alcohol base with flavour, surfactant and a humectant, such as glycerol. A varnish is a polymer-based matrix that slowly releases an agent onto the tooth surface to which it is applied and to saliva (Cummins 1997).

Chlorhexidine

Chlorhexidine is a highly effective antiplaque agent that has become the 'gold standard' against which the effectiveness of other antiplaque agents is

Table 4.3.1. Common ingredients of toothpaste. Data from Forward *et al.* (1997) and Eley (1999).

Ingredient	Purpose	Commonly used substances
Polishing or abrasive agent	Helps remove plaque, removes stained, pellicle, polishes surfaces, restores natural lustre, enhances enamel whiteness	Calcium carbonate, dicalcium phosphate dihydrate, alumina, silicas
Binder or thickener	Controls stability and consistency, affects ease of dispersion in mouth, controls ability to be squeezed from tube and appearance	Carageenates, alginates, sodium carboxymethyl-cellulose, magnesium aluminium silicate, sodium magnesium silicate, colloidal silica
Surfactant	Foaming agent (eases removal of food debris), aids dispersion of product in mouth, antimicrobial properties, solubilizing agent	Sodium lauryl sulphate
Humectant	Reduces loss of moisture, improves texture and feel of product, sweetening agent	Glycerin, sorbitol, polyethylene glycol
Flavour and sweetener	Improves taste and leaves fresh aftertaste	Soluble saccharin, peppermint oil, spearmint oil, winter-green (methyl salicylate)
Therapeutic agents	Anticaries agents, antiplaque agents	Fluoride, triclosan, stannous fluoride, sanguinarine, chlorhexidine salts, cetylpyridinium chloride, enzymes
	Anticalculus agents, antisensitivity agents, tooth-whitening agents	Pyrophosphates, potassium chloride, hydrogen peroxide
Substances to whiten product		Titanium dioxide
Preservatives		Benzoate

measured (Jones 1997). The structure of chlorhexidine is shown in Figure 4.3.3. It is a substituted 1,6-bisguanidohexane, which contains both hydrophobic and hydrophilic groups. Each bisguanido group has an associated proton, which confers two strong positive charges on the molecule. The positive charges are believed to be responsible for the antibacterial effects of chlorhexidine, promoting interaction of the molecule with negative charges on teeth, oral mucosa and bacterial cells. One end of the molecule is thought to bind to the enamel surface (possibly attaching to

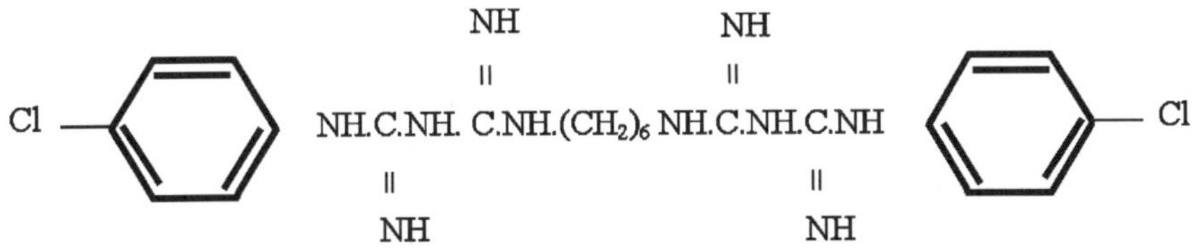

Figure 4.3.3. Structural formula of chlorhexidine. Re-drawn from Cole and Eastoe (1988). Published by Wright, London.

adsorbed proteins in the pellicle) and the other to bacteria. There is specific and strong adsorption to phosphate-containing compounds and, at low concentrations, adsorption to phospholipids in the inner bacterial membrane causes loss of membrane integrity, resulting in increased permeability. In addition, at sub-lethal concentrations chlorhexidine can inhibit acid production in streptococci, inhibit amino acid uptake and catabolism in *Streptococcus sanguis* and inhibit a major protease of *P. gingivalis* (Marsh and Martin 1999). At higher concentrations (depending on the bacterial species), precipitation of bacterial cytoplasm and cell death occurs (Cole and Eastoe 1988; Jones 1997).

Chlorhexidine is effective against Gram-positive and Gram-negative bacteria, aerobes and anaerobes, yeasts, fungi, including *Candida*, and some lipophilic viruses (Hennessy 1973; Emisilon 1977; Jones 1997). It has also been shown to reduce the adherence of *P. gingivalis* to epithelial cells, possibly due to its effect on binding to the bacterial membrane (Eley 1999). The superiority of chlorhexidine over other antiplaque agents is due to its high substantivity: a single rinse can reduce the salivary bacterial counts by over 90% for several hours (Roberts and Addy 1981; Jones 1997). Chlorhexidine is most commonly used as a 0.2% solution of chlorhexidine gluconate in the form of a mouthrinse, but it may also be used in sprays, gels, toothpaste, chewing gum and varnishes. Subgingival irrigation of chlorhexidine solution may be used in periodontitis, and it is also in the sustained-release vehicle PerioChip™ (see below). Its use is recommended for the following applications (Addy and Moran 1997):

- As an adjunct to oral hygiene, particularly in the oral hygiene phase of periodontal treatment.
- Following oral surgery, including periodontal surgery and root planing, gingivectomy and after extraction.
- In patients with intermaxillary fixation to improve oral hygiene and reduce bacterial load in saliva.
- For plaque control amongst physically and mentally handicapped patients.
- In immunocompromised patients predisposed to oral infections.

- In high-risk caries patients.
- In treatment of recurrent minor aphthous[1] oral ulceration.
- In patients with orthodontic appliances where, as a consequence, oral hygiene has deteriorated or with oral ulceration.
- In elderly, long-stay hospital and terminally ill patients.
- To limit bacteraemia in patients at risk of bacterial endocarditis.

Chlorhexidine is reported to be twice as effective as fluoride in controlling the development of caries (Van Rijkom *et al.* 1996), although it may act synergistically with fluoride (Addy and Moran 1997). This may be because chlorhexidine targets the plaque directly, whereas fluoride modifies enamel mineralization and demineralization (Axelsson 1998). Its use in varnishes to reduce caries in children with a high incidence of caries is currently being investigated (Splieth *et al.* 2000).

Plaque is reported to develop more rapidly and in larger amounts around dental implants than natural teeth. The use of chemotherapeutic agents to control plaque around implants, and thus to reduce gingivitis, is increasing. Irrigation of the subgingival tissues around implants with a solution of chlorhexidine, by means of a powered irrigator with a fine tip, has been shown to result in reduction in plaque and gingival indices. The improvement may be because chlorhexidine is poorly retained on implants and irrigation increases its substantivity in the peri-implant subgingival tissues (Felo *et al.* 1997).

Chlorhexidine has some disadvantages. It inhibits plaque formation in a clean mouth but has little effect on pre-existing plaque (Wade and Slayne 1997; Eley 1999). Its activity is reduced in the presence of anionic compounds, including certain components of toothpastes (Jones 1997), and may be inactivated by blood and pus, and by outer membrane vesicles of *P. gingivalis* (Grenier *et al.* 1995; Brecx 1997). Because chlorhexidine is positively charged and binds strongly to any negatively charged surface, it also binds to dietary factors, such as polyphenols and tannins found in some foods, tea, coffee and wine. This leads to staining of the teeth that is very difficult to remove (Figure 4.3.4). Although it is not toxic, chlorhexidine has an unpleasant taste, may cause a burning sensation in the mouth and may interfere with taste perception. It promotes calculus formation and may stain composite and glass-ionomer restorative materials, thus reducing their effective life (Eley 1999). It rarely may cause erosion of the mucosal tissue and swelling of the parotid gland (Addy 1986). For these reasons, chlorhexidine is recommended for only up to 2 weeks, during which time the patient should avoid certain

[1]Aphthous: small chronic ulcers of 10–14 days duration for which there is no known cure.

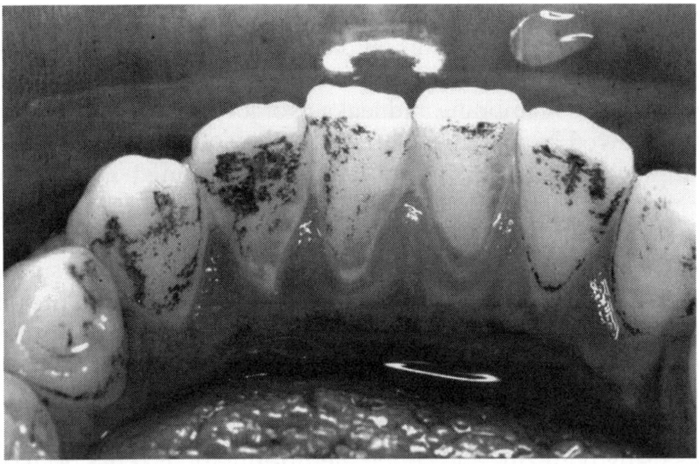

Figure 4.3.4. Teeth stained with chlorhexidine. Photograph courtesy of I.L.C. Chapple and C. Jones, Birmingham School of Dentistry.

foods and drinks. Chlorhexidine gum is also effective at reducing plaque levels, and as it causes less staining could be beneficial for long-term users (Eley 1999).

Essential Oils

Essential (i.e. plant odour-forming) oils are related to phenols and most often contained in mouthrinses. The most well-known member of this family is Listerine®, a formulation that has been in use for over 100 years. Listerine® contains thymol, eucalyptol, menthol, and methyl salicylate in a 26.9 or 21.6% alcohol vehicle (Iacono *et al.* 1998). The essential oils inhibit plaque formation but do not have a significant effect on existing plaque accumulation (Jackson 1997). In comparison with chlorhexidine mouthwash, essential oils are equally effective in reducing gingival inflammation and bleeding, but chlorhexidine mouthwash has been found to be significantly more effective in reducing plaque formation (Overholser *et al.* 1990). However, there was a significant increase in stain and calculus formation with chlorhexidine (Jackson 1997). The alcohol in essential oils may cause a burning sensation that is disagreeable to some patients, and there are concerns that it may cause softening and colour changes of composite resins (Settembrini *et al.* 1995; Jackson 1997).

Triclosan

Triclosan is 2,4,4′-trichloro-2′-hydroxydiphenylether, which is active against both Gram-positive and -negative bacteria with the notable exception of

Pseudomonas species (Gardner and Peel 1991). It acts as a bacteriostatic or bacteriocidal agent, depending on concentration, inhibiting uptake of essential amino acids, uracil and phenylalanine, and causing disorganization of cytoplasmic membrane and leakage of low molecular weight cellular contents (reviewed by Gaffar *et al.* (1997)). Triclosan is moderately effective as an antiplaque agent but is not able to be retained in the mouth for effective periods. However, the addition of zinc ions or a copolymer of methoxyethylene and maleic acid (proprietary name Gantrez®) enhances its affinity for epithelial and tooth surfaces, and thus increases substantivity (Gaffar *et al.* 1997). Structures of triclosan and the copolymer are shown in Figure 4.3.5. The polymer has an attachment and a solubilizing group. The solubilizing group retains triclosan in surfactant micelles, whilst the attachment (COOH) group binds to calcium in the surface layer of liquid on the tooth (Gaffar *et al.* 1997).

In trials using both zinc and copolymer formulations it has been observed that triclosan frequently reduces gingival index scores to a greater extent than plaque index scores, suggesting that it may have anti-inflammatory properties in addition to its antibacterial action (Needleman 1998). It has been shown to reduce a number of inflammatory skin reactions (reviewed by Eley (1999)). The anti-inflammatory effect can be separated from the antimicrobial effect and may be due to the inhibition of cyclo-oxygenase and lipoxygenase pathways and release of products, prostaglandins and leukotrienes. This was demonstrated in gingival fibroblasts stimulated by interleukin 1-β (Gaffar *et al.* 1995, 1997).

Since triclosan is not active against pseudomonads, then it could select for overgrowth of these organisms, but this is more likely to be a problem in its use as a general disinfectant than in the oral cavity. It may be inactivated by non-ionic surfactants. It does not stain the teeth and is non-toxic, except that low levels of contact dermatitis has been reported (Perrenoud *et al.* 1994). Despite its antibacterial activity against a wide range of bacterial species, it has been shown that triclosan and zinc citrate triclosan were more selective against Gram-negative periodontal pathogenic anaerobes than against *S. mutans* in mixed cultures, suggesting that it could be useful in suppressing pathogens in the oral cavity but leaving bacteria that predominate in health (Marsh 1992; Bradshaw *et al.* 1993). In Europe, triclosan has been marketed in toothpastes and mouthwashes since 1992. Clinical trials have shown that triclosan mouthwashes do reduce plaque, but to a much lesser extent than chlorhexidine, whereas the toothpastes are not dramatically more effective than conventional toothpastes at reducing gingival inflammation (Eley 1999). However, in addition to not staining the teeth, triclosan has one clear advantage compared with chlorhexidine, in that in conjunction with zinc citrate it inhibits, rather than promotes, calculus formation and is as effective as pyrophosphates in reducing calculus formation (Iacono *et al.* 1998).

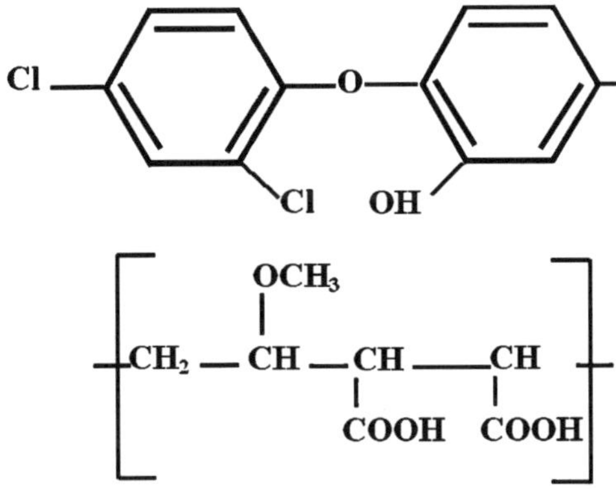

Figure 4.3.5. Structural formulae of triclosan (upper figure) and copolymer of methoxyethylene and maleic acid (lower figure). Re-drawn from Volpe *et al.* (1993), published in the Journal of Clinical Dentistry, reproduced with permission.

Metal Ions: Tin (Stannous Ion) and Zinc

Stannous fluoride has shown some antibacterial effectiveness in toothpastes and mouthrinses. Its activity is due to binding of the stannous (Sn^{2+}) ion to the bacterial cell surface, displacing calcium and altering enzyme function (Ellingsen *et al.* 1980; Boyd *et al.* 1988). Clinical studies of the effectiveness of stannous fluoride in reducing plaque have shown reductions of up to 70%, but results are not consistent and further work is necessary to achieve optimum formulations (Iacono *et al.* 1998). Zinc ions can inhibit sugar transport, acid production and protease activity. Zinc citrate has good adhesive properties and is used in conjunction with triclosan to increase its substantivity (Marsh and Martin 1999). It has been shown that zinc ions can inhibit regrowth of plaque without changing plaque ecology (Ingram *et al.* 1992). Sanguinarine is an alkaloid, extracted from the blood-root plant, with antiplaque activity. It appears to be most effective as an antiplaque agent in conjunction with zinc, which possibly enhances its inhibitory effect on glycolysis (Southard *et al.* 1987; Eisenberg *et al.* 1991; Eley 1999). Copper also has antibacterial activity, but it is not used in oral hygiene products because it causes staining and has an unpleasant taste (Hastings-Drisko 1999). Its main use is in amalgam.

Antibiotics

There is a need to control the systemic use of antibiotics in dentistry to avoid the build up of resistant organisms in the population and adverse drug reactions (Larsen and Fiehn 1996). The second European Workshop on Periodontology recommended the use of systemic antibiotics for severe refractory periodontitis, *P. gingivalis*- and *A. antinomycemcomitans*-associated early onset and juvenile periodontitis, given as an adjunct to mechanical treatment (Lang and Newman 1997; cited by Newman 1998a). The antibiotics most commonly used are those with selective activity against anaerobic organisms, such as minocycline, and combinations of metronidazole–amoxicillin or metroxyzole–ciprofloxin (Steinberg and Friedman 2000). The aim is to bring a shift in the relative proportions of anerobes and aerobes in the subgingival flora, decreasing the numbers of anerobes and thus re-establishing a relative predominance of aerobes, as in healthy plaque (Listgarten 1976).

Simple mouthwashes do not normally penetrate more than 1 mm into the subgingival area (Flotra *et al.* 1972; Mashimo *et al.* 1980) but, in the presence of a periodontal pocket, local delivery of antimicrobial agents may provide a controlled delivery of a standard dose of antibiotic over an extended time period, thus avoiding the problems associated with systemic antibiotics. So called 'sustained-release' vehicles for placement of drugs in the periodontal pocket (reviewed by Killoy (1998) and Steinberg and Friedman (2000)) include tetracycline fibres, a polymer system (for delivery of doxycycline), chlorhexidine-impregnated gelatin chips (PerioChip™; Heasman *et al.* 2001), minocycline ointment and metronidazole gel.

Controlling Plaque by Disrupting the Process of Plaque Accumulation

Delmopinol

Delmopinol is one of the few antiplaque agents that interfere with plaque accumulation. It is an amine alcohol (Figure 4.3.6) and, used in the form of the hydrochloride in 0.1–0.2% solutions, it inhibits plaque formation, growth, gingivitis and calculus formation. Early studies by Simonsson *et al.* (1991) in an artificial mouth system showed that it was able to prevent plaque formation and to dissolve established plaque *in vitro*. It is believed to interfere with plaque matrix formation, causing the plaque to be more loosely adherent to the tooth. In clinical trials it reduced the numbers of dextran-producing streptococci in plaque from delmopinol-treated patients compared with a control group, but there was no major shift in plaque ecology. Side effects of delmopinol include staining of the teeth, transitory numbness of the tongue, taste disturbance and rarely mucosal soreness and erosion, but these were well tolerated. Delmopinol could be a useful antiplaque and antigingivitis agent as a component of mouthwashes or toothpastes (Eley 1999).

Fluoride

The introduction of fluoride dentifrices is the factor that has unequivocally produced the greatest reduction in the frequency of caries in the past 20

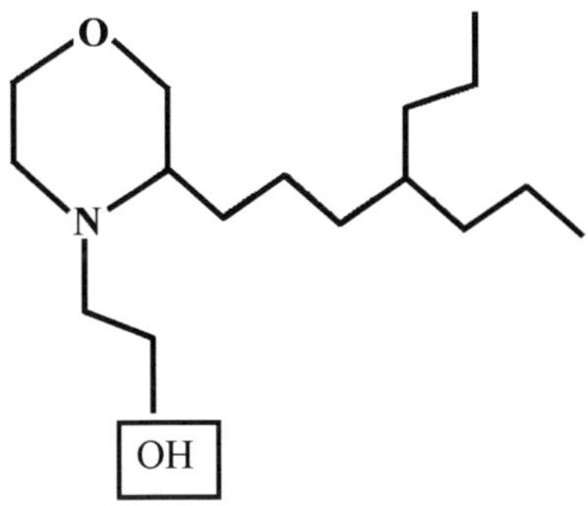

Figure 4.3.6. Structural formula of delmopinol. Re-drawn from Simonsson *et al.* (1991).

years (Needleman 1998). Fluoride promotes re-precipitation of calcium phosphate in small lesions at neutral pH and reduces the solubility of enamel in acid. In addition to its effects on remineralization and demineralization, fluoride has a number of bacteristatic effects on plaque. It inhibits a number of enzymes, including catalases and peroxidases, thus possibly interfering with the redox balance in plaque (Marquis 1995; Marsh and Martin 1999). It also has several effects on bacterial metabolism:

- it reduces glycolysis by inhibition of enolase;
- it indirectly inhibits sugar transport by blocking production of phosphoenolpyruvate;
- it acidifies the interior of cells, thus inhibiting key enzymes;
- it interferes with membrane permeability to ionic transfer;
- it inhibits the synthesis of intracellular storage compounds, especially glycogen (Marsh and Martin 1999).

The antimicrobial effect of fluoride is enhanced if it is combined with a counter-ion that also has antimicrobial activity; thus stannous fluoride is more effective than sodium fluoride (Marsh and Martin 1999).

Fluoride is included in this section because it may also alter the structural integrity and stability of plaque biofilms by inhibiting extracellular polysaccharide production (Marsh and Martin 1999) and by interfering with the formation of bacterial co-aggregates in plaque. Bacteria may associate by means of cationic bridges involving calcium, and the formation of such links is inhibited by fluoride (Rose and Turner 1998a,b), probably reducing the stability of the plaque (Sutherland 1999).

ALTERNATIVE METHODS FOR CONTROLLING PLAQUE

Enhancing the Natural Antimicrobial Function of Saliva

The constant flow of saliva flushes bacteria and food debris away from the oral cavity into the gut. Apart from this mechanical function, saliva contains a number of antimicrobial factors that inhibit bacterial adhesion and protect against harmful bacterial products. These include antibodies, particularly IgA, which inhibit bacterial adhesion, and IgM and IgG, which are thought to enhance bacterial phagocytosis. In addition, saliva contains several enzymes with antimicrobial activities, including peroxidases and lysozyme, lactoferrin, agglutinins, histadine-rich peptides (histatins), proline-rich peptides, cystatins (antiviral activity) and polymorphonuclear leucocytes, responsible for phagocytosis of bacteria (Tenovuo 1998). Peroxidases are responsible for the production of hypothiocyanate, which inactivates a wide range of pathogenic bacteria, viruses and yeasts. Some toothpastes (e.g. Zendium®) contain peroxidase-enzymes, amyloglucosidase and glucose oxidase, and other components to generate hypothiocyanate, and these may be of benefit to people suffering from xerostomia (a dry mouth) and aphthous ulcers (Fridh and Koch 1999).

Sugar Substitutes

One way in which plaque control could conceivably be facilitated is to exploit the flushing effect of saliva by promoting bacterial aggregation. The sugar alcohols, xylitol and mannitol, are used as caries-preventative sweeteners since they decrease fermentation of sugars to acids by oral bacteria, including *S. mutans*. Xylitol cannot be metabolized, but it is accumulated by *S. mutans* as a toxic sugar-phosphate, resulting in growth inhibition (Trahan 1995). It has recently been shown to inhibit protein synthesis, including the expression of heat-shock genes HSP-70 and HSP-60, and thus to impair the survival of wild-type strains of *S. mutans* (Hrimech *et al.* 2000). Repeated culture of *S. mutans*, in the presence of xylitol, unfortunately results in selection for a xylitol-insensitive population in which expression of the *gbpC* gene is elevated. This gene encodes glucan-binding protein C, which is involved in dextran-dependent aggregation and adhesion to dental plaque. The mutant cells showed reduced adhesion to glass when the cultures were shaken, but not when they were static, possibly because the aggregates are easier to dislodge. This could be advantageous from a plaque control point of view for habitual xylitol users, as, following brushing to dislodge aggregates, they might be more readily washed away in saliva (Sato *et al.* 2000).

Targeting Specific Organisms in Plaque: Development of Vaccines

Cariogenic bacteria, such as *S. mutans*, are not primary colonizers in dental plaque but adhere via specific adhesins. This is being exploited in the development of vaccines. In the newborn child, immune defence against cariogenic bacteria is provided by salivary secretory IgA antibodies, which interfere with sucrose-dependent and -independent attachment of *S. mutans* streptococci to tooth surfaces. Current research (reviewed by Russell *et al.* (1999)) is directed towards the development of vaccines to induce mucosal IgA antibodies to surface adhesins and glucosyltransferase. The aim is to induce antibodies directly in the oral cavity, without parenteral administration of antigens. An alternative strategy is to achieve passive immunization using synthetically produced antibodies. Ma *et al.* (1998) have produced a recombinant secretory IgG monoclonal antibody in tobacco plants that recognizes the 185 kDa cell surface adhesion protein of *S. mutans*. The first clinical trials with this vaccine show promising results. It was administered to human volunteers who had previously been treated with chlorhexidine to reduce *S. mutans* to undetectable levels. The vaccine resulted in complete suppression of *S. mutans* recolonization over the 4-month duration of the study, whilst in the control group *S. mutans* recolonized to normal levels. Since the vaccine can be produced economically in plants, it could potentially be incorporated into a dentifrice for home use (Needleman 1998).

Vaccines are also being sought against periodontal pathogens. Page and Houston (1999) have developed a killed cell vaccine against *P. gingivalis* that induces protection against periodontitis in an animal model, even though the disease is caused by several organisms. Further work has shown that a vaccine against a cysteine protease of *P. gingivalis* also conferred protection.

Replacement Therapy

Replacement therapy involves the replacement of a pathogenic organism with a less harmful one. In mixed oral communities, this has the advantage that it is less likely to upset the normal ecology and homeostasis that has developed between the host and the microbial community for maintenance of health. Replacement therapy has been used successfully to replace pathogenic strains of alpha-haemolytic streptococci and *Staphylococcus aureus* with ones of lower virulence (reviewed by Hillman (1999)). In the fight against dental caries, a strain of *S. mutans* has been developed in which the lactate dehydrogenase gene has been deleted, which makes it less cariogenic than wild-type strains. This strain was able to displace normal strains of *S. mutans* in aggressive-displacement and pre-emptive displacement rat models (Hillman *et al.* 2000).

Photodynamic Therapy

When lasers were first used for subgingival debridement in periodontal therapy it was observed that melanin-producing bacteria (which, being darker, absorb more laser light), including *Bacteroides gingivalis* and *Bacteroides intermedius*, were selectively targeted (Midda and Renton Harper 1991). This observation suggested a way of selectively killing pathogens. Initially, however, this was not practical, because it required the use of high-power lasers, which could also damage soft tissue. However, some bacteria, including some oral pathogens, can be killed by red light from a low-power helium–neon laser if they have first been sensitized to light by dyes such as toluidine blue O and methylene blue (Wilson *et al.* 1992, 1993). It is thought that the lethal photosensitization of sensitive bacteria may involve changes in the outer membrane or plasma membrane proteins and DNA damage by singlet oxygen (Bhatti *et al.* 1998). Both Gram-positive (*A. actinomycetemcomitans* and *Fusobacterium nucleatum*) and Gram-negative (*P. gingivalis*) species are sensitive. The treatment also works on biofilms, since, in samples of subgingival plaque from patients with chronic periodontitis, between 90 and 100% reductions in viability of aerobes, anaerobes, black-pigmented anaerobes *P. gingivalis, F. nucleatum* and streptococci were achieved after photosensitization with toluidine blue O and laser treatment (Sarkar and Wilson 1993). A disadvantage of toluidine blue is that it causes dark-blue staining of the oral mucosa. Alternative, non-staining photosensitizers are currently being sought. A potential candidate is chlorin e6, to be used in conjunction with the polycation polylysine, which by virtue of its positive charges should facilitate binding to negative charges on the bacterial cell membrane (Soukos *et al.* 1998).

Ozone

The antimicrobial effect of ozone has led to its use in water purification systems. Baysan *et al.* (1998) have recently shown that exposure to ozonized water for 10–20 s resulted in a significant reduction in numbers of *S. mutans* and *Streptococcus sobrinus* in root lesions and on glass beads. Ozone treatment could thus be a useful alternative to conventional drilling and filling for this type of carious lesion.

CONTROLLING PLAQUE ON RESTORATIVE MATERIALS

The development of secondary caries, pulpal inflammation and gingivitis resulting from bacterial invasion under and around restorations is a major problem that could be combated by the use of restorative materials which release antimicrobial agents (Morrier *et al.* 1998). Amalgam is prepared by mixing alloys of copper, silver, zinc and/or tin with mercury to produce a plastic mixture, which is then compressed into a prepared cavity in the tooth and subsequently hardens (Williams 1990). Amalgams intrinsically have some antibacterial activity. Morrier *et al.* (1998) investigated which elements, phases or alloys were responsible for this effect by comparing growth of *Actinomyces viscosus* and *S. mutans* after 24 h in the presence or absence of amalgams, pure metals, alloys and solutions of metal ions. Mercury, copper, silver–copper alloy, fluoride and zinc all showed antibacterial activity (Hg>Cu>F>Zn). Mercury was the most effective element, as the pure metal or as the chloride in aqueous solution. The mechanism of bacterial growth inhibition by mercury in amalgam is not understood, but it has been suggested that it could inhibit bacterial protein and carbohydrate metabolism (Lyttle and Bowden 1993).

Despite the proven effectiveness and long use of restorative materials, there is concern about the use of amalgams, given the known cytotoxicity of the heavy-metal ions. The potential health risk from mercury toxicity is probably negligible, since even extremely high numbers of amalgam fillings release insufficient mercury to exceed the safety threshold (Mackert and Berglund 1997), and a causal relationship between a range of clinical symptoms attributed to amalgam has not been established (Schuurs *et al.* 2000). However, the grey colour of amalgam is aesthetically unattractive, and alternative, resin-based, tooth-coloured materials are being developed and used clinically. These, however, may not have antibacterial properties and may be susceptible to damage by colonizing bacteria. For example, Willershausen *et al.* (1999) compared the surfaces of two resin-based composite materials and a polyacid-modified composite material after exposure to *A. naeslundii* and *S. mutans*. Whilst the amalgams appeared to be unaffected, there was a significant increase in the surface roughness of the polyacid-modified resin after exposure to *S. mutans* and *A. naeslundii*, which was suggested to be due to ionic disassociation of the organic matrix of the materials with loss of ionic filler particles.

Attempts have been made to prevent plaque accumulation on restorative materials by incorporation of antibacterial agents that are released into the oral environment, such as chlorhexidine (Jedrychowski *et al.* 1983; Ribeiro and Ericson 1991). Although such materials did show antibacterial activity, there was concern that they could exert toxic affects or induce population shifts of plaque microorganisms, or that the mechanical properties of the material would deteriorate over time. Other materials incorporating non-leaching components are currently being investigated. The use of silver ions to inhibit bacterial adhesion to polymeric materials is well known (Berger *et al.* 1976). In an *in vitro* study, Yamamoto *et al.* (1996) demonstrated the antibacterial activity of SiO_2 filler containing silver ions on oral streptococci.

With a similar aim, Imazato *et al.* (1998) have developed and tested the antibacterial effect of a dental resin incorporating the antibacterial monomer 12-methylacryloxydodecylpyridinium bromide (MDPB). So far, it has been shown that growth and plaque formation of *S. mutans* on the surface of MDPB resin was inhibited, there was no elution of any antibacterial component and there were no cytotoxic effects (Imazato *et al.* 1998, 1999). Such results are promising, but further work is required to elucidate the mechanism of the antibacterial effect and to produce acceptable formulations for clinical use.

CONTROLLING PLAQUE BY MODIFICATION OF THE MATERIAL SURFACE TO PREVENT ADHESION

Cracks and crevices on the surface of a tooth or material can aid microbial adherence. On rough surfaces bacteria are protected against shear forces and they have the necessary time to bridge the critical distance to the surface within which van der Waals forces can mediate attachment. Bacterial colonization, therefore, begins in pits and crevices in natural or artificial surfaces (Bollen *et al.* 1997). Both surface roughness and, to a much lesser extent, surface free energy influence initial bacterial adhesion and retention of organisms (Quirynen and Bollen 1995; Quirynen *et al.* 2000).

Fissure Sealants

Since bacteria tend to collect in areas of stagnation, it may be advantageous to modify the contours of the tooth to remove areas where plaque may be retained. This is the purpose of fissure sealants. These consist of plastic, methylmethacrylate film, polymerized *in situ* using UV or chemical catalysis, and are intended to fill in areas of the tooth where stagnation may occur. They have been used with success, particularly in children, to coat caries-susceptible surfaces, but they have to be replaced frequently. Some sealants include fluoride, to promote remineralization of early carious lesions (Cole and Eastoe 1988).

The possibility of changing the surface properties of the enamel to reduce bacterial adhesion and colonization is an attractive possibility, discussed by Wade and Slayne (1997). Bacteria interact with surfaces that are covered with a conditioning film or pellicle of proteins from saliva or crevicular fluid. A balance between the attraction, due to van der Waals forces and the repulsion of negative charges on bacterial and substrate surfaces appears to maintain the bacteria in close proximity to the surface. If, by chance, bacteria are brought within 10 and 20 nm of the surface, in time adhesins on the surface of the bacteria may interact with specific proteins in the pellicle, mediating irreversible attachment (Marsh and Martin 1999). The interaction may be strengthened by the dehydrating effect of hydrophobic interactions between the bacterial cell and protein layer. Substances that interfere with the formation of hydrophobic bonds, such as lithium cations and thiocyanate anions, have been shown to reduce the adherence of *S. sanguis* to saliva-coated hydroxyapatite (Nesbit *et al.* 1982). Non-ionic surfactants have also been used to interfere with surface hydrophobicity, and *in vitro* these were found to be effective in blocking adherence of a range of reference strains of oral streptococci to hydroxyapatite beads. However, they were not effective in reducing plaque *in vivo*, possibly because the oral bacteria possessed specific adhesins that were able to bypass the anti-hydrophobic effect (Moran *et al.* 1995; Wade and Slayne 1997).

In general, to avoid bacterial retention, surfaces should be smooth. However, studies on the influence of surface roughness of dental implant abutments on plaque retention have suggested that there is a threshold surface roughness for bacterial retention (R_a value of 0.2 μm) below which no further reduction in bacterial retention can be expected. Care must be taken when cleaning dental appliances, since polishing of many smooth dental materials (e.g. gold) with abrasive materials and toothpastes can raise the roughness above the threshold, thus increasing the likelihood of plaque retention (Bollen *et al.* 1997). Surface finishing techniques, such as electropolishing and brightening, which are intended to produce a smoother surface to inhibit bacterial attachment, do not necessarily have any effect (Taylor *et al.* 1998).

DISCUSSION AND FUTURE PROSPECTS

Although chemical antiplaque and gingivitis agents continue to be developed, it is unlikely that they will completely supersede mechanical methods of plaque control, since a biofilm needs to be disrupted before it can be eliminated chemically (Needleman 1998). Toothbrushing, in conjunction with toothpaste containing antimicrobial constituents, will probably remain the most popular and effective means of controlling dental plaque for the vast majority of people. Improvements can certainly be made in methods for interdental cleaning, especially for elderly and handicapped people. Safe, non-staining, chemical antiplaque agents for long-term use would be desirable. In developing new techniques and agents for plaque control, the use of biofilm models that closely simulate the environment in the oral cavity is essential. The best current model is possibly the constant-depth film fermenter (Wilson *et al.* 1995). In this, biofilms of defined, reproducible composition, comprising of communities of organisms that can be established on different surfaces. A rotating scraper smears medium (artificial saliva containing mucin) in a thin film over the biofilms, simulating the action of tongue over the teeth and maintaining the biofilms at a constant depth. This device has been used to test the susceptibility of mixed communities of oral bacteria to many antimicrobial agents (Wilson 1996; Pratten *et al.* 1998; Wilson and Pratten 1999). This device should, for the first time, allow an estimation of 'biofilm inhibitory concentration' and 'biofilm killing (or eliminating) concentrations' (Costerton *et al.* 1993), instead of the often misleading minimum inhibitory concentration.

As our knowledge and understanding of oral biofilms increases we are gaining an insight into the complex interactions between microorganisms and host, and it is becoming apparent that homeostasis between the host and oral flora is probably required for the maintenance of our general health. Future developments in the control of oral biofilms must be cognizant of this and be based on a knowledge and understanding of plaque ecology in health and disease. Control programmes should not upset the balance of healthy plaque, but should aim to eliminate or reduce the numbers of pathogens without also eliminating beneficial organisms. As further advances are made in our understanding of the interactions between host and bacterial cytokines, there is considerable scope for advancement in the treatment of periodontal diseases. Rather than eliminating species, plaque control programmes may involve the control of production of bacterial virulence factors or cytokine production or induction (Henderson 1999), with greater emphasis on replacement therapy. Plaque control programmes must be simple and inexpensive, or they will not be conducive to patient compliance. It is also unlikely that improved oral hygiene will affect the incidence of severe periodontal disease, which will continue to affect a small but significant susceptible proportion of the population (Bartold *et al.* 1998). In the case of periodontal disease, since the onset depends on a great variety of host factors, plaque control programmes should be flexible and be tailored to the individual (Newman 1998a,b).

REFERENCES

Abraham, NJ, Ciricione, UK and Glass, RT (1990) Dentists' and dental hygienists' attitudes toward toothbrush replacement and maintenance. *Clinical Preventative Medicine* **12**:28–33.

Addy, M (1986) Chlorhexidine compared with other locally delivered antimicrobials. *Journal of Clinical Periodontology* **13**:957–964.

Addy, M (1998) Antiseptics in periodontal therapy. In: Clinical Periodontology and Implant Dentistry (Eds. Linde, J, Thorring, T and Niklaus, PL), Munksgaard, Copenhagen, pp. 462–487.

Addy, M and Moran, JM (1997) Clinical indications for the use of chemical adjuncts to plaque control: chlorhexidine formulations. *Periodontology 2000* **15**:52–54.

Adriaens, PA, De Boever, JA and Loesche, WJ (1988) Bacterial invasion in root cementum and radicular dentin of periodontally diseased teeth in humans. *Journal of Periodontology* **59**:222–230.

Axelsson, P (1998) Needs related plaque control measures based on risk prediction. In: Proceedings of the European Workshop on Mechanical Plaque Control (Eds. Lang, NP, Attstrom, R and Löe, H), Quintessence Publishing, Chicago, pp. 190–247.

Axelsson, P, Linde, J and Nyström B (1991) On the prevention of caries and periodontal disease. *Journal of Clinical Periodontology* **18**:182–189.

Bartold, PM, Seymour, GJ, Cullinan, MP and Westerman, B (1998) Effect of increased community and professional awareness of plaque control on the management of inflammatory periodontal diseases. *International Dental Journal* **48**(Suppl.):S282–S289.

Baysan, A, Whiley, R and Wright, PS (1998) Use of microwave energy to disinfect a long-term soft lining material contaminated with *Candida albicans* or *Staphylococcus aureus*. *Journal of Prosthetic Dentistry* **79**:454–458.

Beals, D, Ngo, T, Feng, Y, Cook, D, Grau, DG and Weber, DA (2000) Development and laboratory evaluation of a new toothbrush with a novel brush head design. *American Journal of Dentistry* **13**:5A–14A.

Beck, J, Garcia, R, Heiss, G, Vokono, PS and Offenbacher, S (1996) Periodontal disease and cardiovascular disease. *Journal of Periodontology* **67**:1123–1137.

Berger, TJ, Spadaro, JA, Chapin, SE and Becker, RO (1976) Electrically-generated silver ions: quantitative effects on bacterial and mammalian cells. *Antimicrobial Agents Chemotherapy* **9**:357–358.

Bhatti, M, MacRobert, A, Meghji, S, Henderson, B and Wilson, M (1998) A study of the uptake of toluidine blue O by *Porphyromonas gingivalis* and the mechanism of lethal photosensitisation. *Photochemistry and Photobiology* **68**:370–376.

Bollen, CML, Lambrechts, P and Quirynen, M (1997) Comparison of surface roughness of oral hard materials to the threshold surface roughness for bacterial plaque retention: a review of the literature. *Dental Materials* **13**:258–269.

Boyd, RL, Leggott, PJ and Robertson, PB (1988) Effects on gingivitis of two different 0.4% SnF_2 gels. *Journal Dental Research* **67**:503–507.

Bradshaw, DJ, Marsh, PD, Watson, GK and Cummins, D (1993) The effects of triclosan and zinc citrate alone or in combination, on a community of oral bacteria grown *in vitro*. *Journal of Dental Research* **72**:25–30.

Brecx, M (1997) Strategies and agents in supragingival chemical plaque control. *Periodontology 2000* **15**:100–110.

Cancro, LP and Fischman, SL (1995) The expected effect on oral health of dental plaque control through mechanical removal. *Periodontology 2000* **8**:70–74.

Cobb, CM (1999) In: Dental Plaque Revisited. Oral Biofilms in Health and Disease (Eds. Newman, HN and Wilson, M), BioLine, Cardiff, pp. 457–502.

Cole, AS and Eastoe, JE (1988) Biochemistry and Oral Biology, Wright, London, pp. 526–532.

Costerton, JW, Khoury, AE, Ward, KH and Anwar, H (1993) Practical measures to control device-related microbial infections. *International Journal of Artificial Organs* **16**:765–770.

Cummins, D (1991) Routes to chemical plaque control. *Biofouling* **4**:199–207.

Cummins, D (1997) Vehicles: how to deliver the goods. *Periodontology 2000* **15**:84–99.

Daly, CG, Chapple, CC and Cameron, AC (1996) Effect of toothbrush wear on plaque control. *Journal of Clinical Periodontology* **23**:45–49.

de la Rosa, MR, Guerra, JZ, Johnston, DA and Radike, AW (1979) Plaque growth and removal with daily toothbrushing. *Journal of Periodontology* **50**:660–665.

de Stefano, F, Anda, RF, Kahn, HS, Williamson, DF and Russell, CM (1993) Dental disease and risk of coronary heart disease and mortality. *British Medical Journal* **306**:688–691.

Egelberg, J (1964) Gingival exudates measurements for evaluation of inflammatory changes of the gingivae. *Odontologisk Revy* **15**:381–398.

Eisenberg, AD, Young, DA, Fan-Hse, J and Spitz, LM (1991) Interactions of sanguinarine and zinc on oral streptococci and *Actinomyces* species. *Caries Research* **25**:185–190.

Eley, M (1999) Antibacterial agents in the control of supragingival plaque—a review. *British Dental Journal* **186**:286–296.

Ellingsen, JE, Svatun, B and Rølla, G (1980) The effect of stannous and stannic ions on the formation and acidogenicity of dental plaque *in vivo*. *Acta Odontologica Scandinavica* **38**:219–222.

Emisilon, C (1977) Susceptibility of various microorganisms to chlorhexidine. *Scandinavian Journal of Dental Research* **85**:255–265.

Felo, A, Shibly, O, Ciancio, SG, Lauciello, FR and Ho, A (1997) Effects of subgingival chlorhexidine irrigation on peri-implant maintenance. *American Journal of Dentistry* **10**:107–110.

Fischman, SL (1997) The history of oral hygiene products: how far have we come in 6000 years? *Periodontology 2000* **15**:7–14.

Flotra, L, Gjermo, P, Rolla, G and Waerhaug, J (1972) A 4 month study on the effect of chlorhexidine mouthwashes on 50 soldiers. *Scandinavian Journal of Dental Research* **80**:10–16.

Forward, GC, James, AH, Barnett, P and Jackson, RJ (1997) Gum health product formulations: what is in them and why? *Periodontology 2000* **15**:32–39.

Fridh, G and Koch, G (1999) Effect of a mouth rinse containing amyloglucosidase and glucose oxidase on recurrent aphthous ulcers in children and adolescents. *Swedish Dental Journal* **23**:49–57.

Gaffar, A, Scherl, D, Afflito, J and Coleman, EJ (1995) The effect of triclosan on mediators of gingival inflammation. *Journal of Clinical Periodontology* **22**:480–484.

Gaffar, A, Afflitto, J and Nabi, N (1997) Chemical agents for the control of dental plaque and plaque microflora: an overview. *European Journal of Oral Science* **105**:502–507.

Galgut, PN (1991) The need for interdental cleaning. *Dental Health* **30**:8–11.

Gardner, JF and Peel, MM (1991) Introduction to Sterilisation Disinfection and Infection Control, 2nd Edition, Churchill Livingstone, p. 198.

Giuliana, G, Ammatuna, P, Pizzo, G, Capone, F and D'Angelo, M (1997) Occurrence of invading bacteria in radicular dentin of periodontally-diseased teeth: microbiological findings. *Journal of Clinical Periodontology* **24**:478–485.

Glass, RT and Lare, MM (1986) Toothbrush contamination: a potential health risk? *Quintessence International* **17**:39–42.

Glass, RT and Jensen, H (1988) More on the contaminated toothbrush: the viral story. *Quintessence International* **19**:713–716.

Grenier, D, Bertrand, J and Mayrand, D (1995) *Porphyromonas gingivalis* outer membrane vesicles promote bacterial resistance to chlorhexidine. *Oral Microbiology and Immunology* **10**:319–320.

Haffajee, AD, Cugini, MA, Tanner, A, Pollack, RP, Smith, C, Kent, RL Jr and Socransky, SS (1998) Subgingival microbiota in healthy, well-maintained elder and periodontitis subjects. *Journal of Clinical Periodontology* **25**:346–353.

Hastings-Drisko, C (1999) Non-antibiotic plaque chemotherapy. In: Dental Plaque Revisited. Oral Biofilms in Health and Disease (Eds. Newman, HN and Wilson, M), BioLine, Cardiff, pp. 523–548.

Hawkins, BF, Kohout, FJ, Lainson, PA and Heckert, A (1986) Duration of toothbrushing for effective plaque control. *Quintessence International* **17**:361–365.

Heasman, PA, Heasman, L, Stacey, F and McCracken, GI (2001) Local delivery of chlorhexidine gluconate (PerioChip™) in periodontal maintenance patients. *Journal of Clinical Periodontology* **28**:90–95.

Hellstadious, K, Åsman, B and Gustafsson, A (1993) Improved maintenance of plaque control by electrical toothbrushing in patients with low compliance. *Journal of Clinical Periodontology* **20**:235–237.

Henderson, B (1999) Bacteria/host interactions in the periodontal diseases: clues to the development of novel therapeutics for the periodontal diseases. In: Dental Plaque Revisited. Oral Biofilms in Health and Disease (Eds. Newman, HN and Wilson, M), BioLine, Cardiff, pp. 443–456.

Hennessy, T (1973) Some antibacterial properties of chlorhexidine. *Journal of Periodontal Research* **8**(Suppl.):S61–S67.

Hillman, JD (1999) Replacement therapy of dental caries. In: Dental Plaque Revisited. Oral Biofilms in Health and Disease (Eds. Newman, HN and Wilson, M), BioLine, Cardiff, pp. 587–599.

Hillman, JD, Brooks, TA, Michalek, SM, Harmon, CC, Snoep, JL and Vander Weijdden, CC (2000) Construction and characterisation of an effector strain of *Streptococcus mutans* for replacement therapy of dental caries. *Infection and Immunity* **68**:543–549.

Hrimech, M, Mayrand, D, Grenier, D and Trahan, L (2000) Xylitol disturbs protein synthesis, including the expression of HSP-70 and HSP-60, in *Streptococcus mutans*. *Oral Microbiology and Immunology* **15**:249–257.

Iacono, VJ, Aldredge, WA, Lucks, H and Schwartzstein, S (1998) Modern supragingival plaque control. *International Dental Journal* **48**:290–297.

Imazato, S, Imai, T, Russell, RRB, Torii, M and Ebisu, S (1998) Antibacterial activity of cured dental resin incorporating the antibacterial monomer MDPB and an adhesion-promoting monomer. *Journal of Biomedical Materials Research* **39**:511–515.

Imazato, S, Ebi, N, Tarumi, H, Russell, RRB, Kaneko, T and Ebisu, S (1999) Bactericidal activity and cytotoxicity of antibacterial monomer MDPB. *Biomaterials* **20**:899–903.

Ingram, GS, Horay, CP and Stead, WJ (1992) Interaction of zinc with dental mineral. *Caries Research* **26**:248–253.

Jackson, RJ (1997) Metal salts, essential oils and phenols—old or new? *Periodontology 2000* **15**:63–73.

Jedrychowski, JR, Caputo, AA and Kerper, S (1983) Antibacterial and mechanical properties of restorative materials combined with chlorhexidines. *Journal of Oral Rehabilitation* **10**:373–381.

Jepsen, S (1998) The role of manual toothbrushes in effective plaque control: advantages and limitations. In: Proceedings of the European Workshop on Mechanical Plaque Control (Eds. Lang, NP, Attstrom, R and Löe, H), Quintessence Publishing, Chicago, pp. 121–137.

Johnson, BD and McInnes, C (1994) Clinical evaluation of the efficacy and safety of a new sonic toothbrush. *Journal of Periodontology* **65**:692–697.

Jones, CG (1997) Chlorhexidine: is it still the gold standard? *Periodontology 2000* **15**:55–62.

Killoy, WJ (1998) Chemical treatment of periodontitis: local delivery of antimicrobials. *International Dental Journal* **48**(Suppl.):S305–S315.

Kozai, K, Iwai, T and Miura, K (1989) Residual contamination of toothbrushes by micro-organisms. *Journal of Dentistry for Children* **56**:210–214.

Lang, NP (1998) Assessment of periodontal health and disease. In: Proceedings of the European Workshop on Mechanical Plaque Control (Eds. Lang, NP, Attstrom, R and Löe, H), Quintessence Publishing, Chicago, pp. 50–71.

Lang, NP and Newman, HN (1997) Consensus report of session II. In: Proceedings of the Second European Workshop on Periodontology. Chemicals in Periodontics (Eds. Lang, NP and Newman, HN), Quintessence Publishing, Berlin, pp. 192–195.

Larsen, T and Fiehn, NE (1996) Resistance of *Streptococcus sanguis* biofilms to antimicrobial agents. *APMIS* **104**:280–284.

Lindhe, J, Hamp, S-E and Löe, H (1975) Experimental periodontitis in the beagle dog. *Journal of Periodontal Research* **8**:1–10.

Listgarten, MA (1976) Structure of the microbial flora associated with periodontal disease and health in man. A light and electron microscope study. *Journal of Periodontology* **47**:1–18.

Listgarten, MA (1999) Formation of dental plaque and other oral biofilms. In: Dental Plaque Revisited. Oral Biofilms in Health and Disease (Eds. Newman, HN and Wilson, M), BioLine, Cardiff, pp. 187–210.

Löe, H, Thielde, E and Jensen, SB (1970) Experimental gingivitis in man. *Journal of Periodontology* **36**:177–187.

Lyttle, HA and Bowden, GH (1993) The level of mercury in human dental plaque and interaction of *in vitro* between biofilms of *Streptococcus mutans* and dental amalgams. *Journal of Dental Research* **72**:1320–1324.

Ma, J K-C, Hikmat, BY, Wycoff, K, Vine, ND, Chargelegue, LY, Hein, MB and Lehner, T (1998) Characterization of a recombinant plant monoclonal secretory antibody and preventative immunotherapy in humans. *Nature Medicine* **4**:601–606.

Mackert, JR and Berglund, A (1997) Mercury exposure from dental amalgam fillings: absorbed dose and the potential for adverse health effects. *Critical Reviews in Oral Biology and Medicine* **8**:410–436.

Marquis, RE (1995) Oxygen metabolism, oxidative stress and acid-base physiology of dental plaque biofilms. *Journal of Industrial Microbiology* **15**:198–207.

Marsh, PD (1992) Microbiological aspects of the chemical control of plaque and gingivitis. *Journal of Dental Research* **71**:1431–1438.

Marsh, P and Martin, MV (1999) Oral Microbiology, Wright, Oxford, UK, pp. 96–103.

Mashimo, PA, Umemoto, T, Slots, J, Glenco, RJ and Ellison, SA (1980) Pathogenicity testing of *Macaca arctoides* subgingival plaque following chlorhexidine treatment. *Journal of Periodontology* **51**:190–199.

McInnes, C, Engel, D, Moncia, B and Martin, R (1992) Reduction in adherence of *Actinomyces viscosus* after exposure to low-frequency acoustic energy. *Oral Microbiology and Immunology* **7**:171–176.

Meyer, DH and FivesTaylor, PH (1998) Oral pathogens: from dental plaque to cardiac disease. *Current Opinion in Microbiology* **1**:88–95.

Midda, M and Renton Harper, P (1991) Lasers in dentistry. *British Dental Journal* **170**:343–346.

Moran, J, Addy, M, Newcombe, R and Warren, P (1995) The comparative effects on plaque regrowth of phenolic, chlorhexidine and anti-adhesive mouthrinses. *Journal of Clinical Periodontology* **22**:929–934.

Morrier, JJ, Suchett-Kaye, G, Nguyen, D, Rocca, J-P, Blanc-Benon, J and Barsotti, O (1998) Antimicrobial activity of amalgams, alloys and their elements and phases. *Dental Materials* **14**:150–157.

Needleman, IG (1998) Oral hygiene. Today's view. *International Dental Journal* **48**(Suppl. 1):S495–S500.

Nesbit, WE, Doyle, RJ and Taylor, KG (1982) Hydrophobic interactions and the adherence of *Streptococcus sanguis* to hydroxylapatite. *Infection and Immunity* **38**:637–644.

Newman, HN (1998a) Periodontal therapeutics—a viable option? *International Dental Journal* **48**:173–179.

Newman, HN (1998b) The rationale for chemical adjuncts in plaque control. *International Dental Journal* **48**:298–304.

Offenbacher, S, Williams, RC, Champagne, CME, Madianos, PN, Chung, HJ, Liu, Y, Geva, S and Beck, JD (1999) Oral infection and systemic disease: initial evidence for systemic invasion of oral pathogens. In: Dental Plaque Revisited. Oral Biofilms in Health and Disease (Eds. Newman, HN and Wilson, M), BioLine, Cardiff, pp. 375–385.

O'Leary, R, Sved, AM, Davies, EH, Leighton, TG, Wilson, M and Keiser, JB (1997) The bactericidal effects of dental ultrasound on *Actinobacillus actinomycetemcomitans* and *Porphyromonas gingivalis*. An *in vitro* investigation. *Journal of Clinical Periodontology* **24**:432–439.

Overholser, CD, Meiller, TF, Dapaola, LG, Minah, GE and Niehaus, C (1990) Comparative effects of 2 chemotherapeutic mouthrinses on the development of supragingival dental plaque and gingivitis. *Journal of Clinical Periodontology* **17**:575–579.

Page, RC and Houston, LS (1999) Prospects for vaccination against plaque-related oral diseases. In: Dental Plaque Revisited. Oral Biofilms in Health and Disease (Eds. Newman, HN and Wilson, M), BioLine, Cardiff, pp. 563–585.

Papapanou, PN and Lindhe, J (1998) Epidemiology of periodontal disease. In: Clinical Periodontology and Implant Dentistry (Eds. Lindhe, J, Karring, T and Lang, NP), Munksgaard, Copenhagen, pp. 69–101.

Pattison, GL and Pattison, AM (1996) Principles of periodontal instrumentation. In: Clinical Periodontology, 8th edition (Ed. Carranza, FA), WB Saunders Company, Philadelphia, p. 458.

Perrenoud, D, Bircher, A, Hunziker, T, Suter, H, Brucknertuderman, L, Stager, J, Thurlimann, W, Schmid, P, Suard, A and Hunziker, N (1994) Frequency of sensitization to 13 common preservatives in Switzerland. *Contact Dermatitis* **30**:276–279.

Pratten, J, Barnett, P and Wilson, M (1998) Composition and susceptibility to chlorhexidine of multi-species biofilms of oral bacteria. *Applied Environmental Microbiology* **64**:3515–3519.

Quirynen, M and Bollen, CML (1995) The influence of surface roughness and surface-free energy on supra- and subgingival plaque formation in man. A review of the literature. *Journal of Clinical Periodontology* **22**:1–14.

Quirynen, M, Bollen, CML and Papaioanou, W (1996) The influence of titanium abutments surface roughness on plaque accumulation and gingivitis. Short term observations. *International Journal of Oral and Maxillofacial Implants* **11**:169–178.

Ribeiro, J and Ericson, D (1991) *In vitro* antibacterial effect of chlorhexidine added to glass–ionomer cements. *Scandinavian Journal of Dental Research* **99**:533–540.

Roberts, GL, Gardner, P, Longhurst, P, Black, AE and Lucas, VS (2000) Intensity of bacteraemia associated with conservative dental procedures in children. *British Dental Journal* **188**:95–98.

Roberts, WR and Addy, M (1981) Comparison of *in vitro* and *in vivo* antibacterial properties of antiseptic mouthrinses containing chlorhexidine, alexidine, CPC and hexetidine. Relevance to mode of action. *Journal of Clinical Periodontology* **8**:295–310.

Rose, RK and Turner, SJ (1998a) Extracellular volume in streptococcal model biofilms—effects of pH, calcium and fluoride. *Biochimica et Biophysica Acta* **1379**:185–190.

Rose, RK and Turner, SJ (1998b) Fluoride-induced enhancement of diffusion in streptococcal model plaque biofilms. *Caries Research* **32**:227–232.

Russell, MW, Hajishengallis, G, Childers, NK and Michalek, SM (1999) Secretory immunity in defense against cariogenic mutans streptococci. *Caries Research* **33**:4–15.

Rylander, H and Lindhe, J (1998) Cause-related periodontal therapy. In: Clinical Periodontology and Implant Dentistry (Eds. Linde, J, Karring, T and Lang, NP), Munksgaard, Copenhagen, pp. 438–459.

Sarkar, S and Wilson, M (1993) Lethal photosensitisation of bacteria in subgingival plaque from patients with chronic periodontitis. *Journal of Periodontal Research* **28**:204–210.

Sato, Y, Yamamoto, Y and Kizaki, H (2000) Xylitol-induced elevated expression of the *gbpC* gene in a population of *Streptococcus mutans* cells. *European Journal of Oral Science* **108**:538–545.

Schuurs, AHB, Exterkate, RAM and Ten Cate, JM (2000) Biological mercury measurements before and after administration of a chelator (DMPS) and subjective symptoms allegedly due to amalgam. *European Journal of Oral Science* **108**:511–522.

Settembrini, L, Penugonda, B, Scherer, W, Strassler, H and Hillelman, E (1995) Alcohol-containing mouthwashes—effect on composite colour. *Operative Dentistry* **20**:14–17.

Seymour, RA and Steele, JG (1998) Is there a link between periodontal disease and coronary heart disease? *British Dental Journal* **184**:33–38.

Simonsson, T, Hvid, EB, Rundegren, J and Edwardsson, S (1991) Effect of delmopinol on *in vitro* dental plaque formation, bacterial acid production and the number of microorganisms in human saliva. *Oral Microbiology and Immunology* **6**:305–309.

Soukos, N, Ximenez-Fyvie, A, Hamblin, MR, Socransky, SS and Tayyaba, H (1998) Targeted antimicrobial chemotherapy. *Antimicrobial Agents and Chemotherapy* **42**:2595–2601.

Southard, GL, Parsons, LG, Thomas, LG, Boulware, RT, Woodall, IR and Jones, BJB (1987) The relationship of sanguinaria extract concentration and zinc ion to plaque and gingivitis. *Journal of Clinical Periodontology* **14**:315–319.

Splieth, C, Steffen, H, Rosin, M and Welk, A (2000) Caries prevention with chlorhexidine–thymol varnish in high risk schoolchildren. *Community Dental and Oral Epidemiology* **28**:419–423.

Steinberg, D and Friedman, M (2000) Development of sustained release devices for modulation of dental plaque biofilm and treatment of oral infectious diseases. *Drug Development Research* **50**:555–565.

Sutherland, IW (1999) Biofilm matrix polymers—role in adhesion. In: Dental Plaque Revisited. Oral Biofilms in Health and Disease (Eds. Newman, HN and Wilson, M), BioLine, Cardiff, pp. 49–62.

Svanberg, M (1978) Contamination of toothpaste and toothbrush by *Streptococcus mutans*. *Scandinavian Journal of Dental Research* **86**:412–414.

Syed, SA and Loesche, WJ (1978) Bacteriology of human experimental gingivitis: effect of plaque age. *Infection and Immunity* **21**:821–829.

Taylor, R, Maryan, C and Verran, J (1998) Retention of oral microorganisms on cobalt–chromium alloy and dental acrylic resin with different surface finishes. *Journal of Prosthetic Dentistry* **80**:592–597.

Tenovuo, J (1998) Antimicrobial function of human saliva—how important is it for our health? *Acta Odontologica Scandinavica* **56**:250–256.

Trahan, L (1995) Xylitol: a review of its action on mutans streptococci and dental plaque—its clinical significance. *International Dental Journal* **45**:77–95.

Van der Ouderaa, FJG (1991) Anti-plaque agents. Rationale and prospects for prevention of gingivitis and periodontal disease. *Journal of Clinical Periodontology* **18**:447–454.

Van Rijkom, HM, Truin, GJ and van't Hof, MA (1996) A meta-analysis of clinical studies on the caries-inhibiting effect of chlorhexidine treatment. *Journal of Dental Research* **75**:790–795.

Verran, J and Leahy-Gilmartin, AA (1996) Investigation into the microbial contamination of toothbrushes. *Microbios* **85**:231–238.

Verran, J and Maryan, CJ (1997) Retention of *Candida albicans* on acrylic resin and silicone of different surface topography. *Journal of Prosthetic Dentistry* **77**:535–539.

Volpe, AR, Petrone, ME, DeVizio, W and Davies, RM (1993) A review of plaque, gingivitis, calculus and caries clinical efficacy studies with a dentifrice containing triclosan and PVM/MA copolymer. *Journal of Clinical Dentistry* **IV**:31–41.

Wade, WG and Slayne, MA (1997) Controlling plaque by disrupting the process of plaque formation. *Periodontology 2000* **15**:25–31.

Walmsley, AD (1997) The electric toothbrush: a review. *British Dental Journal* **182**:209–218.

Wilkins, EM (1999a) Clinical Practice of the Dental Hygienist, 8th Edition, Lippincott Williams and Wilkins, Philadelphia, London, pp. 298–313.

Wilkins, EM (1999b) Clinical Practice of the Dental Hygienist, 8th Edition, Lippincott Williams and Wilkins, Philadelphia, London, pp. 356–364.

Wilkins, EM (1999c) Clinical Practice of the Dental Hygienist, 8th Edition, Lippincott Williams and Wilkins, Philadelphia, London, pp. 404–405.

Willershausen, B, Callaway, A, Ernst, C-P and Stender, E (1999) The influence of oral bacteria on the surfaces of resin-based dental restorative materials—an *in vitro* study. *International Dental Journal* **49**:231–239.

Williams, D (1990) Dental amalgams. In: Concise Encyclopaedia of Medical and Dental Materials, Pergamon Press, Oxford, pp. 127–134.

Wilson, M (1996) Susceptibility of oral biofilms to antimicrobial agents. *Journal of Medical Microbiology* **44**:79–87.

Wilson, M, Dobson, J and Harvey, W (1992) Sensitization of oral bacteria to killing by low-power laser radiation. *Current Microbiology* **25**:77–81.

Wilson, M, Dobson, J and Sarkar, S (1993) Sensitization of periodontopathic bacteria to killing by light from a low-power laser. *Oral Microbiology and Immunology* **8**:182–187.

Wilson, M, Kpendema, H, Noar, J, Hunt, N and Norden, NJ (1995) Corrosion of intra-oral magnets in the presence and absence of biofilms of *Streptococcus sanguis*. *Biomaterials* **16**:721–725.

Wilson, M and Pratten, J (1999) Laboratory assessment of antimicrobials for plaque-related diseases. In: Dental Plaque Revisited. Oral Biofilms in Health and Disease (Eds. Newman, HN and Wilson, M), BioLine, Cardiff, pp. 503–521.

Yamamoto, K, Shizue, O, Aono, M, Kokubo, T, Yamada, I and Yamauchi, J (1996) Antibacterial activity of silver ions implanted in SiO_2 filler on oral streptococci. *Dental Materials* **12**:227–229.

5 Biofilms Past, Present and Future—New Methods and Control Strategies in Medicine

JAMES T WALKER,[1] SUSANNE SURMAN[2] and JANA JASS[3]
[1]*CAMR, Porton Down, Salisbury, UK*
[2]*Food Safety Microbiology, Central Public Health Laboratory, London, UK*
[3]*Department of Microbiology and Immunology, University of Western Ontario and The Lawson Health Research Institute, London, ON, Canada*

BIOFILMS—THE PAST

A century has passed since the first effect of a surface on the bacterial population was noted (Whipple 1901). Looking back, over the last few years our knowledge of microbial biofilms has increased dramatically, as shown by the increase in the numbers of publications on the subject (Figure 5.1). There are also numerous books and many conferences devoted to biofilms, particularly their visualization, the problems that they may cause and the measures needed to control them. Very little time has been devoted to the beneficial aspects of biofilms.

Looking back over the way our knowledge has accumulated over the years, a few notable steps are highlighted here. First, the recognition of the concept of 'surface-associated microbial activity and colonization', or 'biofilm formation', as a phenomenon that occurs in both natural and man-made environments has become a growing interest in both the medical (Bayston 2000) and the non-medical fields. In reality, not all surface-associated bacteria have been, or still are, generally thought of as biofilms. A prime example of this is in dentistry, where the term 'dental plaque' is used to define a consortium of organisms forming a biofilm.

Medical Biofilms. Edited by J. Jass, S. Surman, J. Walker

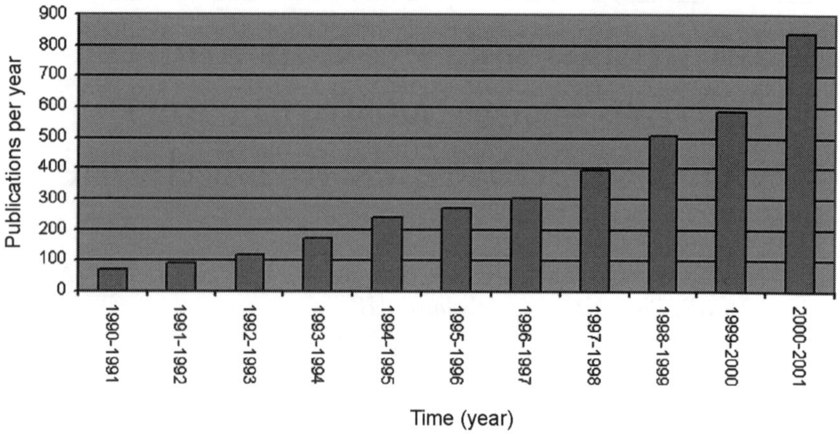

Figure 5.1. Number of publications with the word 'biofilm' cited in PubMed.

Events Leading to Biofilm Formation

By the early 1990s scientists were well on the way to understanding the importance of biofilms, the mechanisms of their formation and their role in microbial survival. The importance of surfaces as sites of increased microbial activity and the principles of adhesion were on the way to being elucidated. The important steps in biofilm formation were appreciated as:

- Surface conditioning (Trulear and Characklis 1982; Allison 1993a) and the mechanisms involved in bacterial adhesion as dependent on both the physiological status of the microorganism (Boyle *et al*. 1991) and on the nature of the substratum.
- The physical and electrochemical nature and relative hydrophobicity of the surface as important factors in the biofilm formation process (Fletcher and Loeb 1979; Dahlbäck *et al*. 1981; Fletcher and Pringle 1983; Konhauser *et al*. 1994), in addition to the importance of receptor interactions in binding to living surfaces. For example, rougher surfaces were preferentially colonized, providing niches protected from the effects of shear stress, turbulent flow and biocide activity (Lytle *et al*. 1989; Konhauser *et al*. 1994; Walker *et al*. 2001).
- Adherence to surfaces in natural and industrial environments—the role of extracellular polysaccharide substances (EPSs), or the glycocalyx, secreted by the cells is thought to be important and to play a role in secondary colonization by different species (Costerton *et al*. 1985). These high molecular weight EPS molecules are believed not to act directly as

adhesins; rather, other factors, possibly low molecular weight polysaccharides, shown to be produced in trace amounts, mediated the initial colonization process, followed by higher molecular weight EPS production as a response to later events (Allison 1993a,b).

- The composition of this glycocalyx was shown to be dynamic, changing as the biofilm developed (Trulear and Characklis 1982). Additionally, the glycocalyx acts as an ionic exchange matrix that is able to trap metal ions (Ferris *et al.* 1987; Geesey *et al.* 1988) and nutrients and thus they may be transported into the cell by highly efficient permeases (Costerton and Geesey 1979). It also plays a role in retention and concentration of digestive enzymes released by the bacteria, thus increasing the metabolic efficiency of the cells (Costerton *et al.* 1978). In some cases, the substrate itself, or its corrosion products, may then be incorporated into the biofilm (Keevil *et al.* 1989; Ellis 1990; Walker *et al.* 1991; Beech and Gaylarde 1991).
- It was also appreciated that biofilms are not homogeneous. Rather, they consist of a consortium of microorganisms that exhibit differing physiological and metabolic properties from their planktonic counterparts in response to the pH, oxygen and nutrient gradients within the EPS matrix (Kepkay *et al.* 1986; Gilbert and Brown 1994). As a result, various niches occur within the biofilm, which may permit the coexistence of microorganisms with differing growth requirements, such as anaerobic and aerobic bacterial populations within the same biofilm (Keevil *et al.* 1994). Metabolic interdependence may occur between species and may be a factor in the increased resistance to physical and chemical stresses exhibited by biofilm members (Caldwell *et al.* 1993).
- Resistance to biocide treatments was shown to be increased in bacteria attached to surfaces (Ridgway and Olsen 1982; Kuchta *et al.* 1985; King *et al.* 1988; Vess *et al.* 1993) and the role of the glycocalyx as a barrier affording various constituents of the biofilm partial protection from antibacterial agents (Costerton *et al.* 1981; Cloete *et al.* 1989) and toxic substrates upon which a biofilm forms, e.g. copper pipes within water distribution systems (Keevil *et al.* 1989). It was still unclear whether this was a phenotypic response of the microbial population to surface growth that played a role in increased resistance (Jass and Lappin-Scott 1994).

Biofilm Heterogeneity

It has long been recognized that the biofilm is not a static entity. Sloughing and erosion processes result in the detachment of portions of a biofilm due to the hydrodynamic conditions or shear forces occurring within a system (Characklis 1981; Taylor *et al.* 1985). The rate of this detachment may be related to the specific bacterial population, since some species have been shown to be more susceptible to shear stresses (Oga *et al.* 1991). Growth-limiting factor(s) may also have an effect on the rate of detachment; for example, Applegate and Bryers (1991) showed differences in both shear and sloughing events between biofilms that were oxygen limited compared with those that were carbon limited. Sloughing or erosion may occur at any time during biofilm development, resulting in the re-suspension of the microorganisms from the biofilm to the planktonic phase (Trulear and Characklis 1982). These may include potential human pathogens such as *Legionella pneumophila*, *Cryptosporidium* spp., *Mycobacterium* spp., *Pseudomonas* spp., *Staphylococcus* spp., *Rotavirus* and *Giardia*, enteroviruses, mycoplasmas and protozoa (Rowbotham 1980; Reasoner 1988; van der Wende *et al.* 1988; Keevil *et al.* 1989; Alary and Joly 1991; Boyle *et al.* 1991; Emde *et al.* 1992). Cells may also actively detach from the surface and subsequently relocate on the substratum, a process termed desorption (Escher and Characklis 1988).

An important breakthrough occurred in 1994, when Nichols suggested that resistance to antimicrobial compounds may not be solely due to the physical impedance of the antimicrobial agent, but that there may be other factors such as absorption or catalytic destruction of the agent by microbes at the biofilm surface (Nichols, 1994). Williams and Stewart (1993) also suggested that glycocalyx formation may be a microbial cooperative response to cell density limitations initiated by bacterial pheromones. The concept of cooperation and signalling between bacterial populations in response to physical/biochemical change is a growing focus of biofilm research.

BIOFILM CONTROL—THE PRESENT

A most important realization over the last few years was that the majority of clinical infections, whether associated with implant or tissue surfaces, are in fact biofilm related and require different strategies for investigation and control. Increased uses of implanted materials, such as catheters (Stickler *et al.* 1998; Crump and Collignon 2000; Fiorina *et al.* 2000; Kunze and Aschoff 2000; Vogel *et al.* 2000; Mermel *et al.* 2001), orthopaedic prostheses (Gracia *et al.* 1997), shunts (Walsh *et al.* 1986; Kockro *et al.* 2000), vascular prostheses (Bergamini *et al.* 1988; Bandyk *et al.* 1991), pacemakers (Marrie and Costerton 1984) and drug delivery systems (Soukos *et al.* 2000; Vyas *et al.* 2000) in medical practice has resulted in a greater number of clinical infections as a result of microbial colonization of the biomaterial surfaces (Finch 1994). Additionally, a new understanding of gastrointestinal infections and infections of normally sterile tissue surfaces have also shown that bacterial attachment and biofilm formation occurs. Infections associated with implanted materials and tissue surfaces are increasingly difficult to treat with antibiotic regimes used to treat the same pathogens successfully where biofilms are not implicated. Consequently, the only treatment often left to the clinician is to remove the implant, followed by a prolonged period of intensive antibiotic therapy. Although the removal of some implants may not be a serious problem medically (i.e. removal of breast implants or catheters), the knock-on effects on the patient's well-being and morale may be severe and hinder recovery significantly, in addition to financial costs. Other colonized materials may require a major operation to treat the infection successfully, such as the removal of an infected prosthetic hip joint, resulting in significant treatment costs and patient recovery time.

In addition to the severe problems with the implant when infected, additional infections and complications can result causing high mortality and morbidity. Approximately 80 000 catheter-related blood-stream infections occur in US intensive care units each year at a cost of $296 million to $2.3 billion and are associated with 2400 to 20 000 deaths per year (Mermel 2000). Prosthetic devices, such as heart valves or joints, inserted deep within the body, run the risk of becoming infected during the surgical procedure or soon after via drains or anachoresis (spread by the blood) from venous catheter sources (Mermel 2001; Mermel *et al.* 2001).

The oral cavity presents us with an environment where most individuals should be able to keep their teeth free from disease-causing plaque or biofilms (Bartold *et al.* 1998). Even so, severe periodontitis affects approximately 10% of most populations; and, despite the dramatic increase in the use of oral hygiene aids, efforts by the dental profession in oral hygiene instruction, and the associated general improvement in oral hygiene levels, the incidence of severe chronic inflammatory periodontal disease has remained the same. It may not be until the adoption of a more specific approach to the control of specific pathogens, which inhabit subgingival biofilms, that major changes in the general incidence of the severe inflammatory periodontal diseases will be seen (Marsh and Martin 1999). Over 20 million patient visits were made to dentists in the UK in 1998, resulting in a cost to the National Health Service that is greater than any other single treatment. It is clear that if greater control of caries and periodontal diseases due to plaque biofilm were available, then there would be significant financial gain to the National Health Service.

In recent years, periodontology has shifted away from surgery and towards medicine. Although surgery, particularly regenerative surgery and the placement of implants (Felo *et al.* 1997), continues to form an important part of periodontal treatment, most future periodontics will be based on a physician-type approach. Improved diagnostics based on more precise periodontal disease classification, simplification of mechanical oral hygiene equipment (Beals *et al.* 2000) and procedures, and the development of chemical and physical adjuncts may be expected to reduce the advance rate of common periodontal diseases and result in less complex treatments. The rationale for non-surgical adjunctive therapy (Hastings-Drisko 1999) is extensive, far beyond the usual antimicrobial logic (Larsen and Fiehn 1996). It will also be important to control the oral microflora for systemic reasons (Hillman *et al.* 2000), since strong links are being established between focal infection of oral origin and a range of systemic diseases, including coronary heart disease, stroke, gastrointestinal disorders and low birth weight, apart from severe, overt systemic infections (Meyer and FivesTaylor 1998). These developments are derived from an improved understanding of the ecological nature of the microbial biofilm that is dental plaque, and of its interactions with its human host (Newman 1998).

Control of Surface-associated Medical Infections

One major problem associated with medical infections of tissue surfaces or the colonization of an implanted device is that conventional antimicrobial sensitivity tests are not clinically predictive of the antibiotic concentrations needed to control and eradicate the infection, therefore, treatment failure and relapse are unfortunately common (Foley and Gilbert 1996). Bacteria within a biofilm have been shown to have an increased resistance to antimicrobial compounds, requiring up to a 1000-fold increase in concentration to eradicate biofilm cells in comparison with bacteria growing in suspension (planktonic) (Costerton *et al.* 1993).

The use of antibiotics does not always guarantee eradication of biofilm-associated infection, and relapse often occurs after an apparently successful treatment is stopped, often resulting in chronic infections (Vorachit *et al.* 1993; Brooun *et al.* 2000; Xu *et al.* 2000). Despite considerable advances in our understanding, there have been few new effective treatments against biofilms (Ceri *et al.* 1999). Antibiotics have been used successfully where the infection has been recognized at early onset, before a mature biofilm is developed, thus occasionally negating the requirement for the implant removal (Zimmerli *et al.* 1998). However, the use of long-term prophylactic antibiotics is not recommended, as they have been associated with selection of resistant microorganisms (Dixon 1998), even though it is general practice that prophylactic antibiotics are given in many hospitals at the time of surgery. This broad, non-selective antibiotic treatment has additional consequences, in that it clears the normal microbial flora in the patient, potentially causing gastrointestinal disturbances.

The failure of antimicrobial treatments may be related to a number of reasons, including: inherent insusceptibility of the target cells to the agents employed (Anderl *et al.* 2000); the acquisition of resistance (Vorachit *et al.*

1993); an emergence of a pre-existing but unexpressed resistant phenotype (Walsh 2000); failure of penetration or inactivation of the antibiotic by enzymes produced by neighbouring cells (Xu *et al.* 2000); or diffusion-reaction mechanisms. To aid our search for novel antimicrobial products we need to seek new targets, and this can only be achieved through a greater knowledge of biofilm structure (Stoodley *et al.* 2001) and physiology. In addition, the different mechanisms of resistance developed within biofilm communities must be clearly characterized.

A number of new strategies have been developed by researchers to overcome the problems encountered in biofilm control *in vitro*. Although many of these are still under investigation and are being tested in clinical applications, they provide the key to future control strategies. These include:

(i) Novel combinations of chemical and physical techniques to control biofilms, such as ultrasound or electrical enhancement of antibiotics (Jass and Lappin-Scott 1996; Rediske *et al.* 2000; Wattanakaroon and Stewart 2000).
(ii) Novel antibiotic derivatives with increased antibacterial activities (Cho *et al.* 2001; Ishikawa *et al.* 2001; Springer *et al.* 2001).
(iii) Novel anti-adhesive compounds that prevent or inhibit bacterial binding to either tissue surfaces or implants, such as soluble receptor analogues (Kihlberg *et al.* 1989; Ohlsson *et al.* 2002), antibodies that block adhesion (Flock 1999) or compounds that prevent the expression of bacterial adhesins (Svensson *et al.* 2001).
(iv) Use of probiotic bacteria and fungi to prevent or remove microbial contamination. For example, the use of bacteria to prevent yeast contamination of artificial voice boxes. Such techniques are of great interest in terms of being cost effective and having a low impact on the environment (Busscher *et al.* 1997, 2000; van der Mei *et al.* 2000).
(v) 'Smart surfaces' that reduce or prevent biofilm growth and contamination. So, whilst materials can be developed that reduce fouling, such techniques may only delay and/or decrease contamination by up to 1.0 log CFU cm^{-2} (Hilbert 2001). Such materials may be of interest for applications such as prosthetic hip replacement, where the material surface may present an initial challenge to microbial attachment. However, such small reductions in contamination levels may not merit the production of smart surfaces.

Although not all infections can be prevented, the use of contamination prevention strategies would assist in decreasing the number of implant failures that occur in surgery due to biofilm growth or would delay biofilm formation, allowing longer use of the device. While we consider the importance of implant success to human health, it must also be viewed as an important economic consequence, since medical resources are limited. The replacement of a prosthetic hip requires an additional two to three surgical procedures in removal and replacement of the implant, exposing the patient to additional risks and trauma. Hence, novel strategies consisting of using smart materials in combination with slow-release antibiotics and/or phage therapy may increase success. These alternative strategies may result in a reduced use of antibiotics, hence reducing cost outlay and the potential development of antibiotic-resistant microorganisms.

It is clearly noted throughout this book that not all infections can be prevented and, therefore that novel strategies to remove established biofilms are necessary. The use of contamination prevention strategies would assist in decreasing the number of failures that occur in the medical community due to biofilm growth. This, together with novel strategies and effective combinations of smart materials, antibiotics, antiadhesives, steroids and/or phage therapy, may significantly reduce the morbidity and impact of colonized medical and tissue surfaces, both from a patient and an economic perspective (Chanishvili *et al.* 2001; Sharp 2001). Such strategies may result in reduced use of antibiotics, and thus lower the potential for developing antibiotic-resistant microorganisms and minimizing disruption of commensal microflora.

Bacteriophage Therapy for Biofilm-related Infections

A combined strategy of bacteriophage (phage) therapy to augment the use of antibiotics (Hughes *et al.* 2001) may prove successful. Bacteriophage therapy is the use of bacterial-specific viruses to treat infections by causing bacterial cell lysis and death. This was a major area of interest 80 years ago in the fight to combat bacterial infections (Brunoghe and Maisin 1921; Beckerich and Hauduroy 1922; Davison 1922; da Costa Cruz 1923; Spence and McKinley 1924). Although a considerable degree of success was demonstrated in many of the early studies, the development of penicillin and other antibiotics during the 1940s provided a more efficient and comprehensive approach to the eradication of infection and, in general, led to the cessation of research in this area. However, in Eastern Europe, work continued and resulted in a number of strategies developed to combat infection with bacteriophages (Chanishvili *et al.* 2001).

Recent studies to re-evaluate bacteriophage-based therapies for the treatment of human and veterinary infections have been undertaken over the past 20 years and reviewed recently (Aliisky *et al.* 1998; Barrow and Soothill 1997; Levin and Bull 1996; Sharp 2001). Smith and coworkers (Smith and Huggins 1982, 1983; Smith *et al.* 1987a,b) published a series of studies on the treatment of systemic *Escherichia coli* infections in mice and diarrhoeal disease in young calves. Soothill (1992) examined the use of bacteriophage to control organisms often implicated in the colonization of skin burns and successfully demonstrated the use of phage to protect skin grafts from destruction by *Pseudomonas aeruginosa* and *Acinetobacter baumanii* (Soothill 1994).

Bacteriophage therapy offers advantages over antibiotic therapy, including activity against drug-resistant organisms and an alternative therapy for patients with antibiotic allergies. It may be used as a prophylactic treatment to combat the spread of infection where the source is identified at an early stage, or where outbreaks occur within a relatively closed institution, such as old people's homes or schools. Bacteriophages are highly specific in the destruction of their targets and, unlike antibiotics, they do not interfere with the natural microflora. In addition, they can be formulated as a unique cocktail or in combination with other antimicrobial compounds in order to destroy multiple strains. A number of medical conditions may be suitable for phage therapy, including catheter infections and biofilms on prosthetic devices.

Antibiotic Resistance of Biofilm Cells

In order to provide techniques and methods for biofilm measurement, scientists require an understanding of the fundamental structure and function of biofilms. We must understand clearly what factors determine microbial biofilm growth and how this relates to its structure. We know that the bacteria alone are often sensitive to current treatment, therefore, the biofilm structure and community interactions are what provides the biofilm with survival ability.

The structure of the biofilm may itself play a role in the defence of surface-associated cells against antimicrobial agents (Stewart 1998). There are a number of well-established mechanisms of antibiotic resistance, such as efflux pumps, modifying enzymes and target mutations, that do not appear to be responsible for the protection of biofilm bacteria *per se* (Walsh 2000). Dispersal of biofilm bacteria usually results in clearance by antibiotics and the body immune system (Williams *et al.* 1997; Anwar *et al.* 1989), suggesting that resistance of biofilms cells is not inherent to the individual cells. Hence, our current understanding of the bacterial antibiotic resistance mechanisms does not appear to explain fully most cases of antibiotic resistance in biofilms. If the biofilm structure and physiology is viewed as a whole, it is evident that there are multiple multicellular strategies that may be involved in conferring resistance to the biofilm (Stewart and Costerton 2001). Resistance may be mediated through reaction diffusion (Costerton *et al.* 1987; Hoyle *et al.* 1992; Huang *et al.* 1995), phenotypic variation (Gilbert *et al.* 1990) and/or heterogeneity (Wentland *et al.* 1996). Hence, resistance of microbial biofilms to a wide variety of antimicrobial agents may be associated with a number of factors, including age, physiology, growth rate and spatial organization. Thus, biofilms continue to adapt and select phenotypes that survive exposure to antimicrobial agents (Gilbert *et al.* 2001).

When considering the range of mechanisms that prevent biofilm control, a multiplicity approach is required in research strategies to provide a basis to increase the understanding of microbial resistance. Such studies are currently being undertaken *in vitro*, however, new techniques are being developed to help investigate what is actually occurring *in vivo* at both the genetic and expression levels. Gene and protein arrays only provide a glimpse of what is occurring, therefore, additional verification techniques are needed to obtain the complete picture of the biofilm resistance mechanism *in vivo*.

Microbial Cell Communication

Whether microbial cells need to communicate with each other for growth has been the focus of current debate (Kaprelyants and Kell 1996). It has been understood for several years that tissue cells communicate with each other, but only recently have there been indications that bacteria also send and receive information. The best characterized example of inter-bacterial signalling is autoinduction (or quorum-sensing) of the symbiotic marine bacterium *Vibrio fischeri* in the light organs of certain marine fish. The autoinduction of luminescence in *V. fischeri* was described in the early 1970s by Eberhard (1972) and Nealson *et al.* (1970). *V. fischeri* accumulates in the fish light organ at high cell densities (10^{10}–10^{11} cells ml^{-1}) and produces a small diffusible compound called the autoinducer. At a critical concentration of the autoinducer, the *lux* genes are activated, producing the characteristic luminescence. Autoinduction only occurs where *V. fischeri* reaches high cell densities, such as those that are found in biofilms where the cell-to-cell association is high. In recent years, a growing number of Gram-negative bacteria, such as *P. aeruginosa*, have been demonstrated to have genes similar to the *lux* genes that are regulated by autoinducer molecules produced at high bacterial densities. These genes have been associated with the regulation of a number of virulence factors (Pearson *et al.* 1994; Manefield *et al.* 2001; Erickson *et al.* 2002).

Quorum-sensing is based on chemical signals, and recognition of these signals may not be limited to the same bacterial species but may also be recognized by other bacterial species or even other cells. This may help monitor and control the diversity of species or regulate intergeneric communication. It has been suggested that quorum-sensing may play an important role in the development of biofilms, since this environment is characterized by both high cell densities and the close proximity of different species (Williams and Stewart 1993; Davies *et al.* 1998). These cell signals may also be involved in attenuating host immune response, aiding in persistent infections.

Starting with single cells attaching to a surface, biofilms develop to mature microcolonies with complex structures consisting of stacks and aggregates of microbial cells (Costerton *et al.* 1995). This structure may have a profound importance in the placement of particular species within the biofilm, possibly dependent on nutrient conditions. Anaerobic species such as *Porphyromonas gingivalis* may indeed use quorum-sensing as a mechanism of cell–cell communication to sense anaerobic sites within the biofilm (Hansen *et al.* 2000). Whether these cells track their way through the plaque biofilm until they find a suitable anaerobic region, or whether they are within the aggregates that form part of a biofilm and only start to grow once the region has become sufficiently reduced, is unknown. However, such movements in biofilms have been demonstrated by aerobes (Stoodley *et al.* 2001) and anaerobes such as sulphate-reducing bacteria (Dunsmore 2002). In general, communication signals produced by one bacterium might function as an attractant for a second microorganism, thus leading to the development of mixed cultures functioning cooperatively within a particular ecological niche. This is one way that quorum-sensing may play a role in the control of gene expression within biofilms such as glycocalyx production. Alternatively, quorum-sensing may lead to the induction of other genes essential for the maintenance of the bacterial attachment, removal (sloughing), oxygenation or even reduction in the immediate locality (Williams and Stewart 1993). Other bacteria may use such communication mechanisms to aid transfer of nutrients and may influence the effectiveness of antimicrobial compounds. Hence, blocking or neutralization of the communication mechanism may provide a strategy to interfere with the biofilm formation.

P. aeruginosa can cause serious clinical problems, since they cause biofilm-related infections that are highly resistant to antibiotic therapy. These organisms have been shown to operate a quorum-sensing system that may be involved in the development of structurally complex biofilms (Stoodley *et al.* 1997). The presence of autoinducers has also been detected in *P. aeruginosa* infections *in vivo* and is believed to be involved in regulating the expression of some virulence factors (Singh *et al.* 2000; Erickson *et al.* 2002). The involvement of an intercellular signal molecule in *P. aeruginosa* biofilm formation suggests a possible target for controlling biofilm growth (Davies *et al.* 1998). Kjelleberg and coworkers (Manefield *et al.* 2001) have identified a compound (halogenated furanone) from red algae that mimics the autoinducer. They have produced a number of synthetic analogues that interfere with the quorum-sensing system. Although these compounds appear not to block initial attachment of bacteria, they alter the biofilm architecture and enhance detachment (Hentzer *et al.* 2002). This has relevance in biofilms on catheters, in cystic fibrosis, and also in industrial environments, where *P. aeruginosa* biofilms are a persistent problem.

BIOFILM RESEARCH—THE FUTURE

Researchers have recently suggested that the notion that biofilm cells have greater resistance than do planktonic cells is misplaced. Although they are not disputing that biofilms are not killed by concentrations of bactericides that are lethal to log-phase planktonic cells, they suggest that biofilm cell resistance is based on growth phase. This is supported by findings where stationary-phase *P. aeruginosa* cells were slightly more tolerant to antibiotics than biofilms, when treated with antimicrobial agents targeting slow-growing or stationary-phase cells (Lewis 2001; Spoering and Lewis 2001). This will further fuel the debate about the inherent resistance of biofilms and should fundamentally alter the way that we target biofilm eradication in the future.

The past failure of most available antimicrobial substances to contend adequately with biofilms has stimulated the search for new compounds that have activity directed primarily against the biofilm phenotype (Gilbert and Allison 2000). Although this has had only limited success, with the rapid growth in molecular typing techniques it is likely that the development of 'designer antimicrobial compounds' will increase. It is perhaps no longer correct to use the term antibiotic, as many newer antimicrobial substances are synthetic or synthetic analogues of antibiotics.

The Future of Bacteriophage Therapy

The development of phage therapies with the possibility of treating chronic *P. aeruginosa* biofilm infections, by the application of phage carrying and encoding hydrolytic enzymes to destroy the alginate supporting the biofilm, offers a major therapeutic benefit. The use of phage technology may expand as our understanding of the structural properties and the stability of phages increases. This could lead to the design of suitable delivery and targeting strategies and co-administration of phage with existing drugs using novel delivery vehicles. In the future, we may see increased delivery of phage in combination with other agents designed to reduce the severity of the symptoms of cystic fibrosis and bacterial colonization, such as antibiotics, DNAse or antimicrobial peptides (Hughes *et al.* 2001).

Quorum-Sensing in Biofilm-related Infections

There are a large number of Gram-negative bacteria involved in biofilm formation and infections that may also be controlled by autoinduction and quorum-sensing systems. Gram-positive bacteria and fungi also produce cell-communication and quorum-sensing signals, however, they are different to those found in Gram-negative bacteria. How these chemical signals affect virulence expression and which genes they regulate are not fully understood. It is clear, therefore, that substantial research is required in this area to determine the role of quorum-sensing in biofilm growth, cell–cell and species–species interaction and virulence. We envisage that, with a clearer understanding about the regulatory properties of cell signalling biochemicals (i.e. furanones) and their effect upon bacteria, we may be able to develop strategies to confuse or alter this signalling to our benefit (Hentzer *et al.* 2002). This strategy would allow us not only to destroy or prevent the development of unwanted biofilms, but perhaps also to promote the development of health-promoting biofilms such as probiotic organisms.

Technological Exchange with Industry

A great opportunity exists to facilitate technology transfer from the water and manufacturing industry to the clinical environment. This may be especially important for controlling the transfer of nosocomial infections that cause substantial morbidity and mortality in many hospitals.

Industrial microbiologists have identified many of their problems, such as corrosion (Hamilton 1985; Little *et al.* 1991), oil souring (Bass *et al.* 1993), and soiling and pipe/filter blocking (Daschner *et al.* 1996), to be mediated by microorganisms. Therefore, a basic understanding of the microbiology and genetics of biofilms must be available to a wider range of disciplines. For example, in the food industry, the use of Hazard Analysis and Critical Control Point (HACCP) has brought about a revolution in the use of risk assessment and microbiological analysis for the control of hygiene in the work place and of the clean-in-process (CIP) rinse systems. The benefits of understanding of fouling biofilm formation and their control have led to longer food product shelf life by assisting in the control of food pathogens and spoilage microorganisms and a reduction of food-poisoning incidents. A similar approach may prove beneficial for cross-infection control within the health and medical systems. Thus, training and educating the medical staff and doctors in clinical biofilm characteristics may aid in treatment selection of those infections that are biofilm related.

The majority of biofilms come to light only after a problem has manifested itself. However, pro-active monitoring of biofouling obviously plays an important factor in biofilm and process control. Such procedures have been adopted by a number of commercial companies who have developed monitoring software to detect biofouling and to manage the subsequent problems. This type of forward-thinking analysis has resulted in the development of multi-parameter information systems allowing companies to implement monitoring and control strategies. This style of technology transfer, of the use of optical fouling monitors and rapid ATP (adenosine triphosphate) analysis, is an excellent example of taking a multi-parametric approach to developing monitoring systems. Although many of these techniques cannot be directly transferred, the concepts of monitoring infection or the ability to rapidly detect infectious biofilms within a commensal bacterial population have far-reaching possibilities. The limitations in clinical practice are that one has to use approaches that are least damaging and invasive to the human host.

Traditional techniques are still used for monitoring biofilm formation on implants or catheter surfaces, including the use of direct contact plates to enable the assessment of total bacterial numbers and the identification of infective species. Plating techniques have their limitations, including the long incubation periods for organism growth, failure to detect organisms with special nutrient requirements and conditions (such as intracellular organisms), and the need for qualified persons to interpret results. Although antibody detection has been successful for a number of infective agents, it is often specific to identification and not to treatment susceptibility.

Another important aspect of controlling infection rates in hospitals is to be able to monitor the environment, water and medical equipment for infectious biofilms that are often resistant to antibiotics and may lead to nosocomial infections. To monitor medical equipment, for example endoscope washers, there is a necessity for fast and improved methods for microbial detection. Extending the principles of water testing kits to detect rapidly any microbial cross-infection of catheters by analysing urine, and other clinical samples (blood, serum, sputum, tissue, etc.), may be possible. With the modern advances in nanotechnology and nanomachines, many laboratory-based monitoring systems have now been moved from the laboratory to on-site monitoring. One such technique is the rapid ATP analysis used for monitoring biofouling of surfaces in industry, which may be adapted to monitor biofilm formation on medical equipment and operating theatre surfaces, reducing the possibility of infection transfer. Although ATP technology still has limitations, primarily the detection limit of $10^3\,\mathrm{CFU\,cm^{-2}}$, it is improving as technology advances. Such modifications could be to combine this technology with a polymerase chain reaction approach to identify specific pathogens rapidly. This would also be of particular use where bioterrorism (Hagmann 2001) or food poisoning is thought to have occurred.

Modern Techniques—Genomics, Proteomics and Bioinformatics

It is difficult to predict the future when we look at the rapid rate of development in molecular, biochemical and nanotechniques. Over the last decade there has been increased awareness of the power of genomics, proteomics, microchip technology and bioinformation to study global gene expression during bacterial growth, however, the most rapid growth has occurred in the last few years with respect to understanding clinical infections and biofilms (Miller and Diaz-Torres 1999; Thulasiraman *et al.* 2001). Current micro-array technology provides proteome maps for only a limited number of strains (Pennisi 1998), and very few of these have been grown in the biofilm phase. Most genomic and proteomic studies have been done on homogeneous populations represented by well-designed liquid-medium experiments. However, the analysis of *in vitro* biofilms, representing a heterogeneous population of cells, is more complex and it will be difficult to elucidate the component populations (Schmid *et al.* 2000). Similarly, microbial infections and biofilms *in vivo* are also more complex owing to their heterogeneous populations. Bioinformatics is a powerful tool in the generation of, primarily, predictive proteomic data from analysis of DNA and RNA. Proteomic biofilm studies may include profiling expression patterns in response to medium composition, nutrient limitations, colony ageing, quorum-sensing and environmental conditions (temperatures or pH within different parts of the biofilm or disease state). Such studies may assist in the understanding of biofilm growth, structure, physiology, resistance patterns and comparisons with planktonic grown cells (Steyn *et al.* 2001). To aid our search for novel antimicrobial products we need to seek new targets, and to do that we require a greater knowledge of biofilm structure (Stoodley *et al.* 2001) and physiology and to examine further the mechanisms of resistance within biofilm communities.

SUMMARY

Although researchers have made substantial progress in understanding biofilms, the detection of their presence in medical infections and controlling biofilm formation has been relatively difficult. Research is essential if we are to develop new, more rapid methods for biofilm detection. Preventing unwanted infectious biofilm growth on implants or tissue surfaces is an ultimate goal, however, to control and eradicate infective biofilms selectively without disturbing the body's commensal population is a more realistic aim for the future. To achieve this requires extensive research to provide us with a fundamental understanding of the mechanisms of biofilm growth and survival *in vivo* and to establish the role that communication between bacterial cells and bacterial and host cells plays in pathogenesis. With the current concerns of increasing occurrence of multi-antibiotic-resistant strains, chronic infections, implant-related infections, gastric disturbances and nosocomial infections, strategies at controlling biofilms both *in vivo* and within hospital environments is essential.

ACKNOWLEDGEMENT

We would like to acknowledge the support of Professor Richard Sharp in compiling the section on phage therapy.

REFERENCES

Alary, M and Joly, JR (1991) Risk factors for contamination of domestic hot water systems by *Legionellae. Applied and Environmental Microbiology* **57**:2360–2367.

Aliisky, J, Ickowski, K, Rapoport, A and Troitsky, N (1998) Bacteriophages show promise as antimicrobial agents. *Journal of Infection* **36**:5–15.

Allison, DG (1993a) Biofilm-associated exopolysaccharides. *Microbiology Europe* (Nov–Dec):16–19.

Allison, DG (1993b) Exopolysaccharide synthesis in bacterial biofilms. In: Bacterial Biofilms and Their Control in Medicine and Industry (Eds. Wimpenny, J, Nichols, W, Stickler, D and Lappin-Scott, H), BioLine, Cardiff, pp. 25–30.

Anderl, JN, Franklin, MJ and Stewart, PS (2000) Role of antibiotic penetration limitation in *Klebsiella pneumoniae* biofilm resistance to ampicillin and ciprofloxacin. *Antimicrobial Agents and Chemotherapy* **44**:1818–1824.

Anwar, H, van Biesen, T, Dasgupta, M, Lam, K and Costerton, JW (1989) Interaction of biofilm bacteria with antibiotics in a novel *in vitro* chemostat system. *Antimicrobial Agents and Chemotherapy* **33**:1824–1826.

Applegate, B and Bryers, JD (1991) Effects of carbon and oxygen limitations and calcium concentrations on biofilm removal processes. *Biotechnology and Bioengineering* **37**:17–25.

Bandyk, DF, Bergamini, TM, Kinney, EV, Seabrook, GR and Towne, JB (1991) *In situ* replacement of vascular prostheses infected by bacterial biofilms. *Journal of Vascular Surgery* **13**:575–583.

Barrow, PA and Soothill, JS (1997) Bacteriophage therapy and prophylaxis: rediscovery and renewed assessment of potential. *Trends in Microbiology* **5**: 268–271.

Bartold, PM, Seymour, GJ, Cullinan, MP and Westerman, B (1998) Effect of increased community and professional awareness of plaque control on the management of inflammatory periodontal diseases. *International Dental Journal* **48**:282–289.

Bass, CJ, Lappin-Scott, HM and Sanders, PF (1993) Bacteria that sour reservoirs: new concept for the mechanisms of reservoir souring by sulphide generating bacteria. *Journal of Offshore Technology* **1**:31–37.

Bayston, R (2000) Medical problems due to biofilms: clincial impact, aetiology, molecular pathogenesis, treatment and prevention. In: Dental Plaque Revisited:

Oral Biofilms in Health and Disease (Eds. Newman, HN and Wislon, M), BioLine, Cardiff, pp. 111–124.

Beals, D, Ngo, T, Feng, Y, Cook, D, Grau, D and Weber, D (2000) Development and laboratory evaluation of a new toothbrush with a novel brush head design. *American Journal of Dentistry* **13**:5–14.

Beckerich, A and Hauduroy, P (1922) The bacteriolysant therapy of bacillary dysentery in children. Therapeutic application of bacteriolysants; d'Herelle's phenomenon. *Comptes Rendues de la Societe de Biologie* **86**:168–175.

Beech, IB and Gaylarde, CC (1991) Microbial polysaccharides and corrosion. *International Biodeterioration* **27**:95–107.

Bergamini, TM, Bandyk, DF, Govostis, D, Kaebnick, HW and Towne, JB (1988) Infection of vascular prostheses caused by bacterial biofilms. *Journal of Vascular Surgery* **7**:21–30.

Boyle, M, Ford, T, Maki, JS and Mitchell, R (1991) Biofilms and the survival of opportunistic pathogens in recycled water. *Waste Management and Research* **9**: 465–470.

Brooun, A, Liu, S and Lewis, K (2000) A dose–response study of antibiotic resistance in *Pseudomonas aeruginosa* biofilms. *Antimicrobial Agents and Chemotherapy* **44**: 640–646.

Brunoghe, R and Maisin, J (1921) Essais de therapeutique au moyen du bacteriophage du Staphylocoque. *Comptes Rendues de la Societe de Biologie* **85**:1029–1121.

Busscher, HJ, Free, RH, Van Weissenbruch, R, Albers, FW and van der Mei, HC (2000) Preliminary observations on influence of dairy products on biofilm removal from silicone rubber voice prostheses *in vitro*. *Journal of Dairy Science* **83**:641–647.

Busscher, HJ, van Hoogmoed, CG, Geertsema-Doornbusch, GI, van der Kuijl-Booij, M and van der Mei, HC (1997) *Streptococcus thermophilus* and its biosurfactants inhibit adhesion by *Candida* spp. on silicone rubber. *Applied and Environmental Microbiology* **63**:3810–3817.

Caldwell, DE, Korber, DR and Lawrence, JR (1993) Analysis of biofilm formation using 2D and 3D imaging. In: Microbial Cell Envelopes: Interactions and Biofilms (Eds. Quesnel, B, Gilbert, P and Handley, PS), Society for Applied Bacteriology Symposium Series no. 22, Blackwell, Oxford, pp. 52S–66S.

Ceri, H, Olson, ME, Stremick, C, Read, RR, Morck, D and Buret, A (1999) The Calgary biofilm device: new technology for rapid determination of antibiotic susceptibilities of bacterial biofilms. *Journal of Clinical Microbiology* **37**:1771–1776.

Chanishvili, N, Chanishvili, T, Tediashvili, M and Barrow, PA (2001) Phages and their application against drug-resistant bacteria. *Journal of Chemical Technology and Biotechnology* **76**:689–699.

Characklis, WG (1981) Bioengineering report, fouling biofilm development: a process analysis. *Biotechnology and Bioengineering* **23**:1923–1960.

Cho, A, Glinka, TW, Ludwikow, M, Fan, AT, Wang, M and Hecker, SJ (2001) New anti-MRSA cephalosporins with a basic aminopyridine at the C-7 position. *Bioorganic Medical Chemistry Letters* **11**:137–140.

Cloete, TE, Brozel, VS and Pressly, J (1989) A bacterial population structure study of water cooling systems in South Africa. *Water SA* **15**:37–42.

Costerton, JW, Geesey, GG and Cheng, K-J (1978) How bacteria stick. *Scientific American* **238**:86–95.

Costerton, JW and Geesey, GG (1979) Which populations of aquatic bacteria should we enumerate. In: Native Aquatic Bacteria. Enumeration, Activity and Ecology,

ASTM STP 695 (Eds. Costerton, JW and Colwell, RR), American Society for Testing and Materials, pp. 7–18.

Costerton, JW, Irvin, RT and Cheng, KJ (1981) The bacterial glycocalyx in nature and disease. *Annual Review of Microbiology* **35**:299–324.

Costerton, JW, Marrie, TJ and Cheng, K-J (1985) Phenomena of bacterial adhesion. In: Bacterial Adhesion (Eds. Savage, DC and Fletcher, M), Plenum Press, London, pp. 3–40.

Costerton, JW, Cheng, KJ, Geesey, GG, Ladd, TI, Nickel, JC, Dasgupta, M and Marrie, TJ (1987) Bacterial biofilms in nature and disease. *Annual Reviews in Microbiology* **41**:435–464.

Costerton, JW, Khoury, AE, Ward, KH and Anwar, H (1993) Practical measures to control device-related bacterial infections. *International Journal of Artificial Organs* **16**:765–770.

Costerton, JW, Lewandowski, Z, Caldwell, DE, Korber, DR and Lappin-Scott, HM (1995) Microbial biofilms. *Annual Reviews in Microbiology* **49**:711–745.

Crump, JA and Collignon, PJ (2000) Intravascular catheter-associated infections. *European Journal of Clinical and Microbiological Infectious Diseases* **19**:1–8.

Da Costa Cruz, J (1923) Bacteriophago en therapeutica. *Brazil Medicine* **37**:298.

Dahlbäck, B, Hermannsson, M, Kjelleberg, S and Norkrans, B (1981) The hydrophobicity of bacteria—an important factor in their initial adhesion at the air–water interface. *Archives of Microbiology* **128**:267–270.

Daschner, FD, Ruden, H, Simon, R and Clotten, J (1996) Microbiological contamination of drinking water in a commercial household water filter system. *European Journal of Clinical Microbiology and Infectious Diseases* **15**:233–237.

Davies, DG, Parsek, MR, Pearson, JP, Iglewski, BH, Costerton, JW and Greenberg, EP (1998) The involvement of cell-to-cell signals in the development of a bacterial biofilm. *Science* **280**:295–298.

Davison, WC (1922) Das le bactériophage dans le traitment de la fièvre typhoïde. *American Journal of Diseases in Children* **23**:531–539.

Dixon, B (1998) New and resurgent infections—new and future responses. *Science Progress* **81**:273–285.

Dunsmore, B (2002) Environmental microbiology and the oil industry, PhD thesis, Univeristy of Exeter, Exeter.

Eberhard, A (1972) Inhibition and activation of bacterial luciferase synthesis. *Journal of Bacteriology* **109**:1101–1105.

Ellis, DM (1990) Microorganisms: their control in cooling systems. *The Plant Engineer* (**Jan–Feb**):35–36.

Emde, KM, Mao, EH and Finch, GR (1992) Detection and occurrence of waterborne bacterial and viral pathogens. *Water Environment Research* **64**:641–647.

Erickson, DL, Endersby, R, Kirkham, A, Stuber, K, Vollman, DD, Rabin, HR, Mitchell, I and Storey, DG (2002) *Pseudomonas aeruginosa* quorum-sensing system may control virulence factor expression in the lungs of patients with cystic fibrosis. *Infection and Immunity* **70**:1783–1790.

Escher, AR and Characklis, WG (1988) Microbial colonisation of a smooth substratum: a kinetic analysis using image analysis. *Water Science and Technology* **20**:277–283.

Felo, A, Shibly, O, Ciancio, SG, Lauciello, FR and Ho, A (1997) Effects of subgingival chlorhexidine irrigation on peri-implant maintenance. *American Journal of Dentistry* **10**:107–110.

Ferris, FG, Schultze, S, Witten, TC, Fyfe, WS and Beveridge, TJ (1987) Metal interactions with microbial biofilms in acidic and neutral pH environments. *Applied and Environmental Microbiology* **55**:1249–1257.

Finch, RG (1994) Medical problems associated with biofilms. In: Bacterial Biofilms and Their Control in Medicine and Industry (Eds. Wimpenny, J, Nichols, W, Stickler, D and Lappin-Scott, H), BioLine, Cardiff, pp. 89–92.

Fiorina, JC, Weber, M and Block, JC (2000) Occurrence of lectins and hydrophobicity of bacteria obtained from biofilm of hospital catheters and water pipes. *Journal of Applied Microbiology* **89**:494–500.

Fletcher, M and Loeb, GI (1979) Influence of substratum characteristics on the attachment of a marine Pseudomonad to solid surfaces. *Applied and Environmental Microbiology* **37**:67–72.

Fletcher, M and Pringle, JH (1983) The effect of surface free energy and medium surface tension on bacterial attachment to solid surfaces. *Journal of Colloid and Interface Science* **104**:5–14.

Flock, J-I (1999) Antibodies that block adherence of *Staphylococcus aureus* to fibronectin. *Trends in Microbiology* **7**:140–141.

Foley, I and Gilbert, P (1996) Antibiotic resistance of biofilms. *Biofouling* **10**:331–346.

Geesey, GG, Jang, L, Jolley, JG, Hankins, MR, Iwaoka, T and Griffiths, PR (1988) Binding of metal ions by extracellular polymers of biofilm bacteria. *Water Science and Technology* **20**:161–165.

Gilbert, P, Collier, PJ and Brown, MR (1990) Influence of growth rate on susceptibility to antimicrobial agents: biofilms, cell cycle, dormancy, and stringent response. *Antimicrobial Agents and Chemotherapy* **34**:1865–1868.

Gilbert, P and Brown, MRW (1994) Experimental approaches to the physiological properties of biofilm cells. In: Bacterial Biofilms and Their Control in Medicine and Industry (Eds. Wimpenny, J, Nichols, W, Stickler, D and Lapin-Scott, H), BioLine, Cardiff, pp. 49–52.

Gilbert, P and Allison, D (2000) Biofilms and their resistance towards antimicrobial agents. In: Dental Plaque Revisited: Oral Biofilms in Health and Disease (Eds. Newman, HN and Wilson, M), BioLine, Cardiff, pp. 125–143.

Gilbert, P, Allison, D, Rickard, A, Sufya, N, Whyte, F and McBain, AJ (2001) Do biofilms present a nidus for the evolution of antibacterial resistance. In: Biofilm Community Interactions: Chance or Necessity? (Eds. Gilbert, P, Allison, D, Brading, M, Verran, J and Walker, J), BioLine, Cardiff, pp. 340–351.

Gracia, E, Fernandez, A, Conchello, P, Lacleriga, A, Paniagua, L, Seral, F and Amorena, B (1997) Adherence of *Staphylococcus aureus* slime-producing strain variants to biomaterials used in orthopaedic surgery. *International Orthopaedics* **21**:46–51.

Hagmann, M (2001) WHO helps countries prepare for bioterror attacks. *Bulletin of the World Health Organisation* **79**:11.

Hamilton, WA (1985) Sulphate-reducing bacteria and anaerobic corrosion. *Annual Reviews in Microbiology* **39**:195–217.

Hansen, MC, Palmer, RJ and White, DC (2000) Flowcell culture of *Porphyromonas gingivalis* biofilms under anaerobic conditions. *Journal of Microbiological Methods* **40**:233–239.

Hastings-Drisko, C (1999) Non-antibiotic plaque chemotherapy. In: Dental Plaque Revisited. Oral Biofilms in Health and Disease (Eds. Newman, HN and Wilson, M), BioLine, Cardiff, pp. 523–548.

Hentzer, M, Riedel, K, Rasmussen, TB, Heydorn, A, Andersen, JB, Parsek, MR, Rice, SA, Eberl, L, Molin, S, Høiby, N, Kjelleberg, S and Givskov, M (2002) Inhibition of quorum sensing in *Pseudomonas aeruginosa* biofilm bacteria by halogenated furanone compound. *Microbiology* **148**:87–102.

Hilbert, L (2001) Integrated strategy to combat biofouling in the food industry. Presented at the Biofouling of Materials Conference (Ed. Sand, W), Frankfurt Dechema, Frankfurt, p. 11.

Hillman, JD, Brooks, TA, Michalek, SM, Harmon, CC, Snoep, JL and Vander Weijdden, CC (2000) Construction and characterisation of an effector strain of *Streptococcus mutans* for replacement therapy of dental caries. *Infection and Immunity* **68**:543–549.

Hoyle, BD, Alcantara, J and Costerton, JW (1992) *Pseudomonas aeruginosa* biofilm as a diffusion barrier to piperacillin. *Antimicrobial Agents and Chemotherapy* **36**: 2054–2056.

Huang, CT, Yu, FP, McFeters, GA and Stewart, PS (1995) Nonuniform spatial patterns of respiratory activity within biofilms during disinfection. *Applied and Environmental Microbiology* **61**:2252–2256.

Hughes, G, North, J, Walker, J and Sharp, R (2001) Biofilms, bacteriophage interactions and bacteriophage therapy. In: Biofilm Community Interactions: Chance or Necessity? (Eds. Gilbert, P, Allison, D, Brading, M, Verran, J and Walker, J), BioLine, Cardiff, pp. 325–331.

Ishikawa, T, Kamiyama, K, Nakayama, Y, Iizawa, Y, Okonogi, K and Miyake, A (2001) Studies on anti-MRSA parenteral cephalosporins. III. Synthesis and antibacterial activity of 7-beta-[2-(5-amino-1,2,4-thiadiazol-3-yl)-2(Z)-alkoxyimino-acetamido-3. *Journal of Antibiotics (Tokyo)* **54**:257–277.

Jass, J and Lappin-Scott, H (1996) The efficacy of antibiotics enhanced by electrical currents against *Pseudomonas aeruginosa* biofilms. *Journal of Antimicrobial Chemotherapy* **38**:987–1000.

Jass, J and Lappin-Scott, H (1994) Sensitivity testing for antimicrobial agents against biofilms. In: Bacterial Biofilms and Their Control in Medicine and Industry (Eds. Wimpenny, J, Nichols, W, Stickler, W and Lappin-Scott, H), BioLine, Cardiff. p. 73–76.

Kaprelyants, AS and Kell, DB (1996) Do bacteria need to communicate with each other for growth. *Trends in Microbiology* **4**:237–242.

Keevil, CW, Walker, JT, McEvoy, J and Colbourne, JS (1989) Detection of biofilms associated with pitting corrosion of copper pipework in Scottish hospitals. *Proceedings of the Biodeterioration Society, Corrosion* **5**:99–117.

Keevil, CW, Dowsett, AB and Rogers, J (1994) *Legionella* biofilms and their control. In: Microbial Biofilms: Formation and Control, Society for Applied Bacteriology, pp. 201–215.

Kepkay, PE, Schwinghamer, P, Willar, T and Bowen, AJ (1986) Metabolism and metal binding by surface colonising bacteria: results of microgradient measurements. *Applied and Environmental Microbiology* **51**:163–170.

Kihlberg, J, Hultgren, SJ, Normark, S and Magnusson, G (1989) Mapping the PapG-adhesin receptor site of uropathogenic *E. coli* by inhibition of hemagglutination using synthetic analogues of galabiose. *Journal of the American Chemical Society* **111**:6364–6368.

King, CH, Shotts, EB, Wooley, RE and Porter, KG (1988) Survival of coliforms and bacterial pathogens within protozoa during chlorination. *Applied and Environmental Microbiology* **54**:3023–3033.

Kockro, RA, Hampl, JA, Jansen, B, Peters, G, Scheihing, M, Giacomelli, R, Kunze, S, and Aschoff, A (2000) Use of scanning electron microscopy to investigate the prophylactic efficacy of rifampin-impregnated CSF shunt catheters. *Journal of Medical Microbiology* **49**:441–450.

Konhauser, KO, Schulze-Lam, S, Ferris, FG, Fyfe WS, Longstaffe, FJ and Beveridge, TJ (1994) Mineral precipitation by epilithic biofilms in the Speed River Ontario, Canada. *Applied and Environmental Microbiology* **60**:549–553.

Kuchta, JM, States, SJ, McGlaughlin, JE, Overmeyer, JH, Wadowsky, RM, McNamara, AM, Wolford, RS and Yee, RB (1985) Enhanced chlorine resistance of tap water-adapted *Legionella pneumophila* as compared with agar medium passaged strains. *Applied and Environmental Microbiology* **50**:21–26.

Kunze, S and Aschoff, A (2000) Use of scanning electron microscopy to investigate the prophylactic efficacy of rifampin-impregnated CSF shunt catheters. *Journal of Medical Microbiology* **49**:441–450.

Larsen, T and Fiehn, NE (1996) Resistance of *Streptococcus sanguis* biofilms to antimicrobial agents. *Acta Pathologica Microbiologica et Immunologica Scandinavica* **104**:280–284.

Levin, BR and Bull, JJ (1996) Phage therapy revisited: the population biology of a bacterial infection and its treatment with bacteriophage and antibodies. *American Naturalist* **147**:881–898.

Lewis, K (2001) Riddle of biofilm resistance. *Antimicrobial Agents and Chemotherapy* **45**:999–1007.

Little, B, Wagner, P and Mansfield, F (1991) Microbiologically influenced corrosion of metal and alloys. *International Material Reviews* **36**:253–272.

Lytle, MS, Adams, JC, Dickman, DG and Bressler, WR (1989) Use of nutrient response techniques to assess the effectiveness of chlorination of rapid sand filter gravel. *Applied and Environmental Microbiology* **55**:29–32.

Manefield, M, Welch, M, Givskov, M, Salmond, GPC and Kjelleberg, S (2001) Halogenated furanones from red alga, *Delisea pulchra*, inhibit carbapenem antibiotic synthesis and exoenzyme virulence factor production in the phytopathogen *Erwinia carotovara*. *FEMS Microbiology Letters* **205**:131–138.

Marrie, TJ and Costerton, JW (1984) Morphology of bacterial attachment to cardiac pacemaker leads and power packs. *Journal of Clinical Microbiology* **19**:911–914.

Marsh, PD and Martin, MV (1999) Oral Microbiology, 4th Edition, Wright, Oxford.

Mermel, LA (2001) New technologies to prevent intravascular catheter-related bloodstream infections. *Emerging Infectious Diseases* **7**:197–199.

Mermel, LA (2000) Prevention of intravascular catheter-related infections. *Annals of International Medicine* **132**:391–402.

Mermel, LA, Farr, BM, Sherertz, RJ, Raad, II, O'Grady, N, Harris, JS and Craven, DE (2001) Guidelines for the management of intravascular catheter-related infections. *Clinical Infectious Diseases* **32**:1249–1272.

Meyer, DH and FivesTaylor, PH (1998) Oral pathogens: from dental plaque to cardiac disease. *Current Opinion in Microbiology* **1**:88–95.

Miller, BS and Diaz-Torres, MR (1999) Proteome analysis of biofilms: growth of *Bacillus subtilis* on solid medium as model. *Methods in Enzymology* **310**:433–441.

Nealson, KH, Platt, T and Hastings, JW (1970) Cellular control of the synthesis and activity of the bacterial luminescent system. *Journal of Bacteriology* **104**:313–322.

Newman, HN (1998) Periodontal therapeutics—a viable option? *International Dental Journal* **48**:173–179.

Nichols, W (1994) Biofilm permeability to antibacterial agents. In: Bacterial Biofilms and Their Control in Medicine and Industry (Eds. Wimpenny, J, Nichols, W, Stickler, D and Lappin-Scott, H), BioLine, Cardiff, pp. 141–150.

Oga, T, Suthersan, S and Ganczarczyk, JJ (1991) Some properties of aerobic biofilms. *Environmental Technology* **12**:431–440.

Ohlsson, J, Jass, J, Uhlin, BE, Kihlberg, J and Nilsson, UJ (2002) Discovery of potent inhibitors of PapG adhesins from uropathogenic *Escherichia coli* through synthesis and evaluation of galabiose derivatives. *ChemBiochem* **3**:101–108.

Pearson, JP, Gray, KM, Passador, L, Tucker, KD, Eberhard, A, Iglewski, BH and Greenberg, EP (1994) Structure of the autoinducer required for expression of *Pseudomonas aeruginosa* virulence genes. *Proceedings of the National Academy of Science of the United States of America* **91**:197–201.

Pennisi, E (1998) Genome data shake tree of life. *Science* **280**:672–674.

Reasoner, DJ (1988) Drinking water microbiology research in the United States: an overview of the past decade. *Water Science and Technology* **20**:101–107.

Rediske, AM, Roeder, BL, Nelson, JL, Robison, RL, Schaalje, GB, Robison, RA and Pitt, WG (2000) Pulsed ultrasound enhances the killing of *Escherichia coli* biofilms by aminoglycoside antibiotics *in vivo. Antimicrobial Agents and Chemotherapy* **44**:771–772.

Ridgway, HF and Olson, BH (1982) Scanning electron microscope evidence for bacterial colonisation of a drinking-water distribution system. *Applied and Environmental Microbiology* **41**:274–287.

Rowbotham, TJ (1980) Preliminary report on the pathogenicity of *Legionella pneumophila* for freshwater and soil amoebae. *Journal of Clinical Pathology* **33**:1179–1183.

Schmid, M, Twachtmann, U, Klein, M, Strous, M, Juretschko, S, Jetten, M, Metzger, JW, Schleifer, KH and Wagner, M (2000) Molecular evidence for genus level diversity of bacteria capable of catalyzing anaerobic ammonium oxidation. *Systematic and Applied Microbiology* **23**:93–106.

Sharp, RJ (2001) Bacteriophages: biology & history. *Journal of Chemical Technology and Biotechnology* **76**:667–672.

Singh, PK, Schaefer, AL, Parsek, MR, Moninger, TO, Welsh, MJ and Greenberg, EP (2000) Quorum-sensing signals indicate that cystic fibrosis lungs are infected with bacterial biofilms. *Nature* **407**:762–764.

Smith, HW and Huggins, MB (1983) Effectiveness of phages in treating experimental *Escherichia coli* diarrhoea in calves, piglets and lambs. *Journal of General Microbiology* **129**:2659–2675.

Smith, HW and Huggins, MB (1982) Successful treatment of experimental *Escherichia coli* infections in mice using phage: its general superiority over antibiotics. *Journal of General Microbiology* **128**:307–318.

Smith, HW, Huggins, MB and Shaw, KM (1987a) The control of experimental *Escherichia coli* diarrhoea in calves by means of bacteriophages. *Journal of General Microbiology* **133**:1111–1126.

Smith, HW, Huggins, MB and Shaw, KM (1987b) Factors influencing the survival and multiplication of bacteriophages in calves and in their environment. *Journal of General Microbiology* **133**:1127–1135.

Soothill, JS (1994) Bacteriophage prevents destruction of skin grafts by *Pseudomonas aeruginosa. Burns* **20**:209–211.

Soothill, JS (1992) Treatment of experimental infections of mice with bacteriophages. *Journal of Medical Microbiology* **37**:258–261.

Soukos, NS, Socransky, SS, Mulholland, SE, Lee, S and Doukas, AG (2000) Photomechanical drug delivery into bacterial biofilms. *Pharmacological Research* **17**:405–409.

Spence, RC and McKinley, EB (1924) The therapeutic value of bacteriophage in the treatment of bacillary dysentery. *Southern Medical Journal* **17**:563–568.

Spoering, AL and Lewis, K (2001) Biofilms and planktonic cells of *Pseudomonas aeruginosa* have similar resistance to killing by antimicrobials. *Journal of Bacteriology* **183**:6746–6751.

Springer, DM, Luh, BY and Bronson, JJ (2001) Anti-MRSA cephems. Part 1: C-3 substituted thiopyridinium derivatives. *Bioorganic and Medicinal Chemistry* **11**: 797–801.

Stewart, PS (1998) A review of experimental measurements of effective diffusive permeabilities and effective diffusion coefficients in biofilms. *Biotechnology and Bioengineering* **59**:261–272.

Stewart, PS and Costerton, JW (2001) Antibiotic resistance of bacteria in biofilms. *Lancet* **358**:135–138.

Steyn, B, Oosthuizen, MC, MacDonald, R, Theron, J and Brozel, VS (2001) The use of glass wool as an attachment surface for studying phenotypic changes in *Pseudomonas aeruginosa* biofilms by two-dimensional gel electrophoresis. *Proteomics* **1**:871–879.

Stickler, D, Morris, N, Moreno, MC and Sabbuba, N (1998) Studies on the formation of crystalline bacterial biofilms on urethral catheters. *European Journal of Clinical Microbiology and Infectious Diseases* **17**:649–652.

Stoodley, P, deBeer, D and Lappin-Scott, HM (1997) Influence of electric fields and pH on biofilm structure as related to the bioelectric effect. *Antimicrobial Agents and Chemotherapy* **41**:1876–1879.

Stoodley, P, Hall-Stoodley, L and Lappin-Scott, HM (2001) Detachment, surface migration and other dynamic behaviour in bacterial biofilms revealed by digital time-lapse imaging. *Methods in Enzymology* **337**:306–319.

Svensson, A, Larsson, A, Emtenäs, H, Hedenström, M, Fex, T, Hultgren, SJ, Pinker, JS, Almqvist, F and Kihlberg, J (2001) Design and evaluation of pilicides: potential novel antibacterial agents directed against uropathogenic *Escherichia coli*. *ChemBiochem* **12**:915–918.

Taylor Eighmy, T and Bishop, PL (1985) Effect of reactor turbulence on the binding protein-mediated aspartate transport system in thin wastewater biofilms. *Applied and Environmental Microbiology* **50**:120–124.

Thulasiraman, V, McCutchen-Maloney, SL, Motin, VL and Garcia, E (2001) Detection and identification of virulence factors in *Yersinia pestis* using SELDI Protein Chip system. *Biotechniques* **30**:428–432.

Trulear, MG and Characklis, WG (1982) Dynamics of biofilm processes. *Journal of Water Pollution Control Federation* **54**:1288–1301.

Van der Mei, HC, Free, RH, Elving, GJ, Van Weissenbruch, R, Albers, FW and Busscher, HJ (2000) Effect of probiotic bacteria on prevalence of yeasts in oropharyngeal biofilms on silicone rubber voice prostheses *in vitro*. *Journal of Medical Microbiology* **49**:713–718.

van der Wende, E, Characklis, WG and Grochowski, J (1988) Bacterial growth in water distribution systems. *Water Science and Technology* **20**:521–524.

Vess, RW, Anderson, RL, Carr, JH, Bond, WW and Favero, MS (1993) The colonisation of solid PVC surfaces and the aquisition of resistance to germicides by water microorganisms. *Journal Applied Bacteriology* **74**:215–221.

Vogel, L, Sloos, JH, Spaargaren, J, Suiker, I and Dijkshoorn, L (2000) Biofilm production by *Staphylococcus epidermidis* isolates associated with catheter related bacteremia. *Diagnostic Microbiology and Infectious Diseases* **36**:139–141.

Vorachit, M, Lam, K, Jayanetra, P and Costerton, JW (1993) Resistance of *Pseudomonas pseudomallei* growing as a biofilm on silastic discs to ceftazidime and co-trimoxazole. *Antimicrobial Agents and Chemotherapy* **37**:2000–2002.

Vyas, SP, Sihorkar, V and Mishra, V (2000) Controlled and targeted drug delivery strategies towards intraperiodontal pocket diseases. *Journal of Clinical Pharmacology and Therapeutics* **25**:21–42.

Walker, JT, Dowsett, AB, Dennis, PJL and Keevil, CW (1991) Continuous culture studies of biofilm associated with copper corrosion. *International Biodeterioration* **27**:121–134.

Walker, JT, Verran, J, Boyd, RD and Percival, S (2001) Microscopy methods to investigate structure of potable water biofilms. *Methods in Enzymology* **337**: 243–255.

Walsh, C (2000) Molecular mechanisms that confer antibacterial drug resistance. *Nature* **406**:775–781.

Walsh, TJ, Schlegel, R, Moody, MM, Costerton, JW and Salcman, M (1986) Ventriculoatrial shunt infection due to *Cryptococcus neoformans*: an ultrastructural and quantitative microbiological study. *Neurosurgery* **18**:373–375.

Wattanakaroon, W and Stewart, PS (2000) Electrical enhancement of *Streptococcus gordonii* biofilm killing by gentamicin. *Archives of Oral Biology* **45**:167–171.

Wentland, EJ, Stewart, PS, Huang, CT and McFeters, GA (1996) Spatial variations in growth rate within *Klebsiella pneumoniae* colonies and biofilm. *Biotechnology Progress* **12**:316–321.

Whipple, GC (1901) Changes that take place in the bacterial contents of waters during transportation. *Technical Quarterly* **14**:21–29.

Williams, P and Stewart, GSAB (1993) Cell density dependent control of gene expression in bacteria—implications for biofilm development and control. In: Bacterial Biofilms and Their Control in Medicine and Industry (Eds. Wimpenny, J, Nichols, W, Stickler, D and Lappin-Scott, H), BioLine, Cardiff, pp. 9–12.

Williams, I, Venables, WA, Lloyd, D, Paul, F and Critchley, I (1997) The effects of adherence to silicone surfaces on antibiotic susceptibility in *Staphylococcus aureus. Microbiology* **143**:2407–2413.

Xu, KD, McFeters, GA and Stewart, PS (2000) Biofilm resistance to antimicrobial agents. *Microbiology* **146**:547–549.

Zimmerli, W, Widmer, A, Blattter, M, Frie, R and Ochsner, P (1998) Role of rifampicin for treatment of orthopedic implant-related staphylococcal infections—a randomised controlled trial. *Journal of the American Medical Association* **279**: 1537–1541.

Index

Note: page references in **bold** refer to figures, those in *italics* refer to tables. Abbreviations used in the index are: IUDs = intrauterine devices; MBEC = minimum biofilm eradication concentration; PCR = polymerase chain reaction.